MONOGRAPHIAE BIOLOGICAE

VOLUME 67

Series Editors

H.J. Dumont and M.J.A. Werger

APPLIED POPULATION BIOLOGY

Applied Population Biology

Edited by

S. K. JAIN

and

L. W. BOTSFORD

KLUWER ACADEMIC PUBLISHERS
DORDRECHT / BOSTON / LONDON

Library of Congress Cataloging-in-Publication Data

Applied population biology / edited by S.K. Jain and L.W. Botsford.
p. cm. -- (Monographiae biologicae ; v. 67)
"This book arose out of a series of seminars at the University of California, Davis, held in the spring of 1986 ...sponsored by the Alfred P. Sloane Foundation"--Foreword.
Includes bibliographical references (p.) and index.
ISBN 0-7923-1425-5 (HB : acid-free paper)
1. Population biology--Congresses. 2. Biology, Economic--Congresses. I. Jain, S. K. (Subodh Kumar), 1954- . II. Botsford, Louis W. III. Alfred P. Sloan Foundation. IV. Series.
QP1.P37 vol. 67
[QH352]
574 s--dc20
[574.5'248] 91-29384

Published by Kluwer Academic Publishers,
P.O. Box 17, 3300 AA Dordrecht, The Netherlands.

Kluwer Academic Publishers incorporates
the publishing programmes of
D. Reidel, Martinus Nijhoff, Dr W. Junk and MTP Press.

Sold and distributed in the U.S.A. and Canada
by Kluwer Academic Publishers,
101 Philip Drive, Norwell, MA 02061, U.S.A.

In all other countries, sold and distributed
by Kluwer Academic Publishers Group,
P.O. Box 322, 3300 AH Dordrecht, The Netherlands.

printed on acid-free paper

Printed in The Netherlands

Contents

Foreword

This book arose out of a series of seminars at the University of California, Davis, held in the spring of 1986. This series on applied population biology was one of three sponsored by the Alfred P. Sloan Foundation, the other two being in human ecology and theoretical ecology. The support of the Sloan Foundation is gratefully acknowledged.

The purpose of the series was to critically examine the *process* of application of the concepts of population biology to practical problems. We focused specifically on the process in the hope that we could uncover some reasons for the perceived limited success in solving practical problems. To do this we tried to choose speakers who were engaged in innovative, applied research in a wide variety of fields in population biology. We attempted to include examples of both plants and animals, population dynamics and genetics, and domestic as well as wild species. (Note that our efforts toward diversity were aided by our backgrounds, being a plant ecologist who works on both domestic and wild species and a population ecologist who works primarily on wild animals). No particular philosophy or orientation such as reductionist versus holistic, or mathematical versus descriptive, was necessarily emphasized.

In addition to the invited seminar on the specific application, a discussion among a small group of faculty, students and the invited speaker was held the next day. At the beginning of each of these, the questions of how the speaker applied the fundamental concepts of population biology to their specific practical problem, and how useful they found them to be in solving that problem, were posed to the speaker. The ensuing discussion usually began with an attempt to first answer the nontrivial question of what the fundamental concepts of population biology are (as well as to define "population biology," itself). Both were interpreted in the widest possible sense. In the remainder of these discussions the responses of the speakers were amazingly similar throughout the series. Several key ideas emerged early on and were repeated by workers from the various fields throughout the series. In its simplest form, the common thread was that only a very limited number of the "principles of population biology" were of direct use in application, and thus

S. K. Jain and L. W. Botsford (eds), Applied Population Biology, vii–ix.

we are not in the position of having established a general body of knowledge from which we can draw predictions about how ecological systems will react to various human activities (a position in which we perceive the "harder" physical sciences to be). Ecologists do not have a set of general laws that we can routinely apply to a variety of ecological and environmental problems to produce successful solutions. This fact should be more widely realized by practitioners and users alike.

This lack of general laws is problematic for applied population biologists. We currently operate as though we do have a set of general principles that can be drawn upon to solve practical problems. The outcome of an approach is not often tested since whether our prescriptions worked is either: (1) not monitored, (2) obscured by variability, or (3) impossible to tell because of the models used. We can continue to do this, and pay little attention to whether our methods work, or when they are shown to have failed, we can ascribe bad predictions to some inherent natural variability. An alternative approach is to treat each application as a unique case and reinvent the wheel for each, independent of past and future applications. The former approach would lead to continued poor performance, while the latter would preclude development of any general principles. We propose here that neither approach be followed, but rather that both the application of population biology and the development of general principles of population biology would be best served by greater integration of the two. We develop and approach that does not presume a general theory, but rather integrates applications into the empirical process of establishing a general body of knowledge.

To improve applications, instead of formulating a solution to each practical problem, then applying it with little further involvement (such as testing results, observing outcomes), we should treat the application of the solution as a continuing part of the empirical process, by monitoring and responding to the results. Taking advantage of continuing feedback of results in practical applications of population biology requires an entirely different approach, as well as substantial differences in actual execution of applications. Viewing applications as continuing experiments rather than final products will lead to different recommended solutions. Adoption of this approach therefore requires a critical evaluation of current research methods and analyses. Our approach to practical problems must be restructured so that we can learn from our mistakes rather than sweeping them under a rug of "unavoidable biological variability".

Treating applications as experiments (albeit imperfect ones) and using results to further develop the field in general, will also ensure rapid progress in population biology itself. The expensive experimentation necessary to understand the natural world can be partially avoided by trying to understand the results of perturbations made for other reasons, and incorporating them into ecological theory (cf., Margalef 1968, p. 45).

These conclusions seemed to be common to all of the applied fields covered in the seminar series. Because their further discussion and dissemination seemed valuable, we decided to publish articles based on these seminars under one cover, along with an introductory chapter describing our approach *ab initio* and an epilogue containing our *a posteriori* conclusions from the discussion groups. Thus, the examples presented here are not just innovative applications of the principles of population biology to a variety of fields, but are also particular examples of a common theme whose recognition could be of considerable value to population biology in general. The organization of the book is similar to the organization of the seminar series. The basic motivation for the series and the questions asked at the beginning of the discussion sections are put forth, along with various definitions, in Chapter 1. They are followed by written versions of each seminar. The book concludes with a chapter describing our conclusions from the series in a more complete fashion than was sketched here.

This book is intended primarily for practitioners of population biology, but some of our conclusions have implications for ecologists in general (e.g., theoreticians). The chapters are written by population biologists for population biologists. Because of the esoteric nature of some applied fields, we requested of each author a brief summary of important issues and recent developments in their field, as well as any helpful definitions. Because of this the book should be quite accessable to graduate students in ecology and could form the basis for a reading/seminar course. We recognize, of course, that no single volume in this fascinating and rapidly expanding field could cover all the issues and outstanding examples; however, we hope that it will provide stimulus for many such works.

We would like to express our appreciation to those who took the time to referee the chapters herein. Their efforts contributed significantly to the quality of this book. We are grateful to Dolores Dumont, Eileen O'Farrell and Marilee Schmidt for their cheerful work on processing the mansucripts through several iterations.

Lastly, we would like to express our sincere apologies to chapter authors who sent us their Chapters and responded to reviews in a timely fashion, yet have had to wait so long for publications; a few others were received as early as 1988.

Literature cited

Hastings, A. (ed.). 1988. Community Ecology. Lecture Notes in Biomathematics, No. 77. Springer-Verlag, New York.

Margalef, R. 1968. Perspectives in Ecological Theory. Univ. of Chicago Press, Chicago.

Davis, California

Subodh K. Jain
Louis W. Botsford

1. Population biology and its application to practical problems

LOUIS W. BOTSFORD and SUBODH K. JAIN

Abstract

The principles of population biology have been successfully applied to a wide variety of practical problems. However, there is a widely held view that the practical application of the principles of population biology has often not lived up to its perceived potential. Possible reasons for this include: (1) inadequate principles (or theory), (2) overwhelming inherent variability in populations, (3) a flawed approach to applications, and (4) other exogenous political and social factors, outside the scope of this book. Biological ecology has long sought general principles in the development of a predictive science, but there is considerable dissatisfaction with progress. Ecologists have sought to characterize the general nature of the organization and dynamic behavior of communities and populations, dealing with issues such as the degree of inter-connectedness within them and their susceptibility to random environmental events. Several specific theories regarding limited aspects of their behavior (e.g., genetic structure, species coexistence) have been developed. There has been a relatively recent shift to an experimental, manipulative approach to field observations. Population biologists have also begun to pay closer attention to the philosophical underpinnings of the relationship between theory and observation. The few rules that have guided the development of theoretical principles have led to a class of simple (non-mechanistic), general (vague or unspecific) models which are central to current population and community theory. The approach to genetic theory has been based more on mechanistic models, even though the mechanisms themselves are simpler than the actual systems. We characterize the several kinds of principles of population genetics and population dynamics. We then introduce the applications covered in the following chapters in the context of the questions posed and the background review provided in this chapter.

Introduction

Historically, much of the activity in population biology has involved practical applications in agricultural, natural, and semi-natural populations of crop plants, domesticated animals, pest species, fish, wildlife, forestry, and others. Currently, we are especially sensitive to the needs for application of population ecology and genetics to pressing problems in biological control of weeds and pests in agricultural systems, on the one hand, and conservation of rare and endangered wild species and habitats, on the other. In most cases a

S. K. Jain and L. W. Botsford (eds), Applied Population Biology, 1—24.

population is the natural unit of study or concern in problems involving manipulation, harvest, monitoring or other intervention. However, both community and ecosystem level issues enter into many of these projects. As examples, success of introduced natural enemies in a biological control program often depends on various multispecies interactions, conservation of an endangered species might very well be more prudent in terms of habitat restoration, and management of rangeland requires both species and community level responses to be integrated. Furthermore, applied problems in population biology typically involve both *population dynamics*, the dynamics of birth, death, migration or persistence of a target or indicator species, and *population genetics*, the population's natural or manipulated genetic composition in the context of evolutionary responses to varying environments.

Several key ideas in population genetics have found useful applications in the fields of plant and animal breeding, disease and pest control, forestry and range management, and more recently, conservation of biodiversity. Breeding methods involve a variety of controlled matings among selected stocks, followed by different selection schemes which depend on the species, breeding objectives, and the nature of commercial variety or genetic stock (Wricke and Weber 1986; Robertson 1980). For example, maximizing hybrid vigor in the pasture legumes (e.g., alfalfa) involves interpopulation crosses and pollination management; however, in most pasture species, hybrid selection and mating designs often aim at general combining ability and the most economical methods of varietal development as well (McIvor and Bray 1983). On the other hand, some form of inbreeding is used for selection involving low heritability traits, so that evolution of inbreeding tolerance might be valuable *per se*. Several recurrent selection methods also consider intermittent matings among replicated selection lines, which resembles Wright's three-phase model of evolution in subdivided populations (Wright 1963; Frey 1981; Hill 1986). Forestry breeders are often concerned with the optimal levels of intermixing different gene pools (essentially ecotypes) so as to avoid inbreeding and to maintain genetic variation within stands (Cheliak 1985; Namkoong 1979). Much of the genetic manipulation of grassland species is based on the role of natural selection in evolving persistence and performance of introduced species (Marten *et al.* 1989). And finally, breeding for disease resistance involves epidemiological aspects of host-pathogen population cycles on one hand and genetic models on the other, which help in predicting durable resistance (due to evolutionary constraints on the pathogen) of heterogeneous populations (varietal mixtures, composite crosses, etc.) (Burdon 1987; Denno and McClure 1983; Jayakar and Zonta 1990).

Studies of genetic variation in natural populations have also suggested some useful applications of population genetics. One example is the simple notion, supported by empirical results, that weeds and pests whose populations are genetically homogeneous, are readily subject to biological control. In conservation biology, small populations raise serious concern for their vulnerability to harmful inbreeding effects and to overall losses of adaptive capacity for response to changing environments (Soulè 1987). Population genetics provides tools for estimating effective population size under a variety of real world situations involving different sex ratios, social systems, fluctuations in population size, highly variable dispersion, recruitment and reproductive patterns, and habitat fragmentation (Chepoko-Sade and Halpin 1987). Moreover, such estimates of population sizes allow us to develop predictions of minimum viable size or area required for conservation. Restoration ecology of disturbed or derelict mined areas, on the other hand, seeks information on the relative success of revegetation efforts based on the presence of genotypes with heavy metal tolerance (Bradshaw 1987). Many of these potential applications of population genetics are cited in the following chapters of this volume.

Overall, an understanding of evolutionary processes can help us design, even on a microevolutionary time and space scale, more efficient and predictable outcomes of genetic manipulation of populations for domestication of new species, genetic resource conservation, increases in agricultural productivity and better environmental management of ecosystems (e.g., Libby 1973; Burdon 1988; Jana and Acharya 1981; Chapman 1989; Bradshaw and McNeilly 1982; Jordan *et al.* 1987). We can certainly expect much more in the next decade, especially with the newly emerging field applications of molecular genetics and biotechnology along with an awareness of their ecological consequences.

Population dynamics has also occupied a central place in the solution to many practical problems in population biology, but in a slightly different way than population genetics. A greater variety of models is used and fewer general rules to guide applications have emerged. Population models are involved in many applications, for example in fields such as conservation biology, pest management, and fisheries. These models vary widely in realism and complexity, even within each field (e.g., in fisheries, from the logistic to age- and stage-structured models, and in conservation biology, from birth-death processes to age-structured models with environmental and catastrophic randomness). In addition to the lack of uniformity of approaches (or possibly because of them), application of population dynamics has not led to the wealth of general rules or guidelines for applications that population genetics has. There are a few, but these are certainly not cast in stone. A couple of examples are: (1) the idea that the optimal harvest is obtained by maintaining population sizes at half the carrying capacity, and (2) the notion that optimal harvest in a random environment involves fixed escapement, rather than fixed harvest. There are less specific rules, such as the approach to pest management and control of vector-borne diseases based on using models and data to search for the most vulnerable life history stage. And there have been some less fortunate applications of "general" rules, such as the idea that one need not be concerned about the number of fish killed by power plants and irrigation projects because the populations will always compensate through density-dependent vital rates. However, on the whole the nature of the message that population dynamics brings to practical applications has not been as uniform, nor as general as that of population genetics, perhaps because there is much more variety and/or uncertainty in population dynamic mechanisms than in genetic mechanisms.

There have also been a few instances in which the effects of both population genetics and population dynamics have been investigated concurrently. In some instances these are combined, as for example in the growing interest in evaluating the effects of harvest as a selective force (e.g., Rowell *et al.* 1989) and the use of introduced range legumes for pasture improvement (Marten *et al.* 1989). In other instances, both are included, but they are not closely combined. Examples include evaluation of the evolution of pesticide tolerance in insect pest management (Mangel and Plant 1983; NRC 1986), and evaluation of both genetic and demographic effects in small, endangered populations (e.g., Soulè 1987; Lande 1988; Jain 1991), and recent attention to the genetic selective forces in fishing (e.g., Law and Grey 1989; Rowell *et al.* 1989).

Effectiveness of applications

There is considerable concern that the principles of population biology are not being applied as successfully as they could be (e.g., NRC 1986; Slobodkin 1988). Some practical applications of population biology have long histories (e.g., in fisheries and pest control), while others are the results of the recent (1960s and 1970s) increased public attention to ecological problems (e.g. environmental impact statements, revegetation of derelict lands, conservation biology). However, there is a widely-held perception that even the older of these applications have not been very successful (e.g. in fisheries see May 1984; Royce 1989; in pest management see Kogan 1986), and in some cases success is purposefully not determined (the outcome of recommended solutions is often not monitored [cf. Larkin 1984]). Because of this lack of success, or at least uncertainty regarding predicted results, the utility of population biology is often legitimately questioned. The potential benefits are probably undervalued, and the number of attempted applications and their funding levels thereby reduced.

There are several possible reasons for limited success in the application of the *principles of ecology* to practical problems. One of them would be the principles themselves. Perhaps population biologists are not developing principles that are well suited to the solution of practical problems. Although entertainment of this view may be merely cathartic, it is still considered here. A second possible reason is the amount of variability inherent in biological processes. Perhaps the dynamics of the biological systems that we manipulate and whose behavior we attempt to describe cannot be predicted with any more accuracy, so that we might be doing as well as possible. A third possible reason is that we are taking the wrong approach to practical problems, thereby setting unrealistic expectations. In our approach to practical problems, we tend to adopt a model based on our perception of applied physics and engineering. We assume that we have assembled a number of principles or laws about how populations work, and that we can use them to confidently predict the outcome of manipulation of these populations. This belief is held by a number of population ecologists and is inherent in the widely desired goal of making population ecology "a general, predictive science", so that an application of population biology typically consists of predicting population behavior under various alternatives, then choosing the best option, and would typically end there.

The fourth and final major possible reason for perceived failure is actually a suite of reasons from fields outside of the realm of population biology proper. Many practical problems receive good biological advice, but are not successfully solved because that advice is subverted by political, economic, social and other forces. Examples are many, and most applied population biologists have experienced the frustration of these effects. However, although this problem is ubiquitous and particularly acute in population ecology, the subject is too broad for more than a few words here. Slobodkin (1988) has compared applied ecology to medicine, because the natural systems to be described and manipulated have similar inherent levels of uncertainty and intractability. However, the differences between these fields illustrate the importance of external influences on progress. Few members of the general public are against medical research on heroic procedures to save a limited number of lives, yet most have little heartfelt objection to habitat destruction and extinction of species when they are "necessary for economic progress" (e.g., provide employment, housing, water, power, etc.). Our species is much more easily motivated by (and consequently spends much more money on) immediate threats to the physical health of the individual than it is by long term threats to our physical health through the general deterioration of our environment (e.g., depletion of the ozone layer and skin cancer rates). The expenditure of substantial funds to understand the impact of our increasing population requires a crisis atmosphere with evidence of an immediate threat to our health and survival (e.g., current political attention to, and scientific funding for, global warming).

Investigation of how the principles of population biology have developed and the process of applying them to practical problems, may lead to useful conclusions as to how we might better develop and apply these principles. In this chapter, we first operationally define population biology, then provide a brief historical sketch of the development of the principles of population biology, and finally pose the essential problem of applying these principles to practical problems. In doing so, we realize that in some cases, because they have developed together over a long period, it is often difficult to separate population biology from its applications, while in other cases (e.g., plant or animal breeding, fisheries, and epidemiology) applications have essentially developed independently of mainstream genetics or theoretical ecology

(Simmonds 1979; McIntosh 1985). We then provide a brief guide to the applications described in the ensuing chapters and preview the implications that each of them holds for the central problem of how one should best go about applying the principles of population biology to practical problems.

For our purposes, *population biology* is defined in the broadest sense of current usage; this is usually understood to include biological endeavors in which the population is the primary focus of study. Following the standard definition, a population is taken to be a group of individuals of the same species living together (sharing common space, resources, community of other species) and potentially interbreeding. In terms of ecological levels of organization our definition of population biology may include studies at the individual and the community levels in addition to the population level. In this chapter and the last chapter of this book we will also discuss some results that have been derived from ecosystem level studies. Our working definition of population biology is thus only slightly narrower than the common definition of biological ecology.

Since its beginnings (e.g., Haeckel coining the word "oekologie" in 1866 in a book relating animal morphology to Darwin's theory of natural selection) biological ecology has been related to evolution, although ecological thought did not begin to incorporate genetic variation and natural selection until the 1950s. As such, ecological thought has increasingly involved genetic concepts, both molecular and quantitative, but often both structure and function are studied in terms of phenotypic variation and its adaptive role in evolution. Because the field of genetics has developed somewhat independently of population dynamics, we discuss the development of each of them separately here.

Population dynamics

Throughout its development, population biology has been motivated by a desire to formulate general principles that would unify the various specific descriptions of natural phenomena. In most instances the goal of this formulation was a general, predictive science. However, although ecologists have long been urged to raise biological ecology above the level of a random assortment of unrelated facts of natural history (McIntosh 1980, 1985), dissatisfaction with progress in this direction continues strongly into recent times. To wit Lawton's (1974) statement, "Ecology suffers from a surfeit of fascinating but apparently unrelated observations, superimposed upon an acute shortage of general theories", is but one of several similar recent statements (cf. Slobodkin 1962; Cragg 1966; Watt 1971).

The search for unity and generality has given rise to a number of ways of viewing the natural biological world. The first was the balance of nature concept, the vague notion that populations are somehow designed to maintain a constancy or stability through their inter-relationships with other populations and relationships between individuals within each population (Egerton 1973). This oldest of ecological theories has roots in antiquity. It developed as much on a theological basis as an ecological one, having such attractive explanatory appeal that it was often proposed as evidence for the existence of a divine being (Egerton 1973; McIntosh 1980).

In the late nineteenth and early twentieth century, ecology began to develop as a distinct field and more attention was paid to dynamic rather than static descriptions. This was in part a response to new interpretations of the geological record and the writings of Charles Darwin (McIntosh 1985). The next major ecological theory of some generality arose out of Clements' description of "dynamic ecology", the view that communities progressed through a series of successional states that culminated in an "association" of species that was integrated enough to be characterized as behaving like an individual organism with internal homeostatic controls (Clements' supraorganism in current parlance). Although the utility of this view was challenged on several accounts (e.g., see Egerton 1973; McIntosh 1980, 1985) a vestige of it remains in current theories (Simberloff 1980; Egerton 1983).

Probably the most consequential criticism of this view was Gleason's (1939) contention that the association was more a fortuitous result of certain species being within migrating distance of a location with the right physical environment at the right time (Sheail 1987, p. 62; McIntosh 1980, 1985). His view involved more randomness, fewer inter-connections, and less order. Although our view of biological communities has developed considerably since then, this essential tension between the view that communities have a higher degree of randomness, fewer inter-connections between species and greater variability, and the view that they are deterministic, less variable, highly inter-connected has remained and survives into current debates in community ecology (e.g., Hairston *et al.* 1960; Murdoch 1966; Ehrlich and Birch 1967; Slobodkin *et al.* 1967; Strong *et al.* 1979; Salt *et al.* 1984; Diamond and Case 1988). The issues are currently much more complex, of course, spanning a space defined by variability in several more descriptors (e.g., patterned versus unpatterned, equilibrial versus non-equilibrial, stochastic versus deterministic, etc., see Schoener 1986).

In parallel to this controversy at the community level, a similar one developed at the population level, involving the relative effects of the (random) environment and density-dependence on population numbers. Andrewartha and Birch (1954) were proponents of the view that the abiotic environment was most influential, while others (e.g., Lack 1966) argued the opposite point of view (see the 1957 Cold Spring Harbor Symposia on Quantitative Biology and the summary in the appendix of Lack 1966). The issues have developed far beyond this simple dichotomy, but the essential choice between randomness and deterministic control is still present. Recent developments include the idea of "density vagueness", the proposition that density-dependent relationships contain some inherent, unresolvable variability, hence are not detectable at intermediate population levels, but have a clear effect at lower and higher levels (e.g., Strong 1986a, b, 1987; Lomnicki 1987). Another recent, complicating development in this area is the observation that even the deterministic models can lead to behavior that is virtually indistinguishable from random behavior (May and Oster 1976).

At the same time as these conceptual developments, and not independently, a tradition of the use of mathematical models in population biology was developing (Hutchinson 1978; Kingsland 1985; McIntosh 1985). It began with Verhulst and Pearl's "correction" to Malthus' concept of exponential growth, the logistic equation, near 1840 (Hutchinson 1978). This equation was "rediscovered" in the early 1900s when it was used to analyze behavior of single populations, and used in a number of different combinations reflecting competition and predation (Lotka 1925; Volterra 1926). The approach of assembling systems of simple models of single populations of identical individuals, to represent communities became more popular in the 1960s and continues to be present in recent writings (e.g., Levins 1968; May 1973).

The logistic model was criticized early on, in favor of models that more realistically reflected mechanisms within populations (e.g., Nicholson 1933; Nicholson and Bailey 1935). Other models that more realistically accounted for differences between individuals due to factors such as age, size sex, etc., were also developed (Sharpe and Lotka 1911; M'Kendrick 1927; Leslie 1945) and are currently developing more rapidly (e.g., Nisbet and Gurney 1982; Metz and Diekmann 1987; Ebenman and Persson 1989; Caswell 1989). Various authors have lobbied for and against their greater use in applied population biology (e.g., Ludwig and Walters 1985; Botsford 1981).

Mathematical models have been employed as an attempt to answer some of the above — mentioned general questions regarding community dynamics. Because models of individual populations could easily be unstable, or at least highly variable, or even go extinct easily, modelers sought reasons for their relative constancy or persistence in communities. Various possibilities, such as increasing the number of species (May 1973), increasing trophic interactions between them, and including a number of spatially distinct, but connected populations have been evaluated (see DeAngelis and Waterhouse 1987 for a review). The concept that communities exist in some sort of equilibrium has received considerable attention (e.g., Caswell 1978; Connell and Sousa 1983; O'Neill *et al.* 1986), although this issue is confused by the fact that it is mathematically convenient to assume an equilibrium because stability analysis of the linearized system about that equilibrium is much simpler (cf. Berryman 1988). To avoid that dependence on mathematical convenience, and perhaps approach a biologically more realistic view, new concepts such as boundedness and persistence have replaced simple local stability (e.g., Botkin and Sobel 1974; Chesson 1978; also see Chapter 9, this volume). There has also been more recent attention paid to transient, rather than long term, behavior (DeAngelis and Waterhouse 1987).

Another general class of models, ecosystem models, developed more recently, beginning in the 1960s but having roots in work by Lindeman (1942) (McIntosh 1985, Chapter 6). These were extremely complex, detailed mechanistic models that were in large part based on energy and mass flows (Patten 1975; Odum 1977; O'Neill 1976). The developers of these models constituted a distinct group, separate from the developers of ecological theory described above, with different ideas about mathematical models and ecology (e.g., Levin 1976, 1989; also see O'Neill *et al.*'s [1986] discussion of the "population-community" approach versus the "process-functional" approach).

During the 1950s and 1960s ecologists' approach to field studies became more quantitative, with much greater attention to method. The quantitative approach involved not only an emphasis on more extensive and careful measurement, but also an awareness of the benefits of an experimental/manipulative approach in the field and careful attention to statistical analysis (e.g., Connell 1974; Hurlbert 1984; Underwood 1981). Current topical issues include the validity of assumed replication in field experiments (i.e., the presence of pseudoreplication), and evaluation of different means of manipulating communities (albeit with simple models) (Bender *et al.*, 1984). Diamond (1986) provides a recent summary of relationships between laboratory experiments, field experiments, and natural experiments in the study of community ecology.

Both field biologists and theorists have developed a concern for the formal philosophical basis of the connection between theory and the observed biological world. As in other issues described thus far, considerable controversy has developed, most often based on acceptance or refutation of Popper's (1935) version of the hypothetico-deductive method (Fretwell 1975; Rosensweig 1976; Caswell 1976). According to Popper and other philosophers, science should progress through a process that involves falsifying alternative hypotheses. Popper differs from the others in his specification that an unfalsified hypothesis can be corroborated, but not confirmed. A strictly Popperian approach is not acceptable to many population biologists, in part because of its implications and ideas with which it has been associated (e.g., the importance of testing null hypotheses [Strong 1984; Simberloff 1984]). He is also unpopular with evolutionary ecologists because he proposed a definition of what is and what is not science, which led him to characterize the study of evolution as being metaphysical, rather than scientific (Ruse 1977; Halstead 1980; Little 1980). But many prominent ecologists simply do not accept his strong falsificationist point of view, preferring to conclude more from instances in which hypotheses or theories are not falsified (positive outcomes) and to allow hypotheses of questionable falsifiability. In their criticisms, ecologists have noted (in a sense, correctly) that Popper's views are passé among modern philosophers (e.g., Diamond 1988; May 1981) and that his description fits neither the way that science is done nor the way we establish our beliefs in everyday life (May 1981; Roughgarden 1984). Others have identified real practical problems that arise from trying to follow a strictly Popperian program (e.g., Quinn and Dunham 1984).

In addition to the more comprehensive, general views of population biology, a number of specific theories have also developed in ecology. The theory of the niche has grown out of the competitive exclusion principle (Hutchinson 1957; Jackson 1981) and is the accepted basis for the interpretation of the effects of competition on community structure. Predation theory describes the possible relationships between predator and prey numbers or density (Holling 1959; Ivlev 1961). Island biogeography describes the relative number of species to be found on islands of various sizes (MacArthur and Wilson 1967). Life history theory and its simpler companion, r and K selection, describe how certain schedules of reproduction, mortality, and growth rates might be selected for (Cole 1954; Stearns 1976, 1980). Optimal foraging theory describes the choice of predatory behavior and prey that optimize various criteria (Werner and Mittelbach 1981; Pierce and Olleson 1987; Stearns and Schmid-Hempel 1987; Mangel and Clark 1988).

To attempt to extract the principles of population biology from these various developments and critically analyze how they are applied to practical problems, we directly examine the purpose that motivated development of these principles and the rules that guided that development. The first observation is that in the search for simple, general principles there was a tendency to phrase issues as a choice between two mutually exclusive points of view (i.e., all populations (communities) are either x or y and one submits that they are all y). Several examples were given earlier (i.e., density-dependent versus density independent, stochastic versus deterministic, etc.). There has been little criticism of the crudeness of description inherent in posing questions as a simple choice between two alternatives, let alone the usual leap to generality from a few examples (but see Schoener 1988).

Rules for the development of principles, or mathematical models at least, were proposed in one instance by Levins (1966), who characterized models as having a mix of three characteristics: realism, generality, and precision. Each model could have only two of these three, and different combinations were most useful for various modeling goals. For example, models designed for practical applications were supposed to be precise (exact predictions were needed) and realistic (they needed to mimic the specific system as closely as possible), but were not required to be general (they need apply only to the specific system of interest). A similar view was held by Holling (1964, 1966), who added holism as a fourth characteristic. Levins' views were criticized by several researchers (e.g., May 1974 suggested most models satisfied none of the three characteristics), but nonetheless they form the basis for the idea that any specific model need not be all things to all people. From this derives the notion that simple models, with much biological detail omitted, are justified because they lead more easily to general conclusions than specific, realistic models which are usually very complex (e.g., May

1974). Use of the former type has been termed the strategic approach while the latter has been termed the tactical approach (Holling 1964, 1966).

This relaxation of realism is an accepted mode of analysis in modern population biology, both in theoretical and practical endeavors. Models need not mimic any specific system exactly, but may be simplified to obtain a model that because of its lack of detail, applies to a wider class of populations or species. These simpler models are then often used in the solutions to practical problems. Little attention has been paid to the fact that in this process of "generalizing", the baby (i.e., the essential biological mechanism) may be thrown out with the bath water (i.e., the complicating details). An exception is Levandowsky's (1976) comment regarding the Lotka-Volterra model,

"Few ecologists are interested now in these misleading equations but mathematicians apparently dote on them and are always trying to foist them on us — a classic case of the drunkard who loses his watch in the dark but looks for it under the lamp post because that's where the light is ".

It is paradoxical that it is often the more mathematically capable ecologists, who tend to be the strongest supporters of simpler models which make the mathematics easier (e.g., the logistic or other ordinary differential equations rather than age or size structured models via partial differential equations). There has also been some work on the implications of some kinds of simplification (i.e., the reduction of dimensionality) on results (Schaeffer 1981).

In addition to the criticism of the looseness with which mathematical models describe populations and communities, a similar dissatisfaction has grown with regard to similar looseness in verbal models of communities. The more *mechanistic* models, which are based on the behavior, physiology and morphology of individuals, have led to more realistic description of communities over the last ten years (e.g., Schoener 1987).

Population genetics

Very early in the history of genetics, plant and animal breeders on the one hand, and Darwinian evolutionists on the other, became interested in the genetic variability present in both wild and domesticated species. The first theoretical development, the Hardy-Weinberg rule, very clearly established the basic tenets of variability studies in random mating species. Soon after, quantitative genetic descriptors of variation and various departures from panmixia were recognized. By the early 1930s the classic body of theoretical, and nonetheless biologically rich, work of S. Wright, R. A. Fisher, and J. B. S. Haldane had laid down the foundation of a profound population genetic treatment of evolution. There followed a long period of discoveries of the nature and amounts of genetic variation, various kinds of selective forces, the interplay of selection and random drift, and eventually, a linkup with some ecological concepts of population structure, frequency-dependent selection, host-pathogen coevolution and competitive interactions. Population biologists in the 1960's began to make numerous explicit attempts to develop a juxtaposition of genetic and ecological models and empirical findings. In the meantime, plant and animal breeders also developed management models based on population structure, and used these, in particular, for improvement of economic traits under selection and manipulation of mating system.

In contrast to the early development of population ecology around various polemics, population genetics seemed more harmonious in developing a synthetic theory of evolution and some simple applications in plant and animal breeding. We illustrate the developments of models by following the inclusion of progressively more complex, detailed mechanisms such as mating structure, genetic makeup of an evolving trait, and more general descriptors of selection, migration, etc. along three main themes: (1) overdominance for maintaining stable gene frequency equilibria, (2) variation levels and patterns as assayed with increasingly larger samples of loci or molecular variants, and (3) evolution of multilocus and genome properties.

Early findings on major gene polymorphisms such as melanism in moths, body color in mammals, or pubescence in plants, led to the development of single locus models in which relative fitnesses (constant selective values) could be estimated from observed gene frequency patterns. Fisher (1922) showed overdominance (i.e., heterozygotes more fit than either homozygote) to provide for the maintenance of such polymorphisms. Numerous simply inherited traits began to yield simple theoretical models of evolutionary changes (see Ford 1964 for a review). Theodosius Dobzhansky and his associates used inversion polymorphisms in *Drosophila* for numerous experimental studies to test those single locus models as well as to estimate parameters of migration, deviation from panaxia, and selection (for a review, see Wright 1978). Human geneticists also used numerous models for measuring the relative roles of migration, drift, and inbreeding in conjunction with genetic loads due to mutation (Mourant *et al.* 1976; Wright 1979). Besides overdominance, various models of selection soon entered into these analyses: directional or disruptive selection, gametic versus zygotic selection, frequency-dependency, etc. Parallel treatment of these models for simplified descriptions of quantitative genetic variation provided models on the role of modifiers, gene interaction, and rates of evolutionary change (Weir *et al.* 1989; Karlin and Nevo 1986). Overall, these ideas in the early history of ecological genetics are now seen more as elementary building blocks, to be eventually replaced by more comprehensive developments in dealing with multilocus variation, environmental factors and numerous details of population structure (Ewens 1980; Barton and Clark 1990; Hartl and Clark 1989).

Since the introduction of electrophoretic assays of isozyme variation in 1966, numerous workers have reported on the patterns and levels of variation in attempts to compare groups of organisms, related species living under different environments, and so on. The discovery of as much as 30 to 40 percent polymorphic loci raised new questions about the forces maintaining this variation. Lewontin (1974) provided an elegant summary of such queries and noted that adaptive or ecological questions were not answered here. On the other hand, the so-called non-Darwinian or neutrality models (Kimura 1964) developed, in which essentially mutation-migration-drift balance is the primary focus. Newer molecular assays of nucleotide sequence variation and restriction fragment length polymorphisms have continued to enrich our information base on the genetic variation and phylogenetic analyses (Nei 1987; Dover and Flavell 1982). We have a powerful battery of genetic tools to analyze both the historical and the current forces governing population structure, mating patterns and random drift effects. But now, molecular and morphological evolutionary findings seem to be telling us different *stories*. Ecologists and most biosystematists have taken greater interest in quantitative genetic variation, assuming that many adaptive and morphogenetic-behavioral issues might be better resolved here (Loeschcke 1987; Michod and Levin 1988). From the perspective of applied biology, both molecular and morphological variants have their different methodological advantages. Since very little is known about the kinds and relative magnitudes of selective forces in natural populations, theoretical developments dealing with patterns of variation still appear rather precocious.

Population geneticists were criticized by Mayr (1963) for emphasizing what he called bean-bag genetics dealing with single locus events, and he made a strong plea for dealing with gene interactions and integrative genome-level evolutionary properties. In fact, the potential role of linked gene systems had already been recognized by Fisher, Haldane, Wright, and Mather, among others, in various contexts. During the early 1960's models and computer simulation rapidly expanded the theory of multilocus systems. There have been continued advances along several lines: response to selection involving multiple correlated traits, evolution of breeding systems and sex; the role of random drift in multilocus genetic shifts (e.g., hitchhiking); and the evolution of supergenes. Experimental work has lagged behind due to a lack of appropriate genetic stocks with some very notable exceptions (Clegg 1985). On the other hand, molecular studies have raised new questions on the role of different kinds of mutations, transposable elements, gene conversion and concerted evolution of multigene families in relation to the evolution of proteins and their genomic basis. Phylogeny of cultivated wheat, for example, analyzed earlier through chromosome pairing and C-banding of chromosomes in different ancestral genomes, can now be analyzed using molecular variation. Here again, we see advances in understanding evolutionary relatedness, but not much else from the evolutionary ecologist's viewpoint. To sum up, population genetics theory and experiments have made rapid advances in mapping genes and genomes and considerable progress toward providing potentially useful materials for field studies in the near future. Many controversies and debates essentially come down to the gap between simplifying (but not unnatural) assumptions of theoretical work and the increased awareness of the complex genetic and environmental espects of population-level processes in both natural and agricultural populations.

Principles of population biology — population dynamics and genetics

With these brief historical sketches as background we can now give some idea of what should be included in the "principles" of population biology. We again resort to a broad, somewhat fuzzy definition. We make no attempt to provide a list of principles, but rather suggest a definition, list several potential sources of principles, discuss the generic differences between different kinds of principles, and give a few examples. In fact, a concrete, definitive scheme of principles would be impossible in light of the observation that there is not even agreement among population biologists as to the differences in definition between a theory, a law, a hypothesis, a concept, a rule, a paradigm, and a principle (McIntosh 1980).

Our general notion of a principle here will be a broad, functional one: we seek statements that answer the question, what do we as population biologists know that is of potential use in many practical applications. This would be a statement of some generality that has not been shown to be untrue, and is of some utility. This is probably closest to what most ecologists mean by the term "law". As applied population biologists we would like to have a set of these that could be applied with confidence to the solution of practical problems.

There are several potential sources of principles in the ecological literature. Some ecologists have assembled lists of the principles of ecology (e.g., Odum 1971; Watt 1973), hence these could be considered. Because, as discussed above, one of the avowed purposes of theory is to develop general laws or principles, another prime source would be the theory of ecology (e.g., May 1980; Roughgarden *et al.* 1989). Some idea of the principles of ecology can be gleaned from the first eight chapters in a recent book that presents the state of ecological knowledge and case studies of how it is applied (NRC 1986). There are also several recent descriptions of how the principles of ecology are applied to specific fields. An example is the book on the use of ecology in environmental assessment (Westman 1985) which attempts to show how the principles of ecology are useful. Another is the book by Kogan (1986) on integrated pest management, which includes several initial chapters on theory. Sainsbury (1988) provides a useful summary of how various views of community ecology might be used in fishery management, and likewise, Getz and Guiterrez (1982) relate various branches of ecological theory to applications in pest management.

The various principles of population biology can be usefully categorized into one of several types. There are generalizations that arise empirically such as: (1) Bergmann's rule, (2) Lindemann's 10 percent rule, and (3) the

3/2 thinning rule. There are also generalizations that arise from models, such as: (1) biological pest control requires an equilibrium at a low pest level, or (2) fish populations must be harvested at one-half the carrying capacity. General observations will tend to have "averaged out" the specific details, and general models have the details removed *a priori*. Various principles have resulted from the central core of theoretical community and population ecology (see Kareiva 1989 for a list of recent ones), as well as the specific theories mentioned above (e.g., (1) from life history theory, species with higher fecundities will be found in more variable environments; (2) from island biogeography, larger areas will contain more species).

In addition to these classes of principles, there are other kinds which we will not include in our definition of the principles of population biology. The first is the principles involved in the techniques and methods that ecologists use (e.g., principles used in modeling populations). The second is the principles involved in organizing knowledge about the natural history of organisms. Although these are of great importance and are certainly valuable tools for population biologists, they are actually laws about methods, not populations.

When we focus more restrictively on genetics, rather than population biology in general, there appears to be, for some reason, greater agreement on principles (e.g., Hartl and Clark 1989). The Hardy-Weinberg principle gives equilibrium gene and genotypic frequency expectations in a panmictic population. Its corollaries allow us to predict the behavior of various deviants from an idealized population (e.g., sex linkage, multiple alleles, inbreeding) and some simple evolutionary processes. This principle has served population genetics well in establishing a solid link with Mendelian genetics on the one hand, and with Darwinism on the other.

There are also several principles about the behavior of genetic variance in small populations, that have been established empirically. For example the rates of loss of genetic variation at neutral loci due to random drift in relation to the substructure of a meta-population depend on migration models, historicity, and mutations. The so-called Haldane-Muller principle states that the effect of mutation on population fitness depends on the mutation rate (and not on selective value of the mutant). Likewise, the rate of gene substitution (evolutionary change, in classical terms) is independent of population size and depends only on the mutation rate. Accordingly, Lynch and Hill (1986) have recently shown that the neutral rate of phenotypic divergence depends on the rate of mutations affecting the phenotype. Theories of the evolution of sex and of various molecular features of genetic diversity rely on several such ideas which might be thought of as principles.

We also have several general principles regarding the outcome of natural selection. Fisher's Fundamental Theorem, in spite of many restrictions, states that the rate of change in population fitness under viability selection models is proportional to the additive genetic variance. In other words, an allele will increase in frequency provided its marginal fitness exceeds mean population fitness. This theorem, which is similar to the Hardy-Weinberg principle, generates useful ideas in our attempts to find the consequences of its limitations in the real world. The most general model of response to artificial selection directly follows from this theorem, as do the models of life history evolution in their most general form. A principle of interdeme selection shows that selection between demes overrides selection within demes only when the benefit to the group, relative to the cost per individual, is greater than twice the average number of migrants among demes per generation (Barton and Clark 1990). Likewise, the so-called Hamilton rule states that a gene for altruism toward a related individual will be favored if the ratio of the loss in fitness of the altruist to the gain in fitness of the relative is less than the coefficient of relationship. On sex ratio evolution, Fisher's principle states that sex ratio at reproductive age evolves to the point at which parental expenditure is the same for both sexes of the offspring generation. An interesting result reported by Lewontin (1974) on the behavior of multi-locus systems under selection (not yet christened as a named principle) shows how interactive linked gene systems evolve (e.g., supergenes, coadaptation, genome organization, hitchhiking effects) in terms of associative gene effects.

These ideas and others from genetic theory have proven useful both empirically and analytically in teaching and understanding evolution, biosystematics, and various applications in agriculture. However, we might explore further the sense in which they form general principles. First, they appear to reflect the fundamental nature of genetic systems, in that though the theory may evolve further in the future, it is doubtful that these principles will be replaced, but rather likely that they will remain in the general structure in altered form. Second, they appear to have a record of proven usefulness in the field, for example in predicting patterns of genetic variation. Examining examples, we often find that colonizing species that have recently expanded their range show lower levels of genetic variation which can be explained by recent population size bottlenecks. We also find that many endangered and threatened species have lost a significant fraction of their genetic variation. However, such outcomes vary widely among different species and often we cannot explain them by a simple model. Clearly, a great deal of uncertainty arises from many unknown and uncontrolled variables not treated in such models. Likewise, response to artificial selection under a wide variety of breeding schemes in both plants and animals is indeed proportional to the amount of variation. However, there is wide variation in the actual pattern of response (progress under breeding), so that various refined breeding methods remain largely trial-and-error projects.

These characteristics of providing good *post hoc* explanations, with limits on *a priori* quantitative predictions also characterize proposed principles of population ecology. We might also ask therefore, whether there is any fundamental difference between the principles of population dynamics and the principles of genetics. In population genetics Mendelian rules and mating systems provide a solid deductive basis for the formulation of laws regarding inheritance (transmission genetics) and for statistical treatment of variation.

Most principles follow from these robust "givens", but we soon run into uncertainty regarding definition of spatio-temporal units, the exact form or rate of mutation, mechanistic details of segregation and genetic recombination (most now being challenged by molecular genetic findings), and modes of selection as well as the genotype-phenotype relationships. The same kinds of problems arise, possibly even more readily in population ecology. The only clearly identifiable difference is that mechanisms appear to be fundamentally more general in genetics. The study of genetics of birds probably tells one more about certain comparable genetic features of lizards, than the study of the ecology of birds tells one about the ecology of lizards.

Applications of population biology

The principles of population biology are used to varying degrees in applied fields of population biology: fisheries and wildlife management, forestry, biological control of agricultural pests, weeds, and disease vectors, epidemiology, conservation, land reclamation, design of reserves, to name a few. The following chapters are examples of some of these applications, presented for the purposes of examining the process of application of general principles, as much as to present the subject matter itself (Table 1).

In Chapter 2 on land restoration, Bradshaw describes recent progress in what was a unique challenge for population biology: a chance to see whether we understood plant communities well enough to cause them to develop and persist in a "foreign" environment: heavy metal mine tailings and other disturbed lands. In Chapter 3, Hedrick describes how the principles of genetics are used in a highly controlled application with much good information: captive breeding and zoos. As the number of extant species in the wild and the amount of natural habitat declines, this field becomes more critical. In Chapter 4, Kesseli further develops the role of genetic variation and population substructure in planning conservation strategies for rare plants. In Chapter 5, Barrett reviews the population biology of weeds, primarily in terms of genetic variation and changes in breeding system under colonization episodes. Both rare and colonizing plants present similar problems of population regulation to an applied population biologist. In Chapter 6, Jain reviews numerous applications of population genetics and gene pool management in the breeding of agricultural plants and in genetic resource conservation. Natural and agricultural populations clearly show many evolutionary processes in common; the role of human intervention changes only the rates and fitness criteria. In Chapters 7 and 8, Dobson and Hudson, and Smith, respectively, describe applications of modeling in parasitology, a field that had been estranged from mainstream population biology, but in which there has been much recent modeling activity (e.g., Anderson and May 1978). In Chapter 9 Murdoch describes his ongoing evaluation of existing theory (and "general" conclusions drawn therefrom) in solution to a problem in pest

Table 1. List of applications covered in the Chapters 2—11.

Chapter	Author	Area	Specific problem	Methods, concepts, principles	Conclusions, solutions	Methodological implications*
2	A. D. Bradshaw	Land restoration	Restoration ecology of mine tailings, derelict lands, etc.	Soil analysis, plant requirements (nutrients, tolerance), densities for self-sustaining populations, choice of genotypes	Successful solutions are problem-specific	Highly empirical, trial and error approach. Physiological facts as much as ecological principles of adaptation
3	P. Hedrick	Biological conservation	Genetic management of endangered animal species	Effective population size, breeding records; inbreeding effects	Genetic variation monitoring provides guidance to managers	Theory robust but more specific models and empirical details are needed.
4	R. Kesseli	Biological conservation	Design of reserves for *Limnanthes* spp.	Genetic surveys, effective population size	Maintain variability. Minimize losses due to drift	Genetic principles lead to some design guidelines, not predictable outcomes
5	S. Barret	Agriculture	Weed control in rice fields	Weak points in life history. Genetic explanations (mimicry, bottleneck). Breeding systems	Asexual weeds are more controllable. Bottlenecks in pas as well as current ecology limit evolutionary change	Genetic principles provide explanation, not prediction; new research focus is developed
6	S. Jain	Agriculture	Improvement of crop plants	New mating designs, selections methods, and synthesis of variable gene pools	Selection and ecological genetics led to new population management methods	Long-term and new experiments are needed on response to selection

Table 1. (Continued)

Chapter	Author	Area	Specific problem	Methods, concepts, principles	Conclusions, solutions	Methodological implications*
7	A. Dobson and P. Hudsen	Wildlife	Cause of cyclic variability in harvest	Models of populations, host-parasite interaction	Parasite may be cause of cycles	Successful use of models and data on individuals
8	G. Smith	Agriculture	Parasites in domestic animals	Population models including host parasite interactions	Treatment regimes evaluated through models	Models and data on individuals useful in projecting success
9	W. Murdoch	Agriculture	Pest control	Realistic host parasitoid model, stability analysis	Control possible, conclusions using simple models not correct	More realistic data-based models needed
10	G. Robinson and J. Quinn	Conservation biology	Fragmentation	Island biogeography (questioned). More reliable models	Earlier models inadequate, more realism needed	Need more realistic theory
11	C. Waters	Fisheries	Harvest management optimization	Models of population dynamics needed to understand dynamics	Experimental management for information	Highly empirical and iterative approach in improved applications

* These refer to the need for new concepts or experimental methods for answering the general questions raised in each of the chapter topics.

control. In Chapter 10, Robinson and Quinn also question recent theories regarding reserve size, based on MacArthur and Wilson's (1967) theory of island biogeography. Walters' area of application in Chapter 11 is fisheries, whose theory has developed somewhat independently of the theory of population biology, but his message about how to treat management decisions as experiments, is of much more general applicability, and is in fact central to the theme of this book.

Literature cited

Anderson, R. M. and R. May. 1978. Regulation and stability of host-parasite population interactions. I. Regulatory processes. J. Animal Evol. 47: 219—249.

Andrewartha, H. G. and L. C. Birch. 1954. The Distribution and Abundance of Animals. University of Chicago Press, Chicago.

Barton, N. and A. Clark. 1990. Population structure and process in evolution. Pp. 115—198 in K. Wohrmann and S. Jain (eds.), Population Biology: Ecological and Evolutionary Viewpoints. Springer-Verlag, Berlin.

Bender, E. A., T. J. Case, and M. E. Gilpin. 1984. Perturbation experiments in community ecology: theory and practice. Ecology 65: 1—13.

Berryman, A. A. 1988. Equilibrium or nonequilibrium: is that the question? Bull. Ecol Soc. XX: 500—502.

Botkin, D. B. and M. J. Sobel. 1975. Stability in time-varying ecosystems. American Naturalist 109: 625—646.

Botsford, L. W. 1981. More realistic fishery models: cycles, collapse, and optimal policy. Pp. 6—20 in T. Vincent and J Skowronski (eds.), Renewable Resource Management, Springer-Verlag, New York.

Bradshaw, A. D. 1987. The reclamation of derelict land and the ecology of ecosystems. Pp. 53—74 in W. R. Jordan, M. E. Gilpin and J. D. Aber (eds.), Restoration Ecology, University Press, Cambridge.

Bradshaw, A. D. and T. McNeilly. 1982. Evolution and Pollution. Arnold, London.

Burdon, J. J. 1987. Diseases and Plant Population Biology. Cambridge U. Press, Cambridge.

Burdon, R. D. 1988. Recruitment for breeding populations: Objectives, genetics and implementation. Pp. 555—572 in B. S. Weir, E. J. Eisen, M. M. Goodman and G. Namkoong (eds.) Quantitative Genetics, Sinauer, Sunderland, Ma.

Caswell, H. 1976. The validation problem. Pp. 313—325 in B. C. Patten (Ed.), Systems Analysis and Simulation in Ecology, Vol. IV. Academic Press, Inc. New York.

Caswell, H. 1989. Matrix Population Models. Sinauer, Sunderland.

Chapman, C. G. D. 1989. Collection strategies for the wild relatives of field crops. Pp. 263—279 in A. H. D. Brown, O. H. Frankel, D. R. Marshall, and J. T. Williams (eds.), The Use of Plant Genetic Resources. Cambridge U. Press, Cambridge.

Cheliak, W. M. 1985. Mating system dynamics in a seed orchard. Pp 107—117 in H. R. Gregorius (ed.) Population Genetics in Forestry, Springer-Verlag, Berlin.

Chepko-Sade, B. D. and Z. T. Halpin (eds.) 1987. Mammalian Dispersal Patterns. Univ. Chicago Press, Chicago.

Chesson, P. L. 1978. Predator-prey theory and variability. Annual Reviews of Ecology and Systematics 9: 323—347.

Clegg, M. T. 1985. Multilocus Systems: a Review. Oxford Surveys in Evolutionary Biology.

Cold Spring Harbor Symposium in Quantitative Biology. 1957. Vol 22.

Cole, L. C. 1954. The population consequences of life history phenomena. Quarterly Review of Biology 29: 103—37.

Connell, J. H. 1975. Some mechanisms producing structure in natural communities: A model and evidence from field experiments. In M. L. Cody and J. M. Diamond (eds.), Ecology and Evolution of Communities. Betknap, Cambridge.
Connell, J. H. and W. P. Sousa. 1983. On the evidence needed to judge ecological stability or persistence. Amer. Nat. 121: 789—824.
Cragg, J. B. 1966 Preface. Advances in Ecological Research 3: vii.
DeAngelis, D. L. and J. C. Waterhouse. 1987. Equilibrium and nonequilibrium concepts in ecological models. Ecological Monographs 57: 1—21.
Denno, R. G. and M. S. McClure (eds.) 1983. Variable Plants and Herbivores in Natural and Managed Systems. Academic Press, New York.
Diamond, J. 1986. Overview: Laboratory experiments, field experiments, and natural experiments. Pp 3—22 in J. Diamond and T. J. Case (eds.), Community Ecology, Harper and Row, New York.
Dover, G. A. and R. B. Flavell (eds.). 1982. Genome Evolution. Academic Press, London.
Ebenman, B. and L. Persson (eds.) 1988. Size-Structured Populations: Ecology and Evolution. Springer-Verlag, Berlin.
Egerton, F. N. 1973. Changing concepts of the balance of nature. Quart. Rev. Biol. 48: 322—350.
Ehrlich, P. R. and L. C. Birch. 1967. The "balance of nature" and "population control". American Naturalist 101: 97—107.
Ewens, W. J. 1980. Mathematical Population Genetics, Springer-Verlag, Berlin.
Fretwell, S. D. 1975. The impact of Robert MacArthur on ecology. Annual Reviews of Ecology and Systematics 6: 1—13.
Frey, K. J. (ed.) 1981. Plant Breeding II. Iowa State Univ. Press, Ames IA.
Getz, W. M. and A. P. Guiterrez. 1982. A perspective on systems analysis in crop production and insect pest management. Ann. Rev. Entom. 27: 447—466.
Gleason, H. A. 1939. The individualistic concept of the plant community. American Midland Naturalist 21: 92—110.
Hartl, D. L. and A. G. Clark. 1988. Principles of Population Genetics. Sinauer, Sunderland.
Hairston, N. G., F. E. Smith, and L. B. Slobodkin. 1960. Community structure, population control, and competition. American Naturalist 94: 421—425.
Halstead, B. 1980. Popper: good philosophy, bad science? New Scientist 4: 215—217.
Hill, W. G. 1986. Population size and design of breeding programmes. Proc. 3rd World Congr. Genet. Appl. Livestock Prod. 12: 245—256.
Holling, C. S. 1959. Some characteristics of simple types of predation and parasitism. Can. Ent. 91: 385—398.
Holling, C. S. 1964. The analysis of complex population processes. Can. Ent. 96: 335—347.
Hurlbert, S. H. 1984. Pseudoreplication and the design of ecological field experiments. Ecological Monographs 54: 187—211.
Hutchinson, G. E. 1957. Concluding remarks. Cold Spring Harbor Symp. Quant. Biol. 22: 415—427.
Hutchinson, G. E. 1978. An Introduction to Population Ecology, Yale University Press, New Haven.
Ivlev, V. S. 1961. Experimental Ecology of the Feeding of Fishes. Yale University Press, New Haven.
Jana, S. and S. N. Acharya. 1981. The value of mass reservoirs in genetic conservation. Pp. 51—57 in Barley Genetics IV, Edinburgh.
Jayakar, S. D. and L. A. Zonta. 1990. Coevolution at two trophic levels. Pp. 349—366 in K. Wohrmann and S. Jain (eds.), Population Biology: Ecological and Evolutionary Viewpoints. Springer-Verlag, Berlin.
Jordan, W. R., M. E. Gilpin, and J. D. Aber (eds.) 1987. Restoration Ecology. Cambridge U. Press.
Kareiva, P. 1989. Renewing the dialogue between theory and experiments in population

ecology. Pp. 68—88 in J. Roughgarden, R. M. May and S. A. Levin (eds.), Perspectives in Ecological Theory. Princeton Univ. Press, Princeton.
Karlin, S. and E. Nevo (eds.). 1976. Evolutionary Processes and Theory. Academic Press, New York.
Kimura, M. 1964. Diffusion Models in Population Genetics. Methuen, London.
Kingsland, S. E. 1985. Modeling Nature. Univ. of Chicago Press, Chicago.
Kogan, M. (ed.) 1986. Ecological Theory and Integrated Pest Management. Wiley Inter-Interscience, New York.
Lack, D. 1954. The Natural Regulation of Animal Numbers. Oxford: Clarendon Press, London.
Lack, D. 1966. Population Studies in Birds. Oxford: Clarendon Press, London.
Lande, R. 1988. Genetics and demography in biological conservation. Science 241: 1455—1460.
Larkin, P. A. 1984. A commentary on environmental impact assessment for large projects affecting lakes and streams. Can. J. Fish. Aquat. Sci. 41: 1121—1127.
Law, R. and D. R. Grey. 1989. Evolution of yield from populations with age-specific cropping. Evol. Ecol. 3: 343—359.
Lawton, J. H. 1974. Review of J. M. Smith's Models in Ecology. Nature 248: 537.
Leslie, P. H. 1945. On the use of matrices in certain population mathematics. Biometrika 33: 183—212.
Levandowsky , M. 1976. The cats in Zanzibar. Quart. Rev. Biol. 51: 417—419.
Levin, S. A. (ed.) 1976. Ecological Theory and Ecosystem Models. The Institute of Ecology, Indianapolis.
Levins, R. 1966. The strategy of model building in population biology. Amer. Sci. 54: 421—431.
Levins, R. 1968. Evolution in Changing Environments. Princeton Univ. Press, Princeton.
Lewontin, R. C. 1974. The Genetic Basis of Evolutionary Change. Columbia University press, New York, NY.
Libby, W. J. 1973. Domestication strategies in forest trees. Canad. J. For. Res. 3: 265—276.
Lindeman, R. L. 1942. The trophic dynamic aspect of ecology. Ecology 23: 399—418.
Little, J. 1980. Evolution: myth, metaphysics, or science? New Scientist 4: 708—709.
Loeschcke, V. (ed.). 1987. Genetic Constraints on Adaptive Evolution. Springer-Verlag, Berlin.
Lomnicki, A. 1987. Density-vague ecology. Trends Ecol. Evol. 2: 76.
Lotka, A. J. 1925. Elements of Physical Biology. Williams and Wilkins, Baltimore.
Ludwig, D. and C. J. Walters. 1985. Are age-structured models appropriate for catch-effort data? Can. J. Fish. Aquat. Sci. 42: 1066—1072.
Lynch, M. and W. G. Hill. 1986. Phenotypic evolution by neutral mutation. Evolution 40: 915—935.
MacArthur, R. H. and E. O. Wilson. 1967. Theory of Island Biogeography. Princeton Univ. Press, Princeton.
Mangel, M. and C. W. Clark. 1988. Dynamic Modeling in Behavioral Ecology. Princeton University Press, Princeton.
Mangel, M. and R. E. Plant. 1983. Multiseasonal management of an agricultural pest: Development of the theory. Ecol. Modelling 20: 1—19.
Marten, G. C., A. G. Matches, R. F. Barnes, R. W. Brougham, R. J. Clements, and G. W. Sheath (eds.) 1989. Persistence of Forage Legumes. Amer. Soc. Agron., Madison.
May, R. M. 1973. Stability and Complexity in Model Ecosystems. Princeton: Princeton University Press.
May, R. M. (ed.) 1984. Exploitation of Marine Communities. Springer-Verlag, New York.
May, R. M. and G. F. Oster. 1976. Bifurcations and dynamic complexity in simple ecological models. American Naturalist 110: 573—599.
Mayr, E. 1963. Animal Species and Evolution. Harvard, Cambridge.
McIntosh, R. P. 1980. The background and some current problems of theoretical ecology. Synthese 43: 195—255.

McIntosh, R. P. 1985. The Background of Ecology: Concept and Theory. Cambridge University Press, Cambridge.
McIvor, J. G. and R. A. Bray (eds.) 1983. Genetic Resources of Forage Plants. C.S.I.R.O., Melbourne.
Metz, J. A. J. and O. Diekmann (eds.) 1986. The Dynamics of Physiologically Structured Populations. Springer, Berlin, Springer Lecture in Biomathematics, vol. 68.
Michod, R. E. and B. R. Levin (eds.). 1988. The Evolution of Sex: an Examination of Current Ideas. Sinauer Press, Sunderland, MA.
M'Kendrick, A. G. 1926. Applications of mathematics to medical problems. Proc. Edinburgh Math. Soc. 44: 98—130.
Mourant, A. E., A. C. Kopec, and K. Domainewska-Sobezak. 1976. The Distribution of the Human Blood Groups and Other Polymorphisms. Oxford Univ. Press.
Murdoch, W. W. 1966. Community structure, population control, and competition-a critique. American Naturalist 100: 219—226.
Namkoong, G. 1979. Introduction to Quantitative Genetics in forestry. U.S.D.A. Forest Service Tech. Bull. 1588.
National Research Council. 1986. Ecological Knowledge and Environmental Problem Solving: Concepts and Case Studies. National Academy Press, Washington, D.C.
Nei, M. 1987. Molecular Evolutionary Genetics. Columbia University Press, New York.
Nicholson, A. J. 1933. The balance of animal populations. J. Animal Ecol. 2: 132—178.
Nicholson, A. J. and V. A. Bailey. 1935. The balance of animal populations. Proceedings of the Zoological Society of London 3: 551—598.
Nisbet, R. M. and W. S. C. Gurney. 1982. Modelling Fluctuating Populations. John Wiley and Sons, New York.
Odum, E. P. 1977. The emergence of ecology as a new integrative discipline. Science 195: 1285—1293.
Odum, H. T. 1971. Environment, Power, and Society. Wiley-Interscience, New York.
O'Neill, R. V., D. L. DeAngelis, J. B. Waide, T. F. H. Allen. 1986. A Heirarchical Concept of Ecosystems. Princeton University Press, Princeton.
Patten, B. C. 1975. Ecosystem modeling in the U.S. International Biological Program. Pp 1—6 in B. C. Patten (ed.) Systems Analysis and Simulation in Ecology, Vol;. 3. Academic Press, New York.
Pierce, G. J. and J. G. Ollason. 1987. Eight reasons why optimal foraging theory is a complete waste of time. Oikos 49: 111—118.
Popper, K. 1935. The Logic of Scientific Discovery. [A translation of Logik der Forschung, 1935] Harper and Row 1959).
Quinn, J. F. and A. E. Dunham. On hypothesis testing in ecology and evolution. Pp. 22—37 in G. W. Salt (ed.) Ecology and Evolutionary Biology: a Round Table on Research. Univ. of Chicago Press, Chicago.
Robertson, A. (ed.). 1980. Pp. 56—68 in Selection Experiments in Laboratory and Domestic Animals. Commonwealth Agric. Bull., Slough, U.K.
Rosensweig, M. L. 1976. Review of: Golley, F. B., K. Petrusekwicz and L. Ryskowski (ed.), Small Mammals. Science 192: 778—779.
Roughgarden, J. 1984. Competition and theory in community ecology. Pp. 3—21 in G. W. Salt (ed.) Ecology and Evolutionary Biology: A Round Table on Research. Univ. of Chicago Press, Chicago.
Rowell, C., K. Stokes and R. Law. 1989. Does fishing generation selection differentials? J. Fish. Biol. 35A; 335—337.
Royce, W. F. 1989. A history of marine fishery management. Rev. Aquatic Sci. 1: 27—44.
Sainsbury, K. J. 1988. The ecological basis of multispecies fisheries, and management of a demorsal fishery in tropical Australia. In J. A. Gulland (ed.), Fish Population Dynamics, Wiley, New York.
Salt, G. W. (ed.) 1984. Ecology and evolutionary biology: A Round Table on Research. U. of Chicago Press, Chicago.

Schaeffer, W. E. 1981. Ecological abstraction: the consequence of reduced dimensionality in ecological models. Ecol. Mono. 51: 383—401.
Schoener, T. W. 1985. Axes of controversy in community ecology. Pp. 8—16 in W. J. Matthews and D. C. Heins, Community and Evolutionary Ecology of North American Stream Fishes. Univ. of Oklahoma Press, Norman.
Schoener, T. W. 1986. Mechanistic approaches to community ecology: A new reductionism. Amer. Zool. 26: 81—106.
Schoener, T. 1988. Overview: Kinds of ecological communities-ecology becomes pluralistic. Pp. 467—479 in J. Diamond and T. J. Case (eds.), Community Ecology. Harper and Row, New York.
Sharpe, F. R. and A. J. Lotka. 1911. A problem in age distribution. Philosophical Magazine 21: 455.
Sheail, J. 1987. Seventy-five Years in Ecology: The British Ecological Society. Blackwell Scientific Publications, London.
Simberloff, D. 1980. A succession of paradigms in ecology: essentialism to materialism and probabalism. Pp. 63—99 in E. Saarinen (ed.) Conceptual Issues in Ecology. D. Reidel Publ. Co., Boston.
Simberloff, D. 1984. Competition theory, hypothesis testing, and other community ecological buzzwords. Pp. 46—55 in G. W. Salt (ed.), Ecology and Evolutionary Biology: A Round Table on Research. U. of Chicago Press, Chicago.
Simmonds, N. W. 1979. Principles of Crop Improvement. Longman, London.
Slobodkin, L. B. 1962. Growth and Regulation of Animal Populations. Holt, Rinehart and Winston, New York.
Slobodkin, L. B. 1988. Intellectual problems of applied ecology. Bioscience 38: 337—342.
Slobodkin, L. B., F. E. Smith, and N. G. Hairston. 1967. Regulation in terrestrial ecosystems, and the implied balance of nature. American Naturalist 101: 109—124.
Soulè, M. E. (ed.) 1987. Viable Populations. Cambridge U. Press, Cambridge.
Stearns, S. C. 1976. Life history tactics: a review of the ideas. Q. Rev. Biol. 51: 3—47.
Stearns, S. C. 1980. A new view of life-history evolution. Oikos 35: 266—281.
Stearns, S. C. and P. Schmid-Hempel. 1987. Evolutionary insights should not be wasted. Oikos 49: 118—125.
Strong, D. R. 1984. Natural variability and the manifold mechanisms of ecological communities. Pp. 56—80 in G. W. Salt (ed.) Ecology and evolutionary biology: A Round Table on Research. U. of Chicago Press, Chicago.
Strong, D. R. 1986. Density-vague population change. Trends Ecol. Evol. 1: 39—42.
Underwood, A. 1981. Techniques of analysis of variance in experimental marine biology and ecology. Ann. Rev. Oceanography and Marine Biol. 19: 513—605.
Volterra, V. 1926. Variazioni e fluttuazioni del numero d'individui in specie animali conviventi, Mem. Acad. Lincei. 2: 31—113.
Watt, K. E. F. 1971. Dynamics of populations: A synthesis. Pp. 568—580 in P. J. den Boer and G. R. Gradwell (eds.) Dynamics of Populations, Centre for Agricultural Publication and documentation, Wageningen.
Watt, K. E. F. 1973. Principles of Environmental Science. McGraw-Hill Inc.
Weir, B. S., E. J. Eisen, M. M. Goodman, and G. Namkoong (eds.). 1988. Proc. Second Intern. Conf. Quant. Genetics, Sinauer, Sunderland.
Werner, E. E. and G. G. Mittlebach. 1981. Optimal foraging: field tests of diet choice and habitat switching. Amer Zool. 21: 813—829.
Westman, W. E. 1985. Ecology, Impact Assessment and Environmental Planning. John Wiley and Sons, New York.
Wricke, G. and W. E. Weber. 1986. Quantitative Genetics and Selection in Plant Breeding. de Gruyter, Berlin.
Wright, S. 1963. Plant and animal improvement in the presence of multiple selection peaks. In W. D. Hanson and H. F. Robinson (eds.) Statistical Genetics and Plant Breeding. NAS-NRC Publ. 982, Washington, D.C.
Wright, S. 1969. Evolution and the Genetics of Populations. Volume 2. Univ. of Chicago, Chicago.

2. The biology of land restoration

A. D. BRADSHAW

Abstract

Disturbance of land is an inevitable consequence of modern civilization which has resulted in a substantial "heritage" of degraded land on which the original ecosystems, plants, animals and soils have been totally destroyed. All that is left is skeletal soil material. The restoration of functional, self-sustaining ecosystems on these materials is a considerable challenge.

The problems of plant habitat degradation that have to be overcome can be separated into three sets: physical, nutritional, and toxicity. Although some of the treatments involve major engineering-oriented manipulations, long term success depends on understanding, copying, and harnessing the ecological and microevolutionary processes that occur in natural succession. This success depends on overcoming all the specific problems that occur within a site, which in turn depends upon a proper appreciation of all that is critical to ecosystem function as well as the demographic and genetic processes of adaptation. Scientific progress in the restoration of derelict lands requires an understanding of successional sequence of colonization, species' autecology, and the dynamics of recruitment, and it is based on the population variability for metal tolerance or similar adaptive genetic criteria; therefore, both community and ecosystem level interactions among species must be investigated. Land restoration is therefore an acid test of our ecological understanding of adaptive processes at all three levels (communities, species, populations) in an integrative applied science.

The inevitability of damage

A large proportion of the material we use to provide ourselves with the comforts of civilized living come out of the ground. It has been so since the early Stone Age; however, now the demand is not just for building materials, but also for fuel, for metals of all sorts, for chemicals and a host of other raw materials. We have learned how to exploit diffuse resources, even though this may mean rejecting 99% of what we excavate, and depositing it on the surrounding land. Whereas surface mining was once limited to how far a wheelbarrow could be taken, and waste heaps were only a few metres across in size, it is now commonplace to strip away covering materials 30 or even 50 m deep, and then excavate down another 100 m, yielding modern waste heaps measured in square kilometers. All this is to satisfy demands that do

S. K. Jain and L. W. Botsford (eds), Applied Population Biology, 25–44.

not stop increasing until the resource is exhausted, and then only move on to some substitute.

It is almost axiomatic that disturbance of land by mining is an inevitable and escalating part of civilization. The extent, for the United States, over the period from 1930 to 1971 is given in Table 1 (Paone *et al.* 1974). But there are other causes of disturbance such as housing, industry, railways, and waste disposal. Many of these are permanent features of the landscape, but under the present changing economic conditions, obsolescence is inevitable and as a result, further disturbed land is produced. In England, careful surveys (Department of the Environment 1984) have shown that the derelict land produced by industrial obsolescence is much more than that due to mining.

Land is a non-renewable resource; any loss of the total capital is therefore a serious loss. At the same time the disturbed land can create important environmental problems, not just as an eyesore but also as a source of pollution and related problems. Since human populations are inevitably associated with the disturbance — in many cases whole towns have developed beside, and because of, the disturbance — the environmental problems are particularly serious (e.g. Fairbrother 1970; Caudhill 1976).

As a result, efforts have been made to restore land after disturbance. How little these were successful up to 1971 in the USA is apparent from Table 1 but now planning and regulatory conditions are tighter. In England, the Government has radically improved the environmental legislation and made strenuous efforts through a system of grants to restore land already derelict. The tide has been stemmed, but the total area of officially derelict land has still increased during the period 1974—82 albeit an area equal to 40% of what existed in 1974 has been reclaimed (Department of the Environment 1984). At least the Red Queen and Alice (Carroll, 1876) managed to run fast enough to stay in the same place.

There is, therefore, in industrialized countries particularly, a serious problem owing to land disturbance. As industrialization spreads and mineral resources are exploited all over the world wherever they are found, there is, however, no area of the world where the problem does not occur. At the same time land can be substantially disturbed and even destroyed by other non-industrial activities such as forestry and agriculture.

The nature of the damage

So far, the somewhat noncommittal word "disturbed" has been used. Perhaps this is somewhat misleading, because it implies that things have been pushed out of place but that nothing has been lost. The truth is of course very different. There has usually been total destruction of the previously existing ecosystem — soil, plants and animals.

The exception is when the surface layers of the soil (topsoil) have been removed, stored and then replaced. This may be a continuous process, as in

Table 1. Land disturbed and reclaimed by mining in the United States, 1920—71 (from Paone *et al.* 1974).

Type of use	Area (ha)							
	Metals		Non-metals		Fossil fuel		Total	
	Disturbed	Reclaimed	Disturbed	Reclaimed	Disturbed	Reclaimed	Disturbed	Reclaimed
Surface mining	58,700	7,040	429,000	102,000	391,000	290,000	879,000	399,000
Overburden and refuse from surface mining	49,800	2,130	118,000	52,000	129,000	108,000	297,000	162,000
Subsidence and disturbance from underground mining	4,940	720	1,850	40	35,600	1,620	42,400	2,400
Refuse from underground mining	8,860	610	840	70	67,200	8,090	76,900	8,770
Wastes from milling and processing	89,400	7,000	81,300	9,430	12,900	2,620	184,000	19,000
Total	212,000	17,500	631,000	164,000	635,000	410,000	1,480,000	591,000

modern strip mining. Because the soil is not lost, final restoration is considerably easier. As a result this is now required by much of the modern legislation, such as the 1977 Federal Surface Mining Act for coal in the USA, and the Town and Country Planning Act (Minerals) 1984 in Britain. However, there are still many situations where the replacement of top soil is not required, and there is also the immense "heritage" of disturbed land from the past. In all these areas the land has been destroyed or at least severely damaged, and it is this which provides the great ecological challenge.

If the original soil is missing, what is found instead? It is possible that there is partly weathered *subsoil*, but it is much more likely that what is exposed is unweathered *overburden* material from deeper layers, or *waste* material, closely associated with the mineral being mined but rejected in the extraction process. These wastes may be coarse *waste rock*, or much finer *tailings* where the original rock has been ground to aid extraction and concentration.

All these are raw skeletal materials where the only connection with developed soils is that they represent materials out of which soils are formed by natural processes over the passage of time. They may consist of fairly inert substances such as silica, or of potentially chemically active but fairly benign substances such as calcium carbonate, or of physically extreme substances such as fine various clay minerals. Any of these may contain traces of substances capable of causing toxicity to plants, such as pyrite (which weathers to give sulfuric acid) or heavy metal ores. But the crucial point is that, whatever the material, there has been no weathering, no structure, no organic matter, small or erratic amounts of available nutrients, and certainly no nitrogen.

The starting point, bearing in mind that the vegetation and animals have been completely destroyed, is therefore, far from the well-developed ecosystem which would have existed before the mining disturbance. If we consider that there are two main dimensions to any ecosystem, *structure*, measured in terms of diversity and complexity, and *function*, measured in terms of biomass productivity and nutrient cycling, we can very readily compare the situation after disturbance with that existing before, on a two dimensional graph (Fig. 1) which is a development of the model of Magnuson *et al.* (1980) and Cairns (1982).

Aims in restoration

With this starting point we might well ask whether any restoration at all is possible. From what we know of natural soil forming processes and natural ecosystem development, the answer is clearly in the affirmative. The timescale, however, for a reasonably well-developed ecosystem starting from very raw materials is of the order of 100 years (Dickson and Crocker 1953; Crocker and Major 1955). On a geological time scale this is a second or so,

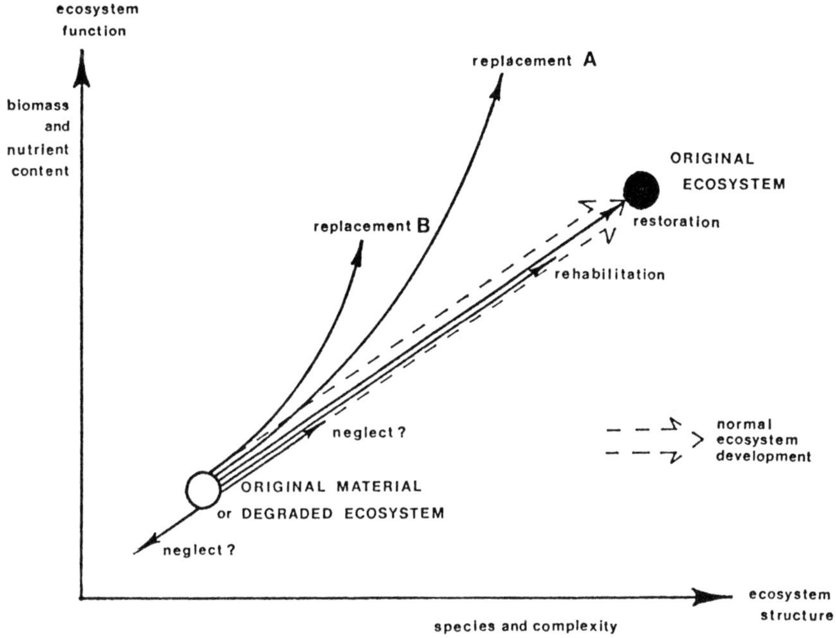

Fig. 1. A development of the model of Magnuson *et al.* (1980) and Cairns (1982).

but to the human beings who live beside and experience the degraded, destroyed land it is nearly two lifetimes. This is not what we should have to offer anyone as a penalty for the comforts of civilization. Indeed we should not have to subject a child during the whole of its childhood, a period of ten years, such a degraded environment.

This is a considerable challenge, and many ecologists and pedologists may think that restoration in a five-year period is impossible. But what is required is not a complete ecosystem fully developed in all its characteristics, but only an ecosystem developed sufficiently in its biological characteristics to be adequate in structure and function. The task is therefore to discover what is required to obtain this.

Before we can consider this, however, we need to look at our aims a little more closely. Are we trying to put back exactly what was there before, true *restoration*, or something else? In practical terms there are in fact at least four alternatives, which are indicated in Fig. 1. (1) True *restoration* is certainly one possibility, and may well be required by legislation, where the ecosystems being disturbed are of considerable value in terms of nature conservation or scenic beauty, as in the case of the Australian mineral sand industry (Brooks 1976). (2) It may sometimes be sufficient to put back an ecosystem which is only partially the same as what was there previously; this can be termed *rehabilitation.* (3) There is, also, the possibility of taking advantage of the disturbance and putting back a new ecosystem, which is perhaps of more "value" to the region than the original; this can be termed *replacement.* The replacement may be more productive, better in terms of function, but simpler and poorer in terms of structure, as when a complex, low yielding, natural woodland is replaced by a simple, high-yielding, planted forest (A in Fig. 1). Equally a natural grassland might be replaced by relatively simpler lower yielding, amenity grassland (B in Fig. 1). (4) Finally, there is the possibility of *neglect*, allowing natural processes to operate. This should lead to slow progression by natural succession back to the original ecosystem. But, depending on the nature of the disturbance, there is also the possibility of further degradation by erosion or equivalent processes. To cite an example of risk here, nearly 20 years ago, a colliery spoil heap perched on a mountainside at Aberfan in South Wales slid into the village school and killed 144 people including 112 children, a whole generation of that village.

Natural succession

Despite the possibility of further degradation, natural regeneration is a valid approach. There are many examples of sites where the pre-existing ecosystems have been totally destroyed, in which a new ecosystem has developed successfully without any external assistance, despite the absence of any sort of soil, plants or animals. The processes which occur come under the category of primary succession. As a result the product may have interesting and valuable characteristics.

In most regions of the world the process of primary succession took place millions of years ago, and the ecosystems now existing are highly developed in both structure and function. However, there are a number of local areas which have been influenced by the very substantial effects of the various glaciations and other effects of the Quaternary period. These ecosystems are much younger, yet most are still 10,000 years old and consequently well developed. In them there is no place for plants adapted to open infertile conditions. The history of the Postglacial period is a progressive loss of these species (Godwin 1975).

Mining and other forms of gross disturbance provide a complete reversal of this situation. Open habitats are created with soils of very low fertility, often extreme in particular characteristics such as high pH or low phosphorus. In these, succession may be slowed down or arrested, allowing specialized plants intolerant of competition, such as members of the Orchidaceae, to find refuge. As a result there are some most interesting, and attractive, floras (Greenwood and Gemmell 1978; Bradshaw 1983). Many of these floras in Britain are of sufficient conservation interest to have been recognized as "Sites of Special Scientific Interest" or adopted as Nature Reserves (Ratcliffe 1974; Usher 1986).

There are many other sites which are less extreme in soil characteristics in which more normal succession occurs. These can harbour a whole range of plant communities belonging to different stages of natural succession, including grassland, scrub and seral woodland, and be important areas for amenity and recreation, particularly where they occur on the outskirts of cities (Teagle 1978). A major attractive public recreation area in London, Hampstead Heath, was once the gravel pit supplying the whole of northern part of the city (Environmental Advisory Unit 1986). Every country and state has its own similar examples. It is very important that we recognize these sites for what they are and preserve them from further "restoration".

Natural succession as a guide to restoration

The slowness of natural succession on degraded land is an interesting phenomenon. If we can understand the causes, this should provide a guide to the problems that have to be overcome to obtain rapid and effective full restoration. The findings of academically directed ecology need to be examined to discover what practical guidance they can provide. There are a number of pertinent studies.

It is perhaps obvious that succession cannot proceed unless the appropriately adapted species are available in the neighborhood to colonize. Where the degraded land is not extreme, these species are likely to be readily available in the surrounding normal ecosystems. But where the land is extreme in certain characteristics, such as the alkaline chemical wastes of N.W. England, appropriately adapted species may not be available nearby, and the whole process of ecosystem development may be held up by the slow and stochastic arrival of these species from long distances (Bradshaw 1983).

Once the species have arrived, their development may be restricted by physical factors, particularly coarseness of texture and lack of organic matter in the substrate. The specific relationships of seed and substrate characteristics are well established ecologically (Harper and Benton 1966). But there is only sparse critical evidence for degraded land. But careful demographic studies certainly indicate that drought can have substantial effects on seedling survival in some years (Park 1982). This can be so important that recruitment of even a large seeded species such as *Ulex europeans* can be restricted on coarse waste material to only those years where the spring is sufficiently wet (Dancer *et al.* 1977). At the same time, as might be expected, there is an interaction between moisture and nutrients. Plants growing slowly because of nutrient are more likely to suffer from surface drought (Humphries 1982). This effect is most likely to be due to inadequate root development (Fitter and Bradshaw 1974).

Nutrients, however, need to be considered more generally, because they are a crucial factor in their own right in succession on degraded land. If we adopt a modern ecosystem approach it becomes obvious that ecosystem development cannot take place without an adequate supply of nutrients, to build up what is contained within the ecosystem and to provide for annual growth in combination with what is cycled within the system. The annual needs can be readily calculated (Bradshaw 1983) (Table 2). These needs will have to be supplied from the soil, or by other inputs such as precipitation or biological fixation. In many of the wastes left by mining the supplies from the soil can initially be very low indeed; this is particularly true for nitrogen which does not occur as a soil mineral.

Observations on natural ecosystem development on degraded land do indeed show that development takes place in parallel with the development of a pool of available plant nutrients in the soil. This is clear for nutrients derived from soil minerals such as phosphorus and potassium on colliery spoil (Knabe 1973), as well as for nitrogen on ironstone spoil banks (Leisman 1957) or on china clay wastes (Dancer *et al.* 1977; Roberts *et al.* 1981). It is difficult from such observations, however, to argue what is cause and what is effect, that is whether nutrient accumulation determines vegetation development or vice versa. However,the levels of available nutrients at any stage in the succession are so low that it would appear that they are indeed controlling, i.e. holding back, succession.

Evidence for this comes particularly from nitrogen in china clay wastes. It is possible to calculate that the size of the soil nitrogen capital stored in organic matter required to provide adequate annual supplies of soluble nitrogen by mineralization is about 1000 kg N ha^{-1} (Bradshaw 1983). This matches the finding that the vegetation in the natural succession only becomes free of nitrogen fixing species, and therefore independent of a special nitrogen supply, when the soil nitrogen capital reaches this same level. Evidence on other natural successions points in the same direction (Marrs *et al.* 1983).

Table 2. Amount of nutrients required annually by ecosystems with different levels of total primary productivity (from Bradshaw 1983).

	Nutrient content assumed (%)	Production level (kg ha^{-1} year^{-1})			
		1,000	5,000	10,000	20,000
N	2.00	20	100	200	400
K	1.10	11	55	110	220
Mg	0.51	5.1	26	51	102
Ca	0.26	2.6	13	26	52
P	0.18	1.8	9.0	18	36
Type of ecosystem		Tundra & desert	Poorly productive temperate ecosystems	Productive temperate & poorly productive tropical	Tropical

The final factor which can restrict natural succession on land degraded by mining and industry is toxicity, particularly that caused by heavy metals. Because heavy metals are stable and not readily leached, metal contaminated soils can remain toxic, and without any vegetation at all, for many decades. As a result the wastes left by metal mining present a particularly extreme environmental hazard, as a constant source of pollution, because of their instability (Bradshaw and Chadwick 1980). But colonization of these sites does occur, and provides us with a most instructive insight into population biology in extreme environments.

It was Prat (1934) who first showed that the species that colonized these sites only did so because they possess local populations specifically adapted to the metal toxicity concerned. We now have a very clear picture of the evolution involved and know that it can be very rapid, but that it is restricted to a limited range of species (Antonovics *et al.* 1971). It appears that the species are limited to those that possess the appropriate variability in their normal populations (Gartside and McNeilly 1974; Bradshaw 1984). Furthermore, there is evidence of considerable variation in metal tolerance among populations of *Agrostis capillaris* L. (Table 3) (e.g. Symeonides *et al.* 1985).

Table 3. Interpopulation variation for tolerance indices to cadmium (Cd), copper (Cu), lead (Pb), nickel (Ni), and zinc (Zn) in *Agrostis capillaris* (after Symeonidis *et al.* 1985).

	% tolerant* individuals to					
Population (Site)	Cd	Cu	Pb	Ni	Zn	Comments
1. Cullingworth	0	0	0	0	0	
2. Denholme	0	0	0	0	1	Field collections
3. Merllyn Gwyn	1	0	5	0	3	in U.K.; non-mined
4. Ruthin	0	8	0	0	1	areas
5. Dwrnudion	4	0	0	0	0	
6. Highland	0	23	0	4	4	American cultivar
7. NZ Browntop	11	3	3	6	0	Cultivar from New Zealand
8. Parys	10	97	44	38	60	Copper tolerant cultivar
9. Goginan	10	70	78	40	93	Pb/Zn tolerant cultivar

* Tolerant individual, identified for more than 50% tolerance index measured in terms of seedling root elongation.

At the same time such mine wastes are usually very deficient in most plant nutrients, particularly phosphorus, and growth is very restricted on this account also (Smith and Bradshaw 1979). As a result the natural floras of metal contaminated areas are very curious and succession is little in evidence. The only places where well developed floras occur is in very old, mainly tropical, areas (Ernst 1974).

One or two other types of toxicity, such as the acidity produced by the presence of weathering pyrite, are less stable. The weathering process which produces the acidity also leaches it away (Knabe 1973). Nevertheless studies of natural colonization of colliery spoil heaps in Britain, which commonly contain pyrite, show that species occurrence is not related to the age of the heap but to pH (Chadwick and Hardiman 1976).

We have therefore an interesting set of processes involved in natural succession on degraded land, firstly those concerned with the colonization process, and secondly those concerned with ecosystem development (Bradshaw 1983) Table 4). The colonization set is strongly deterministic, since only species with particular characteristics are able to colonize. But at the same time it has a strong probablistic component, since chance will play a considerable part in determining which species out of the range of possibilities actually arrive and establish. The development set of processes is partly allogenic due to the effects of plants, particularly those which are N-fixing, in accumulating organic matter and nutrients, and also the effects of soil animals in improving soil structure. There is clear evidence of facilitation, but at the same time the whole process can be held up by special characteristics of the substrate.

Table 4. Natural succession on degraded land.

Starting point:	Extreme skeletal soils and no plants or animals
Colonization: Process is deterministic with a probablistic component resulting in distinctive floras	(1) Immigration of species: only those with appropiate dispersal mechanisms
	(2) Selection of the adapted: only those with the appropriate ecological characteristics
	(3) Selection of the adaptable: only those which can evolve the appropriate characteristics
Ecosystem development: Process is partly allogenic due to external factors and partly autogenic due to plants, etc. (facilitation), resulting in increased biomass and diversity	(1) Accumulation of nutrients: (a) in plants (b) in soil — increased amount and availability
	(2) Changes in soil structure: due to plant and animal activity
	(3) Changes in toxicities: usually reduced

With this background from population biology and ecosystem studies, we are now in a position to make some sense of the methods that are employed in practice in ecosystem restoration (Jain, in press).

Restoration methods in practice

There are a number of publications which describe restoration methods in some detail (Schaller and Sutton 1978, Bradshaw and Chadwick 1980; Coppin and Bradshaw 1982; Williamson *et al.* 1982; Cairns 1988). As a result there is no need to detail them here. However, a general summary is provided in Table 5. What is crucial is that there are a series of specific problems that have to be overcome, which divide themselves, as do the factors restricting ecosystem development, into three groups. At the same time the solutions divide themselves up into those that are short term and those that are long term. The similarity between the long term solutions and what occurs in natural ecosystem development is not accidental, but is because successful long term solutions depend on harnessing natural processes.

It is commonplace among many engineers and landscape architects to believe that the only solution to land degradation is to import top soil. By this method all existing problems are covered up, and if the soil is of ade-

Table 5. The major problems of derelict land and their treatment (adapted from Bradshaw 1983).

Limiting factor	Specific variables	Descriptors	Immediate treatment	Long-term treatment
Physical	Structure	Too compact	Rip or scarify	Vegetation
		Too open	Compact or cover with fine material	Vegetation
	Stability	Unstable	Stabilizer/mulch or nurse	Regrade or vegetation
	Moisture	Too wet	Drain	Drain
		Too dry	Organic Mulch or nurse	Tolerant vegetation
Nutrition	Macronutrients	Nitrogen	Fertilizer	Legume
		Others	Fertilizer + lime	Fertilizer + lime
	Micronutrients	Deficient	Fertilizer	—
Toxicity	pH	Too high	Pyritic waste or organic matter	Weathering
		Too low	Lime	Lime
	Heavy metals	Too high	Organic mulch or metal tolerant cultivar	Inert covering ormetal tolerant cultivar
	Salinity	Too high	Weathering or irrigate	Tolerant species or cultivar

quate quality a satisfactory supply of nutrients will have been provided. This is an attractive solution for small scale problems. But it is prohibitively expensive to bring in top soil from elsewhere to cover a large site, quite apart from the fact that the material may well not be available. It is also common experience in Britain that materials sold as top soil are extremely poor in quality, often no better as a medium for plant growth than the substrates that they are being used to cover (Bloomfield *et al.* 1981). If a cover material is required it may be better to use some other inert, perhaps subsoil, material and upgrade that (Roberts and Roberts 1986).

The normal approach for degraded land is therefore to identify and treat the specific individual problems as they arise on individual sites.

a) *Physical problems*

The texture of degraded land materials is often not particularly extreme. Indeed in some situations, for instance tailings produced after ore extraction can be physically more satisfactory, in terms of available water capacity, that the original soils (Ludeke 1973). But waste rock materials can be too porous, and some overburdens with a high clay content too consolidated especially as a result of earth moving by heavy machinery. All materials will lack organic matter and a properly developed soil structure.

Substantial immediate improvements can be achieved by physical treatments. Coarse waste rock can be broken down and consolidated by the passage of heavy machinery. Excessive compaction of fine materials can be relieved by ripping with tines drawn by a crawler tractor to a depth of 80 cm; winged tine are particularly effective. All this simulates natural weathering. But these treatments are never enough by themselves. A satisfactory soil structure is only achieved by the accumulation of organic matter and the activities of plant roots and soil animals. Adequate growth of vegetation is therefore crucial. This will only occur if nutrient deficiency and other limiting factors are treated. Improvements in soil structure and plant growth therefore take place in concert.

Organic matter has a direct effect on soil structure. But it also has an indirect effect by being the substrate for fungi and bacteria, whose growth and secretions are responsible for crumb formation and the stability of pores and larger structure, and also as the substrate for soil animals. Whose activities create many of the smaller pores. Effectively, all of these organisms will colonize the developing soil without assistance, even the larger soil arthropods (Neumann 1973). But one important soil animal, the earthworm, will not. It is able to disperse at a rate of about 10 m yr^{-1} but no more. It is, therefore advisable to inoculate it into any areas being restored which are large. The results can be spectacular. On reclaimed polders in Holland the bulk density of the soil is lowered, infiltration is increased, and there is rapid disappearance of unincorporated dead material lying on the surface (Hoogerkamp *et al.* 1983). Surface casting by the worms leads to the development of a high grade surface soil and the burying of stones (Bradshaw 1983). All this shows that there is little substitute for understanding and assisting natural processes.

Lack of organic matter will lend to poor water retention at the surface and therefore failure in seed germination and establishment. This can be readily overcome by the application of a surface mulch, of peat, vegetable fibres or even chopped straw. These can be readily applied by machine (Kay 1978). Mulches of this sort are particularly important in hydroseeding, where the seed is sown by spraying onto the soil surface in an aqueous suspension with fertilizer and other materials. Without a mulch the seed is uniquely exposed. The mulch simulates admirably the complex protection to seed afforded by a natural surface. Chemical stabilizers are often also used, but many of these are toxic to germinating seedlings, as are high levels of fertilizer as might be expected. Both must therefore be applied with care (Sheldon and Bradshaw 1977).

The final possible treatment simulates the natural situation very closely. This is to sow another, appropriately chosen, species to act as a nurse. This provides both stability and a moist micro-environment. The use of nurse species is widespread, but the choice of species and its sowing density has been established more by experience than by published experiments. What is required is reliably rapid growing species with a growth habitat that does not lead to excessive competition. It is also essential that the nurse does not persist and take over from the intended species. These requirements commonly lead to the choice of a cereal, such as rye (*Secale cereale*), or a short level grass species, such as ryegrass c.v. Westernwolths (*Lolium italicum*). But larger species such as sorghum are sometimes very valuable. Sowing density must obviously be kept low. Nurse species can be important even for the establishment of wild species such as heather (*Calluna vulgaris*) whose sensitivity to competition in the adult stage is well known (Putwain *et al* 1983; Bradshaw 1987). There is considerable scope for further practically oriented research by population biologists in this area.

b) *Nutrient problems*

In natural situations the accumulation of adequate pools of available nutrients can be very slow. For elements which are in short supply means must be found to supply them artificially. This is usually very easily done by fertilizers and lime. Large amounts of nutrients such as phosphorus (if necessary as much as 400 kg P ha^{-1}) can be applied without difficulty. Doses of soluble elements such as potassium may have to be split to avoid leaching losses, and it may be necessary for elements such as phosphorus which are bound readily by some substrates and therefore become unavailable, to apply a second dose after a few years.

In all cases, of course, the amounts applied must be related to the needs of the ecosystem being established. But it is very rare to find that establishment is best without any nutrient application at all, even species such as *Calluna vulgaris* which have already been mentioned. Not only are degraded or disturbed soils nearly always nutrient deficient, but the seedling phase of most species are quite responsive to enhanced nutrient supply. Such response in terms of increase root and shoot growth increases the resilience of species to physical stresses.

Nitrogen constitutes a special problem. Not only are all its salts very soluble and easily leached, but there has to be a large organic-based store of it (about 1000 kg N ha^{-1}) in the soil to maintain adequate annual supplies, to which reference has already been made. This cannot be provided by fertilizers unless doses of 100—200 kg N ha^{-1} are applied annually for a number of years, which is rather expensive. Nitrogen is therefore commonly provided by biological fixation. Free living N-fixing organisms contribute only small amounts, usually less than 5 kg N ha^{-1} yr^{-1}, in degraded soils. Legumes and other species with symbiotic N-fixing organisms are far more effective (Skeffington and Bradshaw 1980).

With the right choice of species and attention to their other nutrient requirements, annual accumulation rates of 100 kg N ha^{-1} can be obtained even on very degraded materials (Dancer *et al.* 1977). One important aspect of the accumulation process is that the nitrogen is held in organic matter of low C/N ratio, and mineralization rates are correspondingly higher than normal. This means that the nitrogen accumulated is readily recycled. As this process continues, rates of mineralization fall but remain higher than in well established ecosystems where a considerable fraction of the nitrogen is stored in rather recalcitrant material (Skeffington and Bradshaw 1981). In all this the artificial restoration system is certainly copying natural succession, in which it is very clear that nitrogen fixing species play a crucial role. This is well shown on moraines at Glacier Bay, Alaska by *Dryas drummondii*, the major early colonists, and *Alnus crispa* the major late colonist, neither of course Leguminosae but effective nitrogen fixers (Lawrence *et al.* 1967; Crocker and Major 1965).

There is one interesting alternative method of building the necessary organic soil nitrogen pool in one step. This is by application of sewage sludge (Halderson and Zenz 1978; Hall *et al.* 1986). Although, because of its water content, it is not economically practicable to carry sludge long distances, there are many sites in urban regions close to sewage treatment plants. If solidified sludge cake is used,with a 25% dry solids content, a single application of 100—200 tonnes ha^{-1} can provide approximately 1000 kg N in an organic form which mineralizes readily. At the same time a very large quantity of phosphorus is supplied, which indeed could sometimes be excessive (Table 6). A very effective compromise to achieve a productive ecosystem is to reduce the amount of sludge and include a legume which will respond very favorably to the high phosphorus.

c) *Toxicity problems*

In all considerations of problems of toxicity, it is crucial to establish whether or not the toxicity is stable or degradable, and whether it can or cannot, be lost from the ecosystem. Toxins are only likely to be serious and require close attention in degraded land when they are stable and are not lost from the ecosystem.

Acidity produced by the presence of pyrite is a common problem, particularly in the spoils and overburdens produced by coal and metal mining. Spoil pH's can be as low as 2.5. The acidity, which is due to H_2SO_4, is readily leached, although some H^+ may be retained in the soil exchange complex. Weathering of pyrite is usually slow and some pyrite commonly persists to generate further acidity. However, in gold mine tailings in S. Africa the acidity can actually be washed out by systematic leaching and there is insufficient residual pyrite to cause further problems.

In most situations, particularly in colliery spoils, this is not possible, and lime ($CaCO_3$) must be added instead, if any satisfactory ecosystem is to be established. Originally sufficient lime was added to neutralize just the available acidity. Often, as a result, after an original period of success, there could be spectacular degeneration, as the pH fell on the oxidation of further pyrite. Permanent success is now achieved by adding sufficient lime to neutralize not only present acidity but also any further acidity which will be produced in the future. This can involve additions of 150 tonnes $CaCO_3ha^{-1}$ (Costigan *et al.* 1981).

Heavy metal toxicity cannot be dealt with in the same way. The metals cannot be leached or degraded and there is no treatment by which their

Table 6. Nutrients provided in a single application of sewage sludge.

	Sludge applied (wet)					
	25	50	100	200	400	(tonnes or m^3 ha^{-1})
	2.5	5	10	20	40	(depth in mm)
Nitrogen available						
liquid (undig.)	13	25	50	100	200	kg N ha^{-1} (at 0.5 kg m^{-3})
liquid (dig.) or cake	25	50	100	200	400	kg N ha^{-1} (at 1 kg m^{-3})
total						
liquid	50	100	200	400	800	kg N ha^{-1} (at 2 kg m^{-3})
cake	175	350	700	1400	2800	kg N ha^{-1} (at 7 kg m^{-3})
Phosphorus available						
liquid (undig.)	20	40	80	160	320	kg P ha^{-1} (at 0.8 kg m^{-3})
liquid (dig.) or cake	100	200	400	800	1600	kg P ha^{-1} (at 4 kg m^{-3})
total						
liquid	38	75	150	300	600	kg P ha^{-1} (at 1.5 kg m^{-3})
cake	200	400	800	1600	3200	kg P ha^{-1} (at 8 kg m^{-3})

(Concentrations as given in Hall, Daw & Bayas 1986).

effects can be neutralized. It is true that in some materials metal levels may be low enough for them to be complexed by addition of organic matter, farm slurry or sewage sludge (Ludeke 1973; Johnson *et al.* 1976). But in situations where the metal levels are high, although an organic treatment may be effective for a few years, it can then fail disastrously (Goodman *et al.* 1973).

The most interesting alternative treatment is to establish a vegetation cover using the naturally occurring populations of wild species which have evolved metal tolerance. Any nutrient deficiencies will need to be corrected. But tolerant populations can then be remarkably successful (Smith and Bradshaw 1979; McNeilly 1987). It is obviously necessary to look for tolerance among species which are suited to the climate and edaphic conditions of the site. But evolution of metal tolerance, although specific to individual metals and limited to certain species (Bradshaw 1984), is a sufficiently widespread phenomenon for effective material to be found in many different regions of the world (Hill 1977, Cox and Hutchinson 1980).

The major problem with the use of metal tolerant material is that although the surface of the waste is stabilized and pollution substantially controlled, there is still uptake of metal, albeit at a reduced level, into the vegetation, and particles of the waste itself may still become ingested by grazing animals. At the same time fertility must be maintained. Provision of nitrogen particularly has been a difficulty. However, recently metal tolerant populations of legumes, *Lotus purshianus* and *Lupinus bicolor*, have been found, which should be an important step forward with this technique (Wu and Kruckeberg 1985).

Nevertheless for total control of metal pollution the technique does have limitations. The only alternative, then, is to cover the waste with an inert layer of material, the physical characteristics of which ensure that the surface is isolated from the metalliferous wastes beneath. This is most effectively achieved by a layer of coarse material which limits any chance of upward capillary movement of metals (Johnson *et al.* 1977; Parry and Bell 1985).

Salinity is the final type of toxicity in degraded land material, that deserves our attention. High concentrations of salt can be found in overburdens, where they are usually of fossil origin, or in mine waste, where they can be produced by weathering processes, e.g. magnesium sulphate from the interaction of native magnesium carbonate with sulfuric acid from pyrite.

If saline materials are exposed in moist climates the salinity is rapidly removed by natural leaching. Temporary amelioration by additions of gypsum, $CaSO_4$, may be nevertheless be useful. In more arid climates, where the salinity may be a permanent feature, the only method is to follow what occurs in nature, that is to utilize salt tolerant material. Whereas metal tolerance appears to occur almost entirely in local populations, salt tolerance is a property of whole species. The appropriate species may well be a component of the local flora. But it may be easier to use species well known in agriculture which are salt tolerant, such as rhodes grass, *Chloris gayana*.

Conclusions

Degraded land, of whatever origin, is an inevitable concomitant of civilization. It does not, however, have to be an inevitable, permanent feature of civilization. It is to be hoped that, as a result of new legislation in many countries, much of the degradation will be obviated by storage of surface soils and their immediate replacement after the disturbance is over. However, this will not, and indeed cannot, always occur, and we also have an immense amount of existing degraded land produced from past mining and other industrial activity.

This land is an immense challenge to the ecologist. Normally ecologists can pursue investigations of those particular aspects of populations, species and ecosystems which interest them. They do not even have to be critical aspects. Perhaps now, alas, they just have to attract financial support. By this sort of work we hope we obtain a better understanding of ecosystems and the way they function. However, whether or not we have achieved this understanding is usually only tested by logical argument and by experiments which explore only the problem in which we are interested. This is rather imperfect proof, and certainly does not test the completeness of our understanding.

The challenge of restoration of ecosystems on degraded land is that it provides us with an ultimate test of the completeness of our ecological understanding (Bradshaw 1987). If we can create an ecosystem which works, and has the properties we require, and is self-sustaining in the way that natural ecosystems are, then we have tested and proved that understanding. It is a particularly acid test and it follows that in situations where we fail, we shall be forced to address problems that have previously been overlooked or incorrectly understood. The work is therefore an excellent corrective for incomplete understanding.

The situation as it now exists in land restoration, which I hope is apparent from this survey, is that success is now readily achievable. But it does depend on an awareness of the many different factors which can restrict plant and animal colonization and ecosystem development. While some of these factors can only be overcome by what can be termed engineering solutions — major physical or chemical manipulations, most factors can only be overcome, at least in the long term, by a thorough awareness and use of natural processes. Indeed although the economics of land restoration has not been considered, there is no doubt that considerable financial savings in the initial treatment and subsequent maintenance can be achieved if natural processes are harnessed wherever possible.

In a volume where many different aspects of the biology of populations and species have been described and analyzed on account of their own intrinsic interest, it is important to realize that the work that we carry out for its own sake can have important practical significance (Jordan *et al.* 1987). Nature itself provides the model and the guidance for its own restoration.

Literature cited

Antonovics, J., Bradshaw, A. D. and Turner, R. G. 1971. Heavy metal tolerance in plants. *Advances in Ecological Research* 7: 1—85.
Bloomfield, H. E., Handley, J. F. and Bradshaw, A. D. 1981. Top soil quality. *Landscape Design* 135: 32—34.
Bradshaw, A. D. 1983. The reconstruction of ecosystems. *Journal of Applied Ecology* 20: 1—17.
Bradshaw, A. D. 1984. The importance of evolutionary ideas in ecology — and vice versa. In *Evolutionary Ecology* ed. B. Shorrocks, 1—25. Blackwell, Oxford.
Bradshaw, A. D. 1987. Restoration: an ecological acid test. In *Restoration Ecology — A Synthetic Approach to Ecological Research.* eds. W. R. Jordan, M. Gilpin and J. D. Aber. Cambridge University Press, Cambridge.
Bradshaw, A. D. and Chadwick, M. J. 1980. *The Restoration of Land.* Blackwell, Oxford.
Brooks, D. R. 1976. Rehabilitation following mineral sand mining on North Stradbrooke Island, Queensland. In *Landscaping and Land Use Planning as related to Mining Operations.* ed. Australian Institute of Mining and Metallurgy, 93—104. Australian Institute of Mining and Metallurgy, Adelaide.
Cairns, J. 1982. Restoration of damaged ecosystems. In *Research on Fish and Wildlife Habitat.* ed. W. T. Mason and S. Iker, 220—239. U.S. Environmental Protection Agency, Washington.
Carroll, L. 1876. *Alice Through the Looking Glass.* Macmillan, London.
Caudhill, H. M. 1976. *The Watches of the Night.* Atlantic — Little, Brown, New York.
Chadwick, M. J. and Hardiman, K. M. 1976. Vegetating colliery spoil. In *Papers of the Land Reclamation Conference 1976.* ed. J. Essex and P. Higgins, 431—442. Thurrock Borough Council, Grays, Essex.
Copping, N. J. and Bradshaw, A. D. 1982. *Quarry Reclamation.* Mining Journal Books, London.
Costigan, P. A., Bradshaw, A. D. and Gemmell, R. P. 1981. The reclamation of acidic colliery spoil. I. Acid production potential. *Journal of Applied Ecology* 18: 865—878.
Cairns, J., Jr. (ed.). 1988. *Rehabilitating Damaged Ecosystems.* Vols. I, II. C.R.C. Press, Baco Raton.
Cox, R. and Hutchinson, T. C. 1980. Multiple metal tolerance in the grass *Deschampsia caespitosa* (L.) Beauv. from the Sudbury smelting area. *New Phytologist* 84: 631—647.
Crocker, R. L. and Major, J. 1955. Soil development in relation to vegetation and surface age at Glacier Bay, Alaska. *Journal of Ecology* 43: 427—448.
Dancer, W. S., Handley, J. F. and Bradshaw, A. D. 1977. Nitrogen accumulation in kaolin mining wastes in Cornwall. I. Natural communities. *Plant and Soil* 48: 153—167.
Dancer, W. S., Handley, J. F. and Bradshaw, A. D. 1977. Nitrogen accumulation in kaolin mining in Cornwall. II. Forage legumes. *Plant and Soil* 48: 303—314.
Department of the Environment. 1984. *Survey of Derelict land In England 1982.* Department of the Environment, London.
Dickson, B. A. and Crocker, R. L. 1953. A chronosequence of soils and vegetation near Mt. Shasta, California. *Journal of Soil Science* 4: 123—54.
Environmental Advisory Unit, University of Liverpool. 1985. *Transferring our Waste Land — The Way Forward.* Her Majesty's Stationery Office, London.
Ernst, W. 1974. *Schmermetall vegetation der Erde.* Fischer, Stuttgart.
Fairbrother, N. 1970. *New Lives, New Landscapes.* Architectural Press, London.
Fitter, A. H. and Bradshaw, A. D. 1974. Root penetration of *Lolium perenne* on colliery shale in response to reclamation treatments. *Journal of Applied Ecology*, 11: 609—616.
Gartside, D. W. and McNeilly, T. 1974. The potential for evolution of heavy metal tolerance in plants. II. Copper tolerance in normal populations of different plant species. *Heredity* 32: 335—348.

Godwin, H. 1975. *The History of the British Flora* (2nd ed.). Cambridge University Press, London.
Goodman, G. T., Pitcairn, C. E. R. and Gemmell, R. P. 1973. Ecological factors affecting growth on sites contaminated by heavy metals. In *Ecology and Reclamation of Devastated Land.* ed. R. J. Hutnik & G. Davis, 149—174. Gordon and Breach, New York.
Greenwood, E. F. and Gemmell, R. G. 1978. Derelict industrial land as a habitat for rare plants in S. Lancs (V.C. 59) and W. Lancs (V.C. 60). *Watsonia* 12: 33—40.
Halderson, J. L. and Zenz, D. R. 1978. Use of municipal sewage sludge in reclamation of soils. In *Reclamation of Drastically Disturbed Lands.* ed. F. W. Schaller and P. Sutton, 355—378. American Society of Agronomy, Madison.
Hall, J. E., Daw, A. P. and Bayes, C. D. 1986. *The use of Sewage Sludge in Land Reclamation.* WRC Environment, Medmenham, Bucks.
Harper, J. L. and Benton, R. A. 1966. The behavior of seeds in soil, part 2. The germination of seeds on the surface of a water supplying substrate. *Journal of Ecology* 54: 151—166.
Hill, J. C. 1977. Establishment of vegetation on copper —, gold — and nickel-mining wastes in Rhodesia. *Transactions of the Institute of Mining and Metallurgy 86A*: 135—145.
Hoogerkamp, M., Rogaar, H. and Eijsackers, H. J. P. 1983. The effect of earthworms (Lumbricidinae) on grassland on recently reclaimed polder soils in the Netherlands. In *Earthworm Ecology* ed. J. E. Satchell, 85—104. Chapman and Hall, London.
Humphries, R. N. 1982. The establishment of vegetation on quarry materials: physical and chemical constraints. In *Ecology of Quarries*, ed. B. N. K. Davis, 55—61. Institute of Terrestrial Ecology, Cambridge.
Jain, S. K. In press. Restoration ecology: Some basic principles and research needs. *In*: P. Fiedler and S. Jain (eds.) Conservation Biology: Theory and Practice. Chapman and Hall, London.
Johnson, M. S., Bradshaw, A. D. and Handley, J. F. 1976. Revegetation of metalliferous fluorspar mine tailings. *Transaction of the Institute of Mining and Metallurgy* 85A: 32—37.
Johnson, M. S., McNeilly, T. and Putwain, P. D. 1977. Revegetation of metalliferous mine spoil contaminated by lead and zinc. *Environmental Pollution* 12: 261—277.
Jordan, W. R., M. E. Gilpin, and J. D. Aber. (eds.). 1987. *Restoration Ecology.* Cambridge Univ. Press.
Kay, B. L. 1978. Mulch and chemical stabilizers for land reclamation in dry regions. In *Reclamation of Drastically Disturbed Lands.* ed. F. W. Schuller and P. Sutton, 467—483. American Society of Agronomy, Madison.
Knabe, W. 1973. Investigation of soils and tree growth in five deep-mine refuse piles in the hard coal region of the Ruhr. In *Ecological and Reclamation of Devastated Land.* ed. R. J. Hutnik and G. Davis, Vol. 1, 307—324. Gordon and Breach, New York.
Lawrence, D. B., Schoenike, R. E., Quispel, A. and Bond, G. 1967. The role of *Dryas drummondii* in vegetation development following ice recession at Glacier Bay, Alaska, with special reference to its nitrogen fixation by root nodules. *Journal of Ecology* 55: 793—813.
Leisman, G. A. 1957. A vegetation and soil chronosequence on the Mesabi Iron Range spoil banks, Minnesota. *Ecological Monographs* 27: 221—245.
Ludeke, K. L. 1973. Vegetative stabilization of copper mine tailing disposal berms of Pima Mining Company. In *Tailings Disposal Today.* ed. C. L. Aplin and G. O. Argall. 377—408. San Francisco.
Magnuson, J. J., Reigier, H. A., Christie, W. J. and Sonzogi, W. C. 1980. To rehabilitate and restore Great Lakes ecosystems. In *The Recovery Process in Damaged Ecosystems.* ed. J. Cairns, 95—112. Ann Arbor Science, Michigan.
Marrs, R. H., Roberts, R. D., Skeffington, R. A. and Bradshaw, A. D. 1983. Nitrogen and the development of ecosystems. In *Nitrogen as an Ecological Factor.* ed. J. A. Lee, S. McNeill and I. H. Rorison, 113—136. Blackwell, Oxford.
McNeilly, T. 1987. Evolutionary lessons from degraded ecosystems. *In*: *Restoration Ecology: A Synthetic Approach to Ecological Research.* eds. W. R. Jordan III, M. E. Gilpin, and J. D. Aber, 271—286. Cambridge.

Newmann, U. 1973. Succession of soil fauna in afforested spoil banks of the brown-coal district of Cologne. In *Ecology and Reclamation of Devastated Land.* ed. Hutnik, R. J. and Davis, G. Vol. 1: 335—348. Gordon and Breach, New York.

Paone, J., Morning, J. L. and Guorgettil. 1974. *Land Utilization and Reclamation in the Mining Industry, 1930—71.* Information Circular 8642. U.S. Bureau of Mines, Utah.

Park, D. G. 1982. Seedling demography in quarry habitats. In *Ecology of Quarries.* ed. B. N. K. Davis, 32—40, Institute of Terrestrial Ecology, Cambridge.

Parry, G. D. R. and Bell, R. M. 1985. Covering systems. In *Contaminated Land.* ed. M. A. Smith, Chapter 5, Plenum, New York.

Prat, S. 1934. Die Erblichkeit der Resistenz gegen Kupfer. *Bering Deutsch Botanische Gesellschaft* 102: 65—67.

Putwain, P. D., Gillham, D. A. and Holliday, R. J. 1983. Restoration of heather moorland and lowland heathland with special reference to pipelines. *Environmental Conservation* 9: 225—235.

Ratcliffe, D. 1974. Ecological effects of mineral exploitation in the United Kingdom and their significance to nature conservation. *Proceedings of the Royal Society, London* A339: 355—372.

Roberts, R. D., Marrs, R. H., Skeffington, R. A. and Bradshaw, A. D. 1981. Ecosystem development on naturally colonized China clay wastes. I. Vegetation changes and overall accumulation of organic matter and nutrients. *Journal of Ecology* 69: 153—61.

Roberts, R. D. and Roberts, J. M. 1986. The selection and management of soil in landscape schemes. In *Ecological and Design in Landscape.* ed. A. D. Bradshaw, D. A. Goode and E. Thorp, 99—126. Blackwell, Oxford.

Schaller, F. W. and Sutton, P. 1978. *Reclamation of Drastically Disturbed Lands.* American Society of Agronomy, Madison.

Sheldon, J. C. and Bradshaw, A. D. 1977. The development of a hydraulic seeding technique for unstable sand slopes. I. Effects of fertilizers, mulches and stabilizers. *Journal of Applied Ecology* 14: 905—918.

Skeffington, R. A. and Bradshaw, A. D. 1980. Nitrogen fixation by plants grown on reclaimed china clay wastes. *Journal of Applied Ecology* 17: 469—477.

Skeffington, R. A. and Bradshaw, A. D. 1981. Nitrogen accumulation in kaolin mining wastes in Cornwall. IV. Sward quality and the development of a nitrogen cycle. *Plant and Soil* 62: 439—451.

Smith, R. A. H. and Bradshaw, A. D. 1979. The use of metal tolerant plant populations for the reclamation of metalliferous wastes. *Journal of Applied Ecology* 16: 595—612.

Symeonidis, L., T. McNeilly, and A. D. Bradshaw. 1985. Interpopulation variation in metal tolerance to cadmium, copper, lead, nickel, and zinc in nine populations of *Agrostis capillaris* L. New Phytol. 101: 317—324.

Teagle, W. G. 1978. *The Endless Village.* Nature Conservancy Council, Shrewsbury.

Usher, M. B. 1986. (ed.). *Biological Conservation Evaluation.* Chapman and Hall, London.

Williamson, N. A., Johnson, M. S. and Bradshaw, A. D. 1982. *Mine Wastes Reclamation.* Mining Journal Books, London.

Wu, L. and Kruckeberg, A. L. 1985. Copper tolerance in two legume species from a copper mine habitat. *New Phytologist* 99: 565—570.

3. Genetic conservation in captive populations and endangered species

PHILIP W. HEDRICK

Abstract

Many rare and endangered species require genetic conservation strategies involving an *ex situ* management phase prior to their reintroduction in nature. Since small populations are likely to lose genetic variation and suffer from inbreeding depression, these strategies are designed in genetic terms to monitor and avoid these effects as short-term goals, and to retain reestablishment or adaptive potential in nature as the long-term goal. Population genetic theory helps us define the concepts of effective population size, inbreeding and population subdivision, and discuss changes in homozygosis, genetic variance, and fitness in small populations. Estimates of several parameters in these models are presented for three examples (Cheetah, Przewalski's horse, Speke's gazelle). Intermatings among different populations bred in captivity or conserved in natural habitats can be guided (but not exactly prescribed) by the genetic models of migration, mating systems, and drift. Rates of changes in the level of inbreeding can be monitored to avoid drastic fitness losses, and thereby evolving tolerance of certain amount of inbreeding. These *ex situ* methods are designed to supplement and never to substitute for *in situ* conservation.

Introduction

Because of the increase in the human population and expansion of human-related activity in recent decades, many species have become extinct or are on the verge of extinction (e.g., Ehrlich and Ehrlich 1982; Soulè 1986, 1987). Although the preservation of most endangered species depends primarily on the protection of the species and the maintenance of adequate natural habitat, in the past half decade attention has focused on the conservation of genetic variation in endangered species and captive animal populations (e.g., Frankel and Soulè 1981; Schonewald-Cox *et al.* 1983; Ralls and Ballou 1986; Woodruff 1989). This has been possible because some species currently (and probably many more will in the future) exist only in captive or protected situations. Even though much of the concern and planning has focused on genetic conservation in the so-called "charismatic megavertebrates", most of the same principles should apply in general to other endangered vertebrates, invertebrates and plants.

S. K. Jain and L. W. Botsford (eds), Applied Population Biology, 45–68.

The immediate cause of extinction of a species is often due to several factors but human predation, predation by human-associated organisms, and habitat destruction appear to have been important factors in the extinction of nearly all mammal or bird species in the last three centuries (Ziswiller 1967). However, even when there appears to be adequate natural habitat and a species is protected, random factors may cause a species to become extinct (e.g., Shaffer 1981; Soulè and Simberloff 1986). These factors can be classified (see Table 1) into extrinsic factors which includes variation in other species such as predators or pathogens and abiotic variation such as floods, fires, or droughts or into factors intrinsic to the species such as demographic stochasticity, genetic deterioration, or behavior that is maladaptive at low population number.

The extinction of the American heath hen, *Tympanuchus cupido cupido*, illustrates the role of these random factors. Once a relatively common bird in the northeastern United States, the heath hen was reduced to a remnant population on Martha's vineyard at the turn of the century (Fig. 1). In 1907, the remaining birds, less than 100, were protected in a refuge and they increased to nearly 2,000 over the next decade. Then, within a year a fire (abiotic catastrophe) and high predation by goshawsk, *Accipiter gentilis*, (biotic stochasticity) reduced the number to less than 200. The population again rebounded but disease in 1921 (biotic stochasticity) again reduced their numbers. In the final years, a large excess of males (demographic stochasticity), high sterility (perhaps both genetic stochasticity and social dysfunction) contributed to their final extinction. In other words, the extinction of the heath hen, and presumably other species, was probably caused by several random factors including those affecting the genetic variation.

How can we examine the genetic factors that may influence the survival of a captive population of an endangered species? Here I will focus on genetic

Table 1. Random factors that may result in the extinction of a species (after Shaffer 1981).

(a) *Extrinsic factors*
- (1) *Biotic stochasticity* — variation in the populations of competitors, predators, prey, parasites, and pathogens, i.e., species with potential detrimental effects on the endangered species.
- (2) *Abiotic stochasticity or catastrophe* — deleterious changes in the habitat or physical environment, such as floods, droughts, fires, wind storms, or volcanic eruptions.

(b) *Intrinsic factors*
- (1) *Demographic stochasticity* — chance events affecting the survival and reproductive success of a species, such as variation in the number of males and females, birth and death rates, population growth rates and the population age structure.
- (2) *Genetic deterioration* — detrimental genetic changes resulting from chance events such as genetic drift or inbreeding.
- (3) *Social dysfunction* — behaviors that become detrimental to population survival at small population size such as group foraging, breeding, or roosting.

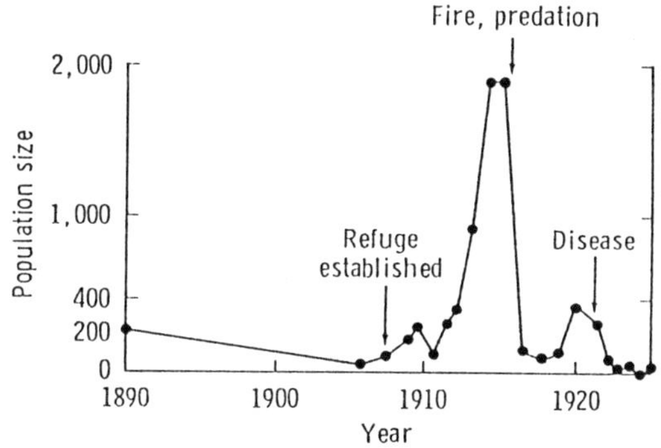

Fig. 1. The population size of the American heath hen and indication of events thought important in its extinction (after Ziswiller 1967).

conservation in a species that, when conditions permit, will hopefully be reintroduced to its natural habitat or a relatively free-living situation. First, before examining these factors, it is useful to separate the short-term and long-term goals important in maintaining a species. The major short-term goal is to insure the vigorous survival of the species for a given but limited time period. In this situation, the prime genetic considerations are the avoidance of both inbreeding depression and the chance fixation fo potentially nonadapted genotypes. On the other hand, the long-term goals involve both the maintenance and survival of the species and as well as retention of the genetic potential for successful expansion or reestablishment in natural populations. In this case, genetic considerations include the avoidance of inbreeding depression, retention of genetic variation so that future adaptation may be possible, and the avoidance of selection for the captive environment. Below I will discuss both the problems of inbreeding depression and retention of genetic variation. Avoidance of selection for a captive environment has been discussed at some length by Frankham *et al.* (1986). Obviously genetic changes affecting behaviors that are advantageous in captive situations, e.g., tameness, would be disadvantageous in a population released to a natural situation having poachers or predators. An example of such a detrimental genetic change caused by artificial rearing conditions occurred in a biological control program for the screwworm, *Cochliomyia hominivorax.* In this case, artificially reared individuals evolved a different form of the flight speed or capacity enzyme α-Gdh and were unable to adequately compete with wild flies when released in infested areas (Bush 1978).

The organization of this review is as follows. First, I will discuss several population genetics concepts — inbreeding depression, effective population size, founder contribution, population subdivision, and retention of genetic variation — that are important in understanding genetic conservation in captive or endangered species. Second, I will briefly relate these concepts in relationship to the genetic variation of several captive and endangered species, in particular the cheetah, Przewalski's horse, and Speke's gazelle. Finally, I will give a brief overall perspective of the role of genetic conservation in captive and endangered species.

Population genetics concepts

Inbreeding depression

Until recent years, some individuals engaged in the breeding of captive animals were not aware of the extent of reduced fitness resulting from inbreeding (e.g., Greig 1979) and considered it to be unimportant. One reason for this misconception is that until recently there were few records (stud books) giving the relationship between breeding animals in zoos. Ralls and her coworkers (e.g., Ralls 1984; Ralls *et al.* 1988), using pedigree information from a number of captive species, documented that the overall mortality of inbred offspring is nearly double that of noninbred offspring. As an example, Table 2 gives the infant mortality for eight mammalian species with relatively large numbers of both noninbred and inbred progeny (from Ralls and Ballou 1986; cheetah data from O'Brien *et al.* 1985). The species are ordered in Table 2 by their juvenile mortality in noninbreds which ranges from 0.106 in the degu to 0.501 in the golden lion tamarin. For all the species, the mortality in inbreds is higher, and in six of the eight species, the difference is significant using the χ^2 test. We should note that indirect (preliminary) evidence of inbreeding depression may be obtained by docu-

Table 2. The proportion of juvenile mortality in noninbred and inbred progeny in eight species (after Ralls and Ballou 1983; O'Brien *et al.* 1985).

Species or subspecies	Juvenile mortality (sample size in parethesis)	
	Noninbred	Inbred
Degu (*Octodontomys degus*)	0.106 (66)	0.235 (51)*
Elephant shrew (*Elephantulus refescens*)	0.153 (144)	0.233 (43)
Pygmy hippopotamus (*Cheoropsis liberiensis*)	0.245 (184)	0.549 (52)*
Cheetah (*Acinonyx jubatus*)	0.263 (194)	0.442 (43)*
Dorcas gazelle (*Gazella dorcas*)	0.280 (50)	0.595 (42)*
Japanese serow (*Capricornis crispus*)	0.288 (73)	0.565 (62)*
Greater galago (*Galago crassicaudatus crassicaudatus*)	0.308 (146)	0.386 (44)
Golden lion tamarin (Leontopithecus *rosalia rosalia*)	0.501 (369)	0.634 (145)*

* $p < 0.05$

menting lower fitness in known inbred populations although inbreeding depression may be confounded with other factors, such as poor habitat (Table 3).

Table 3. Approaches for documenting and reducing inbreeding depression.

(a) *Documentation of inbreeding depression*
   (1) correlation of fitness components, such as juvenile mortality or sterility, with the inbreeding coefficient calculated from pedigrees.
   (2) indirect or preliminary evidence, such as increased juvenile mortality or sterility in populations thought to be inbred.

(b) *Strategies for reducing inbreeding depression*
   (1) minimize inbreeding by mating unrelated or distantly related individuals, moving unrelated between colonies or zoos
   (2) introduce new wild caught individuals into the population
   (3) allow only a small increase in inbreeding per generation, say 1—3% so that lethals are gradually removed from the population
   (4) design a breeding program to purge the population of lethals and then minimize inbreeding
   (5) provide an optimal environment to reduce fitness losses in inbred individuals

What is the cause of reduced fitness in inbred individuals, generally called inbreeding depression? Indications from detailed studies from the vinegar fly, *Drosophila melanogaster*, indicate that homozygosity of recessive lethals in inbred offspring accounts for approximately half of the observed inbreeding depression in viability. The exact cause of the remaining inbreeding depression is not clear, but it appears to be polygenic. To understand the genetic basis of inbreeding depression, let us assume that a detrimental allele A_2 at locus A is maintained in the population by mutation-selection balance where u is the rate of mutation to the detrimental allele and the relative fitnesses of the genotypes A_1A_1, A_1A_2, and A_2A_2 are 1, 1, and $1 - s$, respectively. Using elementary population genetics theory (e.g., Hedrick, 1983), the expected change in allelic frequency is approximately

$$\Delta q = up - spq^2$$

where p and q are the frequencies of alleles A_1 and A_2, respectively, the first term gives the increase from mutation, and second term the decrease from selection. By setting $\Delta q = 0$ we can solve for the equilibrium frequency of genotype A_2A_2 which is

$$q_e^2 \approx \frac{u}{s} .$$

If there is a given amount of inbreeding f (assuming for the moment no inbreeding in previous generations), the frequency of the recessive genotype

A_2A_2 is

$$Q = q_e^2 + f p_e q_e$$
$$= \frac{u}{s} + f\left[\left(\frac{u}{s}\right)^{1/2} - \frac{u}{s}\right]$$

where P_e is the frequency of the dominant allele A_1. Assuming $s = 1$, i.e., A_2 is a recessive lethal, then

$$Q = u + f(u^{1/2} - u)$$
$$= fu^{1/2}.$$

Because u is generally small, say 10^{-5} to 10^{-6}, if there is any inbreeding in the population, most of homozygous A_2A_2 individuals will be the result of inbreeding. For example, using the first equation for Q above if $u = 10^{-6}$ and $f = 0.3$, then $Q = 3.0 \times 10^{-4}$ while if $f = 0$ for the same mutation rate $q^2 = 10^{-6}$, 300-fold less.

Let us assume that there are n loci that can mutate to lethals in the population, all are at equilibrium, and they independently affect viability. In this case, the expected proportion of viable offspring using the lower equation for A (they are not homozygous at any lethal loci) will be

$$v = (2 - Q)^n$$
$$= (1 - fu^{1/2})^n$$

and the expected mortality is $1 - v$. Obviously, if the number of lethal loci is relatively large and there is some inbreeding in the population, then the expected proportion of viable offspring will be significantly reduced. For example, if as above $f = 0.3$ and $u = 10^{-6}$ but there are 2000 such loci in the whole genome, then the viability is reduced to 0.55 that in a non-inbred population.

A general statistical approach to measure the effect of inbreeding on viability is to assume that there are two sources of mortality, one independent of inbreeding, composed of environmental mortality and genetic mortality not due to inbreeding, and another component that increases with inbreeding (e.g., Cavalli-Sforza and Bodmer 1971; Templeton and Read 1984). To examine the inbreeding-dependent component, let B equal the rate of occurrence of recessive lethals over all loci per gamete. For an offspring to survive, it must not be identical by descent (homozygous through inbreeding) for any such lethal, a probability (v_g) that is the zero term in the Poisson probability distribution or

$$v_g = e^{-Bf}.$$

Next, mortality can occur from environmental causes or genetic factors unrelated to inbreeding. Assuming all these causes are independent of each other, then

$$v_e = e^{-A}$$

where if $A = 0$, then there are no deaths unrelated to inbreeding and if A is large, then, for example, environmental causes of mortality are important. The overall viability probability then becomes

$$v = e^{-A - Bf}.$$

B can also be interpreted as the number of lethal equivalents in a haploid genome, i.e., it includes both lethals and detrimentals that when several are homozygous can cause mortality. As an example, let us assume all mortality results from inbreeding ($A = 0$), there are two lethal equivalents per gamete (four per individual on the average) and $f = 0.3$, assumptions that make the expected viability of the offspring 0.55. Note that this probability is the same as found above for 2,000 lethal loci in mutation-selection balance. In other words, given that genetic model there would be an average two lethal equivalents per gamete, similar to the value found in many human populations (e.g., Cavalli-Sforza and Bodmer, 1971).

This approach is useful when matings producing offspring with different inbreeding levels (f_1) are available. If we take the natural logarithms of the expression above, assuming that inbreeding level (i_f) results in viability v_i, then

$$\ln(v_i) = -A - BF_i$$

and B becomes the regression coefficient in a linear regression equation (see Templeton and Read 1984, for more details and an application). In other words, the number of lethal equivalents can be estimated using linear regression analysis.

The above genetic treatment of the extent of inbreeding depression is a static one. Given a certain level of inbreeding for a period of time, the frequency of lethal or detrimental alleles will decline from the level before inbreeding because inbreeding exposes more recessive lethal or detrimental alleles to selection in homozygotes, thereby causing a reduction in their frequency. Assuming that inbreeding continues at a given level for a long time, the new equilibrium allele frequency is

$$q^e = \frac{-f + \left[f^2 + 4(1-f)\dfrac{u}{s} \right]^{1/2}}{2(1-f)}$$

(Haldane 1940). For example, if $s = 1$ and $u = 10^{-6}$ and $f = 0$, then $q_e = 10^{-3}$. However, if $f = 0.10$, then $q_e = 9.9 \times 10^{-6}$, a hundred-fold less.

There are some examples of human populations that have had continual inbreeding for generations and now have little inbreeding depression (e.g., Rao and Inbaraj 1979). In addition, in several endangered or captive species there is little evidence of inbreeding depression. For example, in Pere David's deer, the juvenile mortality in noninbred progeny is 0.118 and in inbred progeny 0.136. Although the full history of this species is not known, all the living animals (around 700 today) descend from 11 animals "discovered" in the hunting compound of the Chinese emperor in 1865. It is thought that this population may have been in captivity more than 300 years, during which it might have undergone inbreeding that reduced the frequency of or eliminated many detrimental or lethal alleles.

In general, species that are outbreeding in nature, when exposed to high inbreeding, have relatively severe inbreeding depression. Obviously these effects would be most harmful when the inbred individuals are placed in a suboptimal environment. For example, although European bison, *Bison bonasus*, are highly inbred (Slatis 1960), they may in part have had a smaller inbreeding depression because they were raised in a semi-natural environment. Other animals inbred in nonnative habitat or stressful captive situations may exhibit more inbreeding depression resulting from this type of genotype-environment interaction.

First, to avoid inbreeding depression, inbreeding should be minimized by mating distantly related or unrelated individuals (Table 3). Second, although it is not possible in many species, the recurrent introduction of new wild-caught individuals into the captive population can also reduce inbreeding depression. Third, extensive inbreeding depression may be avoided by gradually increasing inbreeding over a number of generations. Experience in animal breeding suggests that 1—3% increase in inbreeding per generation can be tolerated by domesticated animals (Frankel and Soulè 1981). However, in many endangered species the inbreeding depression may be higher because domesticated animals may be somewhat inbred already. Finally, an extreme program designed to cope with situations in which these alternatives are not possible was developed for the Spekes gazelle (Templeton and Read 1984; and see discussion below). In this case, animals were deliberately inbred for a few generations to purge the population of lethals and then inbreeding was minimized.

Effective population size

Endangered species or captive populations in almost all cases consist of small, finite populations. An extreme example of such a species is the California condor, *Gymnogyps californianus*, which in 1986 consisted of only 5 wild birds and 21 captive birds. However, it is not correct to say that the evolutionary population size is 26 for several reasons. First, a number of these birds are incapable of breeding because they are too old or are damaged by pesticides. Second, many of these birds are closely related so that they have alleles in common. As a result of these and other considerations, it is important to have an approach that allows quantification of the evolutionarily important population size of such species. In addition, how can this population size be maximized to counter the potential loss of genetic fabrication?

A general approach to this problem is to compare the actual population size to a theoretical one, called the effective population size, with the same evolutionary properties, e.g., the same rate of increase in inbreeding over time. In such a theoretical population, there is an equal probability of reproduction for each parent, resulting in a binomial distribution of family size. Actual populations potentially differ from this ideal in a number of ways, most of which reduce the effective population size below the actual size. We should note that pedigrees are available from stud books for a number of species that exist primarily in captive situations, making different and more comprehensive analyses possible (see below). However, the following discussion is still useful in these cases to focus attention on the major factors affecting the loss of genetic variation in the population.

As we will discuss later, the retention of genetic variation in a population is greater when the effective population size, N_e, is larger. There are several ways in which N_e can be increased or maximized in a population. First, all reproductive individuals should have a chance to contribute to the breeding population unless they are deemed genetically unfit in some manner. For example, many captive golden lion tomarins, *Leontopithecus rosalia*, have a genetic defect in their diaphragm. However, even in such an instance, an effort to include contributions from all founder individuals may result in breeding these individuals and selecting for unaffected progeny.

Second, N_e is a function of the effective number of males (N_{em}) and effective number of females (N_{ef}) as

$$N_e = \frac{4N_{ef}N_{em}}{N_{ef} + N_{em}}$$

For example, if there is only one breeding male, then N_e at maximum value is this expression is $N_e = 4$. On the other hand, the maximum that this expression can be for a given total number of individuals is when the effective numbers of males and females are equal. For monogamous species, as some endangered bird species, the number of breeding males and females should be similar.

If we consider variation in the number of progeny of the two sexes, then we can calculate the effective numbers of males and females (Lande and Barrowclough 1986). If we let k_m and k_f represent the number of progeny, attributed by paternity analyses, to the male and female gene pools, respectively, then

$$N_{em} = \frac{N_m k_m - 1}{k_m - 1 + V_{km}/k_m}, \quad \text{and} \quad N_{ef} = \frac{N_f k_m - 1}{\bar{k}_f - 1 + V_{kf}/\bar{k}_f},$$

where N_m and V_{km} are the number of males and variance of progeny number produced by males (the subscript f indicates the same values for females). Obviously, high variance in offspring number will result in N_{ef} and N_{em} values les than N_f and N_m, a typical result in many natural populations. In many captive species, e.g., Grevy's zebra (*Equus grevyi*) and the Scimitar-horned oryx (*Oryx dammah*), both on unequal sex-ratio (fewer breeding males) in breeding individuals and a high variance of progeny number produced by males have resulted in a N_e/N ratio of 0.2 to 0.4 (e.g., Mace 1986, and citations therein). However, by planned matings or other management practices, the variance may be reduced (family sizes equalized). If the variances approach zero, then N_{ef} approaches $2N_f$ and N_{em} approaches $2N_m$. In other words, the effective size may be made double the actual size when all individuals have equal progeny numbers, a technique used in animal breeding to maximize effective population size.

The effective population size over a number of generations is also important for retention of genetic variation. When there are fluctuations in size in different generations, this effect can be combined into the expression

$$N_e = 1 \Bigg/ \left\{ 2\left(1 - \left[\prod_{i=1}^{t}\left(1 - \frac{1}{2N_{e(i)}}\right)\right]^{1/t}\right)\right\},$$

where there are t different generations and the effective size in the ith generation is $N_{e(i)}$ (Lande and Barrowclough 1986). As is well known and we will illustrate below, generations with a low effective population size (bottlenecks) have a disproportionate effect in this expression. The approximation to this expression using the harmonic mean slightly over estimates N_e, more so when there are large fluctuations in population size.

Lande and Barrowclough (1986) give a useful example that combines all of the factors mentioned above to illustrate the calculation of the effective population size. Table 4 gives the number of males and females having different numbers of progeny in three generations of a given population. For example, there were two females that each had four progeny in generation 1. As in many species, there are fewer breeding males than breeding females in this example and the variance of progeny number is higher in males than females (see Table 4).

Using the effective population size calculated for each generation in this example (Table 5), then the effective population size for the three-generation period is given by

$$N_e = 1 \Bigg/ \left\{ 2\left(1 - \left[\left(1 - \frac{1}{2(4.1)}\right)\left(1 - \frac{1}{2(9.0)}\right)\left(1 - \frac{1}{2(13.1)}\right)^{1/3}\right]\right)\right\} = 6.9$$

In other words, although the effective population size has grown to 13.1 (and

Table 4. The number of males and females in a hypothetical population having different number of offspring in three generations (after Lande and Barrowclough 1987).

Generation	No. males	No. offspring	No. females	No. offspring
1	1	9	2	4
	1	3	1	3
		1	1	
2	1	9	1	5
	1	5	2	4
	2	3	1	3
		2	2	
		2	0	
3	1	12	2	5
	1	9	2	4
	1	7	3	3
	1	5	4	2
	1	3	1	1
	3	0		

Table 5. The values calculated from the hypothetical population given in Table 4.

	Males				Females				
Generation	N_m	k_m	V_{km}	N_{em}	N_f	k_f	V_{kf}	N_{ef}	$N_{e(i)}$
1	3	6.0	18.0	1.4	4	3.0	2.0	4.1	4.1
2	4	5.0	8.0	3.4	8	2.5	3.4	6.6	9.0
3	8	4.5	20.9	5.3	12	3.0	1.6	13.7	13.1

the actual number to 36) by generation 3, the effective population size over the whole period is heavily weighted by the bottleneck in the first generation.

Of course many populations have overlapping generations so that the discrete generation model above is not applicable. In this case, the number of progeny per individual can be calculated using the birth and death rates of the population (see Lande and Barrowclough 1986). Calculating the generation length in the two sexes, T_m and T_f, as the average age at reproduction and assuming $T = 1/2\ (T_m + T_f)$, then the effective population size per average generation is

$$N_e = \frac{4T_mT_fN_{em}N_{ef}}{\overline{T}(N_{ef}T_f + N_{em}T_m)}$$

This equation is maximized for a given T when $T_m = T_f$ (differences in generation length between the sexes have a slightly lower N_e). The effective population size per unit maturation time is

$$N_e^* = N_e\overline{T}.$$

In other words, if T is increased by delaying reproduction or using older offspring for breeding, then the effective population size per unit time is increased.

Founder contribution

Many captive populations have been established from only a few individuals, e.g., Speke's gazelle (from 4 individuals), Przewalski's horse (12 plus a Mongolian domestic horse), Siberian tiger, *Panthera tigris altaria* (25), European bison, *Bison bonasus* (13), and Indian rhino, *Rhinoceros unicornis* (17). These founder numbers represent the number of animals from the wild that successfully reproduced in captivity. Often a high proportion of the wild born animals brought into captivity do not successfully breed, thereby significantly lowering the founder number of the captive population (e.g., Mace 1986). Ideally, if all founders continued to be represented equally, then the expected genetic variation would be maximized.

When there is complete pedigree information from a stud book (see Glatston 1986), then the expected (over all loci) contribution of each founder can be calculated for various generations (e.g., MacCluer *et al.* 1986; Thompson 1986). Different founder lineages can contribute more or less to extant populations for a number of reasons such as variation in reproduction, time when an individual was introduced to the population, or different generation lengths in various lineages. If we define x_i as the expected contribution to the population from founder i, then the effective number of founders can be defined as

$$1/N = \frac{1}{\sum_{i=1}^{N} x_i^2}.$$

When each founder contributes equally, then all $x_i = 1/N_x$ and the effective number of founders equals the actual number, or $N_{ex} = N_x$. We should note that unlike the inbreeding coefficient, the founder proportion is not necessarily a function of population size. For example, populations of different sizes may have the same distribution of founder contributions but a smaller population would increase in inbreeding coefficient faster than a large population. An extreme example illustrates another difference between the inbreeding coefficient and the founder proportion. Assume a population was founded by a given number of mating pairs. If each founding mating pair generated an equal sized noninterbreeding lineage, then the founder contributions would be equalized but inbreeding would not be minimized because each lineage would be descended from a different full-sib mating.

MacCluer *et al.* (1986), using replicated Monte Carlo simulations, has calculated the expected frequency (and its distribution) of a given founder allele for particular pedigrees. From this information, the proportion of a given founder genome that has been lost or is at high risk of being lost (allelic frequency <0.01 or some other criterion) can be estimated. However, there seems to be no simple approach to equalizing founder contributions after a population has been extant for a number of generations. In general, mating preference can be given to individuals containing ancestry from under-represented founders but the effect of particular breeding patterns on the founder proportions probably needs to be determined numerically following procedures suggested by MacCluer *et al.* (1986) or Thompson (1986). Cronquist Jones *et al.* (1985) discuss the related problem of incorporating new founders into a captive population. Below we will discuss founder contribution for Przewalski's horse and Speke's gazelle.

Population subdivision

Natural populations of endangered or captive species may be fragmented because available habitat or reserves are not contiguous or because captive populations may be divided among zoos, respectively. Maintenance of separate populations has the advantage of avoiding epidemics throughout the whole population and subdividing the responsibility and expense of maintenance. However, such a scheme may result in relatively high homozygosity within each subpopulation, perhaps a condition very unlike natural populations.

It has been suggested that exchange of one individual each generation between subpopulations is sufficient to keep the total population as one random-mating population (e.g., Allendorf 1983; Frankel 1983). Such conclusions are based on theoretical considerations showing that at equilibrium, the genetic differentiation among subpopulations is small. For example, for the infinite island model

$$F_{ST} = \frac{1}{4Nm + 1}$$

where N is the size of each subpopulation and m is the rate of gene flow into a subpopulation (e.g., Wright 1951). F_{ST} is a measure of genetic divergence among subpopulations, having a range from 0 for no divergence to 1 when different alleles are fixed in different subpopulations. If the number of migrants per generation is $Nm = 1$, then $F_{ST} = 0.2$, showing that there is relatively little differentiation among populations.

There are a number of assumptions in the application of this logic. First, it assumes that it is possible to successfully move individuals or their gametes between all subpopulations. For example, in natural populations it may be difficult to transfer individuals and have them bred in new subpopulations. Second, it assumes that differentiation is not wanted, i.e., selection and/or genetic drift may be detrimental. Perhaps, it might be advisable to allow different subpopulations to become genetically adapted to their particular environment. Third, this theory is based on genetic properties at equilibrium with a large number of subpopulations. Varvio *et al.* (1986) illustrate numerically that when there is a finite number of islands, the differentiation is less than that for the infinite island model and that it may take some time to approach equilibrium conditions. Finally, Ewens *et al.* (1987) discuss a model in which subpopulations go extinct as the result of catastrophes and are replaced by descendants from another subpopulation. When the rate of subpopulation extinction is relatively high, then the rate of loss of genetic variation may be controlled by this process rather than the finite size of the subpopulation.

With coordinated programs for particular species (see below) exchange between zoos is likely. Detailed pedigree information allows such exchange to bring in unrelated individuals when they are available. Potentially, this regular exchange can be of embryos or of semen when adequate techniques are available.

Retention of genetic variation

The loss of genetic variation in a population can be reduced by maximizing the effective population size, minimizing the inbreeding coefficient, or by equalizing the contribution of founder individuals. If there is no history of inbreeding, in other words, a genetically based breeding program begins from the first generation, the three different approaches to reduce loss of genetic variation lead to similar or identical results. However, when the population has been bred on some other basis and the goal becomes one of these genetic ones, then different schemes may be necessary for different goals.

Let us briefly discuss the effect of finite population size on genetic variation (see Lande and Barrowclough 1986, for an extensive discussion including quantitative genetic variation). First, the expected heterozygosity in generation t at a number of neutral loci where there is no mutation is

$$H = \left(1 - \frac{1}{2N_e}\right)^t H_0$$

where t is the number of generations and H_0 and H_t are the heterozygosities in generations 0 and t, respectively. In other words, the rate of loss of heterozygosity is higher for lower population sizes. If we let $x = H_t/H_0$, then

N_e is approximately

$$N_e = -\frac{1}{2 \ln x}.$$

For example, if we want to retain 90% of the original heterozygosity (x = 0.9), then N_e = 4.7t. If the retention is planned for 10 generations, then N_e must be around 50.

On the other hand, if the inbreeding coefficient in generation t, f_t, is known from pedigree information (Boyce 1983), then

$$H_t = \frac{(1 - f_t)}{(1 - f_0)} H_0 .$$

Obviously if $f_0 = 0$, then $f_t = 1 - x$.

In other words, to maintain 90% of the variation the inbreeding coefficient must be below 0.1 (see discussion of Przewalski's horse below). Furthermore,

$$\left(1 - \frac{1}{2N_e}\right)^t = \frac{(1 - f_t)}{(1 - f_0)},$$

because both sides of this equation are equal to H_t/H_0. If we know f_0, f_t and t as in the Przewalskis horse (discussed below), we can estimate N_e from this formula.

However, it is sometimes more appropriate to determine the effective population size necessary to retain a given amount of genetic variation for a given time period, say 100 or 200 years (see Soulè *et al.* 1986 for a more detailed discussion). Obviously, with such a goal then for short generation animals, such as the striped grass mouse, *Lemniscomys striatus*, with a generation time of less than a year, N_e must be larger than for the white-naped crane, *Grus vipis*, with a generation time of more than a decade. For example, an animal with a generation length (T) of 10 years would have approximately 20 generations in 200 years so that N_e would need to be around 100. For a species with T = 1 (t = 200), on the other hand, N_e would need to be tenfold higher or around 1000.

There has been some suggestion that maintaining as many alleles as possible is important (e.g., Allendorf 1986; Fuerst and Maruyama 1986). The expected number of alleles in generation t is

$$n_t = n_0 \sum_{i=0}^{t-1} \left(1 - \frac{n_i(n_i - 1)}{2N_e}\right)$$

assuming that the population is in equilibrium between mutation and genetic drift (Kimura 1955; Felsenstein 1971). Then the average number of alleles per locus is greater than 1.62, then the rate of loss of alleles is greater than that of heterozygosity while it is less if the number of alleles is lower. For example, if we let n_0 = 4, t = 2, N_e = 50 then n_2 is expected to be 2.90. In this case 27.5 percent of the alleles will be lost while only 2.6 percent of the heterozygosity would be lost illustrating that the loss of alleles may occur faster than the loss of heterozygosity. The distribution of allelic frequencies may not be close to the neutrality equilibrium so that the expression

$$n_{t+1} = n_t - \sum_{i=1}^{t} n(1-p)^{2N}$$

is useful (Denniston 1978). Obviously if there are many rare alleles then the number of alleles is going to decline quickly. For example, given $2N = 10$ and four alleles with frequencies 0.9, 0.06, 0.03, and 0.01 then $n_{t+1} = 2.35$, a reduction of 41 percent compared to a heterozygosity reduction of five percent.

Although these calculations indicate that many rare alleles may be quickly lost in a finite population, it seems that selective breeding for rare alleles would be unwise (Hedrick *et al.* 1986). Perhaps the most important consideration here is that increasing the contribution of individuals with rare alleles may actually decrease the overall amount of genetic variation. In other words, maintaining a rare allele of unknown fitness value that represents only a small portion of the genome could result in an increase in homozygosity in the rest of the genome.

Finally, because potential adaptation is determined by the amount of genetic variance in the population, it is noteworthy that the extent of additive genetic variance for a quantitative trait (given no dominance) is affected by finite population size in a manner identical to heterozygosity (James 1971). In other words, as also noted earlier for $H_{t,\, V_t} = (1 = 1/2N_e)^t \; V_0 = ((1 - f_t)/(1 - f_0))\, V_0$ (Lande and Barrowclough 1987, for detailed discussion).

Case studies in conservation genetics

In recent years, there has been an extensive focus on understanding the biology, including the population genetics, of particular endangered species. For 37 species, a Species Survival Plan based on knowledge of the biology of each species has been developed under the auspices of the American Association of Zoological Parks and Aquariums (Ralls 1984). As examples here of the type of information being obtained on endangered species, I will give an overview of population genetic information known for three endangered species; the cheetah, Przewalski's horse, and Speke's gazelle.

Cheetah

The cheetah, *Acinonyx jubatus*, is restricted to two wild populations in southern and eastern Africa (estimates of the species number vary from

1,500 to 25,000). Examination of 52 enzyme loci has failed to show any polymorphism although other cat species have 8 to 21 percent of their enzyme loci polymorphic (O'Brien *et al.* 1985). In addition, 14 reciprocal skin grafts between unrelated cheetahs were accepted, suggesting that those individuals studied were identical for alleles at the major histocompatibility loci.

The cheetah also suffers from several problems that may affect the potential survival of the species. First, cheetahs have quite high juvenile mortality, 28.6% for all noninbred captive-born individuals (the proportion given in Table 2 is for North American born cheetahs only). Second, spermatozoal concentrations in samples of cheetah semen are one-tenth those in domestic cats and the proportion of abnormal spermatozoa was nearly three times that of domestic cats. Finally, a feline virus infected an Oregon cat colony in 1983 and was fatal to 18 of the 42 cheetahs, while none of the ten lions or three exposed house cat kittens were affected. This observation is consistent with the hypothesis that genetic variation at the major histocompatibility loci (or other loci) is important in conferring resistant to epidemic diseases (e.g., see O'Brien and Evermann 1988).

O'Brien *et al.* (1985) suggest that the low genetic variation is an indication of a past bottleneck that reduced population size and that the high juvenile mortality and low spermatozoal concentrations are indications of inbreeding depression in the species as a whole. We should note that cheetahs that are known to be inbred have a higher juvenile mortality than noninbred progeny (Table 2). This suggests that even species with low genetic variation as measured by enzyme loci may still experience a lowered fitness when inbred, suggesting genetic variation for traits affecting fitness.

Przewalski's horse

The Mongolian wild horse or Przewalski's horse, *Equus przewalskii*, appears to be extinct in the wild. The Przewalski's horses now in captivity descended from 11 animals captured at the turn of the century, a mare caught in 1947, and a Mongolian pony. At the present, there are approximately 400 horses in Europe, Russia and North America.

The largest proportion of Przewalski's horses in the United States are descended from 12 horses imported from Munich between 1956—1963. There is complete pedigree information on all the horses which permits calculation of inbreeding coefficients of all individuals. The average inbreeding coefficient of these imported animals and the following four generations are given in Fig. 2 (from Ryder and Wedemeyer 1982). In addition, the projected inbreeding coefficient in the next several generations is given. Notice that the inbreeding coefficient is projected to be slightly reduced by planning matings between unrelated individuals and rotating head stallion between zoos.

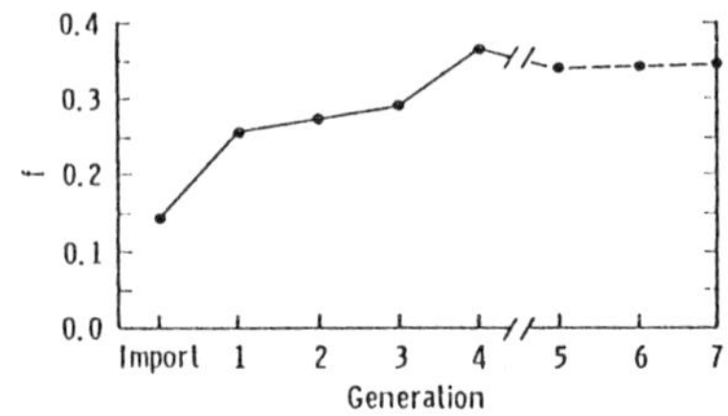

Fig. 2. The inbreeding coefficient of the North American Munich line of Przewalski's horses in the first four generations (solid line) and that projected by avoiding further inbreeding (broken line).

Using the approach given above, we can calculate the effective population size over this time period. With $f_0 - 0.13$ and $f_4 = 0.37$ then $N_e = 6.4$, quite a small value. Using this value and knowing from Ryder *et al.* that $H_4 = 0.049$, we can estimate the value of H_0. The estimated value of H_0 is 0.068, suggesting that the heterozygosity had dropped approximately 28% in the first four generations.

In the top half part of Table 6, the proportion of founder representation,

Table 6. The estimated proportion (x_i) and proportion of founder genome lost or at high risk in (a) Przewalski's horse and (b) Speke's gazelle (from MacCluer *et al.* 1986).

	founder	Estimated proportion of founder genome f_i	% loss	% risk
(a) Przewalski's horse	1	0.038	0.663	0.060
	5	0.019	0.817	0.077
	11	0.126	0.504	0.035
	12	0.063	0.710	0.045
	17	0.061	0.726	0.019
	18	0.061	0.731	0.026
	39	0.110	0.338	0.200
	40	0.098	0.442	0.165
	52	0.019	0.818	0.074
	MDOM	0.063	0.713	0.040
	211	0.067	0.751	0.010
	212	0.201	0.377	0.008
	231	0.074	0.123	0.053
(b) Speke's gazelle	Greenie	0.426	0.0002	0.0014
	Lisa	0.301	0.0	0.0003
	Iodine	0.085	0.3929	0.1711
	Chicago	0.187	0.0	0.0339

the estimated proportion of their genome that has been lost or is at high risk (allele frequency less than 0.01) for the 13 animals is given. In general, there is a negative correlation between the founder proportion and proportion of genome lost, e.g., individual 52 has $x = 0.019$ and proportion lost of 0.818. An exception is individual 231 which is the mare caught in 1947. Both the contribution and proportion lost are small for this animal. Using these x_i values, $N_{ex} = 9.48$, about 73% of the actual 13 contributing animals.

Speke's gazelle

The Speke's gazelle, *Gazella spekei*, is a small gazelle native to the border area between Ethiopia and Somalia. Because of conflict in this area and political difficulties, no new individuals can be obtained and the status in nature of this nearly extinct animal is not known. All of the animals in captivity, around 50, are descendants of one male and three females imported to the United States between 1969 and 1982. As a result of this restricted founder number and the presence of only one male, inbreeding was inevitable in this captive population. Templeton *et al.* (1983) and Templeton and Read (1983, 1984) designed a program to maintain this population.

First, there was evidence of a substantial inbreeding depression in these animals, suggesting the expectation of survival of the population was low. Table 7 gives the birth weight and the 30-day survival of inbred offspring from non-inbred parents. Both survival and birth weight were substantially affected by inbreeding. Using the standard techniques, Templeton and Read (1984) conclude that there were five to seven lethal equivalents per founder animal, similar or slightly larger than the number of lethal equivalents in outbred human populations. Because of these problems, a program was devised to both eliminate lethals from the population by selecting inbred individuals of good viability and to equalize founder contributions.

In the bottom part of Table 6, the founder proportions for the four founders are given. Iodine, who was bred less than the other females, has the lowest x_i value and the highest proportion of her genome lost while Greenie,

Table 7. The 30-day survival and average birth weight of offspring in the Speke's gazelle with different inbreeding coefficients (f) and non-inbred parents (sample size is in parentheses).

f	Birth weight, kg	30-day survival
0	1.67 (6)	0.786 (28)
1/8	1.31 (7)	0.660 (12)
1/4	1.17 (5)	0.333 (21)

the only male founder, has the highest x_i value. The effective number of founders is 3.19, about 80% of the actual founder number of 4. Figure 3 gives the simulated distributions of allelic frequencies for the four founders (from MacCluer *et al.* 1986). From this simulation, it appears that some particular founder alleles have been lost from the population. On the other hand, it seems very unlikely that any loci are fixed (identical by descent) for a given founder allele; the maximum frequency from the simulations for any founder allele is less than 0.6.

Genetic conservation perspective

The application of evolutionary genetics principles to genetic conservation should be of benefit in saving some species from extinction. We must not forget, however, that the fundamental cause of most recent extinctions and near extinctions is human-related activity in terms of habitat destruction, hunting, and human commensals. In other words, loss of genetic diversity may hasten extinction by limiting the future evolutionary potential of a species but if there is not suitable habitat left for natural populations, then problems relating to genetic diversity are moot (Hedrick 1985). Recall that our stated (and perhaps idealistic) goal here is not to preserve species in zoos for Sunday visitors but to maintain an endangered species in captivity for a period until its reintroduction into the wild becomes feasible.

The tentative nature of recommendations by population geneticists con-

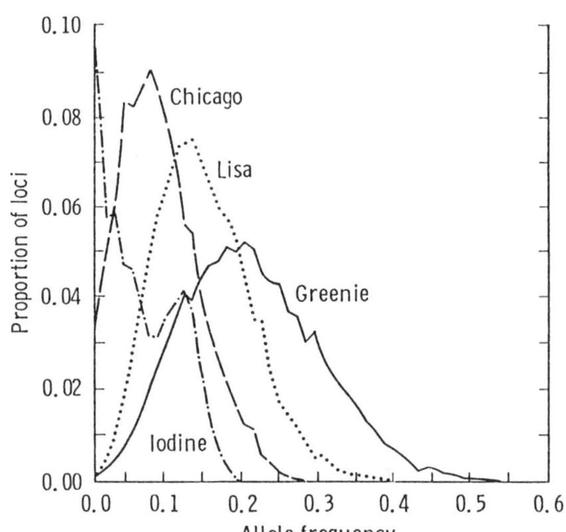

Fig. 3. The proportion of loci having different frequencies for the four founders in the Speke's gazelle.

cerning breeding schemes and management should be understood. Such plans often assume that information is known about the genetics, breeding systems, and kind of genetic variation in the target species, information that is obviously lacking in most endangered or captive species. For example, plans are now being formulated to breed the six captive black-footed ferrets without background information concerning genetic variation or inbreeding depression. In addition, presumptions such as having 25 founders, an effective population size of 50, two percent increase in inbreeding per generation, or exchanging one migrant between populations each generation, ignore the special properties of many species. Before these "magic numbers" are used, their scientific basis and logical use must be understood in order to recommend effective guidelines. If we err, it should be on the up side, suggesting too many founders, too big a population size, avoiding inbreeding too much, or too many migrants because to err on the down side may mean even greater dangers of extinction. Furthermore, as noted by Pimm and Gilpin (1989), the genetic and demographic tradeoffs in managing viable populations require further information on the short- and long-term fitness and population growth variables.

Two additional areas not mentioned earlier merit attention. First, little is known about what genetic factors are important in successful reintroduction or colonization. There are some disturbing indications that introduction success in insects is reduced by the length of time in captivity (e.g., Myers and Sabath 1981). In other words, captive breeding may preclude reintroduction unless careful steps are taken to preserve the natural genetic variation. It is encouraging to note that surplus golden lion tamarins appear to have been successfully reintroduced in Brazil.

Second, it is sometimes difficult to define the evolutionarily important unit that should be conserved (Ryder 1986). Of the 37 taxa designated for Special Survival Plans, 16 are designed as subspecies. For example the black rhino, *Diceros bicornis*, has been divided into seven subspecies, three of which are nearly extinct. When can subspecies lines be ignored and subspecies interbred? Obviously not when populations of a species are significantly differentiated so that hybrid, F_2 or backcross progeny have lowered fitness, a phenomenon sometimes called outbreeding depression (e.g., Endler 1977; Hedrick *et al.* 1978).

A relevant example of outbreeding depression (after Greig 1979) is that of the ibex, *Capia ibex ibex*, which became extinct in Czechoslavakia by overhunting. Ibex were successfully introduced from Austria, but later, to augment the herd, individuals from subspecies in Turkey and Sinai were also introduced. Although the hybrids were fertile, they rutted in early fall instead of winter, thus causing the hybrid offspring to be born in February. The young could not tolerate the cold and the population subsequently went extinct. Templeton and Read (1984) provide a scheme to detect outbreeding depression in species with known pedigrees. In initial applications of this approach, no outbreeding depression was found in the Speke's gazelle and

Ralls and Ballou (personal communication) found no outbreeding depression in several other captive mammals.

It is essential that the scientific community directly responsible for captive and endangered species continues to develop the expertise needed to prevent the extinction of many species. The sophistication of zoo biologists in appreciating genetic factors has increased dramatically in the last decade, more so in the last five years. Furthermore, the application of new reproductive technology, such as semen storage and artificial insemination, in conjunction with population genetic principles may result in long-term maintenance of genetic variation (e.g., Ballou 1984). On the other hand, I worry that some consider conservation genetics as a precisely and routinely definable technological solution to the preservation of species. This might be too heavy a burden for so inexact a discipline, relying on the complex multivariate and probabilistic analyses.

Acknowledgements

I appreciate the comments of Fred Allendorf, Jean MacCluer, and Pekka Pamilo on the manuscript.

Literature cited

Allendorf, F. W. 1983. Isolation, gene flow and genetic differentiation among populations, p. 51—65. *In* C. M. Schonewald-Cox *et al.* (eds.) Genetics and Conservation. Benjamin/Cummings.

Allendorf, F. W. 1986. Genetic drift and the loss of alleles versus heterozygosity. Zoo. Bio. 5: 181—190.

Ballou, J. D. 1984. Strategies for maintaining genetic diversity in captive populations through reproductive technology. Zoo. Biol. 3: 311—323.

Boyce, A. J. 1983. Computation of inbreeding and kinship coefficients on extended pedigree. J. Heredity 74: 400—404.

Bush, G. L. 1978. Planning a rational quality control program for the screwworm fly, p. 37—84. In R. H. Richardson (ed.) The Screwworm Problem. Univ. Texas Press, Austin.

Cavalli-Sforza, C. C., and W. F. Bodmer. 1971. The Genetics of Human Populations. W. H. Freeman, San Francisco.

Cronquist Jones, K. G., N. R. Flesness, and T. J. Foose. 1985. Incorporation of new founders into captive populations. Zool Biol. 4: 153—167.

Denniston, C. 1978. Small population size and genetic diversity. Implications for endangered species, p. 281—290. *In* S. A. Temple (ed.) Endangered Birds: Management Techniques for Preserving Endangered Species. Univ. Wisconsin Press, Madison, Wisconsin.

Ehrlich, P. and A. H. Ehrlich. 1981. Extinction: The Causes and Consequences of the Disappearance of Species. Random House, New York.

Endler, T. A. 1977. Geographic Variation, Speciation and Clines. Princeton Univ. Press, Princeton, NJ.

Ewens, W. J., P. J. Brockwell, J. M. Gani, and S. I. Reswick. 1987. Minimum viable population size in the presence of catastrophes. *In*: M. E. Soulè (ed.) Viable Populations for Conservation. Cambridge Univ. Press, Cambridge.

Felsenstein, J. 1971. The rate of loss of multiple alleles in finite haploid populations. Theoret. Pop. Biol. 2: 391—403.
Frankel, D. H. 1983. The place of management in conservation, p. 1—14. *In* C. M. Schonewald-Cox *et al.* (eds.) Genetics and Conservation. Benjamin/Cummings.
Frankel, D. H. and M. E. Soulè. 1981. Conservation and Evolution. Cambridge Univ. Press, Cambridge.
Frankham, R., H. Hemmer, D. A. Ryder, E. G. Cothran, M. E. Soulè, N. D. Murray, and M. Snyder. 1986. Selection in captive populations. Zoo. Biology 5: 127—138.
Fuerst, P. A., and T. Maruyama. 1986. Considerations on the conservation of alleles and of genic heterozygosity in small managed populations. Zoo. Biol. 5: 171—180.
Glatston, A. R. 1986. Stud books: the basis of breeding programs. Inter. Zoo. Yb. 24/25: 162—167.
Greig, J. C. 1979. Principles of genetic conservation in relation to wildlife management in Southern Africa. S. Afr. J. Wildlife Res. 9: 57—78.
Haldane, J. B. S. 1940. The conflict between selection and mutation of harmful recessive genes. Ann. Eugen. 10: 417—421.
Hedrick, P. W. 1983. Genetics of Populations. Jones and Bartlett, Boston.
Hedrick, P. W. 1985. Genetics and conservation. Evolution 39: 1180—1181.
Hedrick, P. W., P. F. Brussard, F. W. Allendorf, J. A. Beardmore, and S. Drzack. 1986. Protein variation, fitness, and captive propagation. Zool. Bio. 5: 91—99.
Hedrick, P. W., S. Jain, and L. Holden. 1978. Multilocus systems in evolution. Evol. Biol. 11: 101—182.
Kimura, M. 1955. Random genetic drift in multi-allelic locus. Evolution 9: 419—435.
James, J. W. 1971. The founder effect and responses to artificial selection. Genet. Res. 16: 241—250.
Lande, R., and G. R. Barrowclough. 1987. Effective population size, genetic variation, and their use in population management. *In*: M. E. Soulè (ed.) Viable Populations for Conservation. Cambridge Univ. Press, Cambridge.
MacCluer, J. W., J. L. Vandelberg, B. Read, and D. A. Ryder. 1986. Pedigree analysis by computer simulation. Zoo. Biol. 5: 148—160.
Mace, G. M. 1986. Genetic management of small populations. Intern. Zoo. Yearbook 25: 167—174.
Myers, J. H., and M. D. Sabath. 1981. Genetic and phenotypic variability, genetic variance, and the success of establishment of insect introductions for the biological control of weeds, pp. 91—102. *In* E. S. Delfolsse (ed.), Proceedings V International Symposium on the biological Control of Weeds, Brisbane, CSIRO, Canberra, Aust.
O'Brien, S. J. and J. F. Evermann. 1988. Interactive influence of infectious disease and genetic diversity in natural populations. Trends Ecol. Evol. 3: 254—259.
O'Brien, S. J., M. E. Roelke, L. Marker, A. Newman, C. A. Winkler, D. Meltzer, L. Cally, J. F. Everman, M. Bush, and D. E. Wildt. 1985. Genetic basis for species vulnerability in the cheetah. Science 227: 1428—1434.
Pimm, S. L. and M. E. Gilpin. 1989. Theoretical issues in conservation biology. *In*: J. Roughgarden, R. M. May, and S. A. Levin (eds.) Perspectives in Ecological Theory, Princeton Univ. Press, Princeton. pp. 287—304.
Ralls, K. 1984. Genetic management of zoo animals. Biologist 21: 99—102.
Ralls, K., and J. D. Ballou (eds.). 1986. Proceedings of the Workshop on Genetic Management of Captive Populations. Zool Biol. 5(2).
Ralls, K., J. D. Ballou, and A. Templeton. 1988. Estimates of lethal equivalents and the cost of inbreeding in mannals. Conserv. Biol. 2: 185—193.
Rao, P. S. S., and S. G. Inbaraj. 1979. Trends in human reproductive wastage in relation to long-term practice of inbreeding. Ann. Hum. Genet. 42: 401—413.
Ryder, D. A. 1986. Species conservation and systematics: the dilemma of subspecies. Trends Ecol. Evol. 1: 9—10.
Ryder, D. A., P. C. Brisbin, A. T. Bowlking and E. A. Wedemeyer. 1981. Monitoring genetic

variation in endangered species, p. 417—424. *In* G. G. E. Scudder and J. L. Reveal (eds.) Evolution Today, Proc. Sec. Intern. Congr. Syst. Evol. Biol.

Ryder, D. A. and E. A. Wedemeyer. 1982. A cooperative breeding program for the Mongolian wild horse *Equus przewalskii* in the United States. Biol. Cons. 22: 259—271.

Schonewald-Cox, C. M., S. M. Chambers, B. MacBryde, and L. Thomas (eds.). 1983. Genetics and Conservation. Benjamin Cummings, Menlo Park, California.

Shaffer, M. L. 1981. Minimum population sizes for species conservation. Bioscience 31: 131—134.

Slatis, M. H. 1960. An analysis of inbreeding in the European bison. Genetics XXX: 275—287.

Soulè, M. E. (ed.) 1986. Conservation Biology: The Science of Scarcity and Diversity. Sinauer, Sunderland, Mass.

Soulè, M., M. Gilpin, W. Conway, and T. Foose. 1986. The millenium ark: How long a voyage, how many staterooms, how many passengers. Zoo. Biol. 5: 101—114.

Soulè, M. E., and D. Simberloff. 1986. What do genetics and ecology tell us about the design of nature reserves. Biol. Conservation 35: 18—40.

Templeton, A. R., H. Hemmer, G. Mace, U. S. Seal, W. M. Sheilds, and D. S. Woodruff. 1986. Local adaptation, coadaptation, and population boundaries. Zool Biol. 5: 115—125.

Templeton, A. R. and B. Read. 1984. Factors eliminating inbreeding depression in a captive herd of Speke's gazelle. Zool Biology 3: 177—199.

Thompson, E. A. 1986. Ancestry of alleles and extinction of genes in populations with defined pedigrees. Zoo. Biology 5: 161—170.

Varvio, S-L., R. Chakraborty, and M. Nei. 1986. Genetic variation in subdivided populations and conservation genetics. Heredity 57: 189—198.

Woodruff, D. S. 1989. The problems of conserving genes and species. pp. 76—88, eds. D. Western and M. C. Pearl, Conservation for the Twenty-first Century. Oxford Univ. Press.

Wright, S. 1951. The genetical structure of populations. Annals Eugenics 15: 323—354.

Ziswiller, V. 1967. Extinct and Vanishign Animals. Springer-Verlag. New York.

4. Population biology and conservation of rare plants

RICHARD V. KESSELI

Abstract

Conservation of rare and endangered plant taxa and their habitats requires an understanding of the population dynamic factors affecting recruitment, reproduction, dispersal and genetic capacity to evolve. Population size, dispersion and certain genetic structure variables determining effective population size are reviewed in order to emphasize the maintenance of genetic variation in small, subdivided populations. Research on an annual plant genus, *Limnanthes* living in vernal pools as islands and containing several rare taxa, provided excellent examples of many of these genetic and demographic properties. Genetic variation patterns based on allozyme surveys appeared to fit an island model such that rarity based on narrow geographical range and highly localized gene flow clearly affected the patterns. Several other rare plant examples are also reviewed briefly in order to point out the critical population biological research needed for designing nature reserves and their monitoring or management strategies. Current research in conservation biology should advance the *in situ* approaches, and some basic field studies as outlined here; in most cases, even preliminary information on life history, genetic structure, gene flow, breeding system and population size variation would prove highly valuable.

Introduction

Presently we are in the midst of a mass extinction with the present worldwide rate estimated at 40—400 times greater than those of the past. Some 20% of all species will be lost by the year 2000; probably one to five million species in the next 20 years (Myers 1983; Global 2000 Report 1980). In the U.S. more than 20,000 species are endangered particularly in Hawaii, Texas, Florida, and California. Disruption of habitats and overall environmental degradation by humans is the ultimate underlying cause of the accelerated extinction rates. The biological, economic and social consequences of this potential disaster have been clearly stated (Myers 1983; Frankel and Soule 1981; Ehrlich 1983). To abate the losses, even on a local or regional level, we as biologists, resource managers, and citizens must define rarity and characterize endangerment based on our studies of particular ecological systems and the biology of specific taxa. Population genetic principles need

S. K. Jain and L. W. Botsford (eds), Applied Population Biology, 69—90.

to be widely understood and applied in order to estimate the minimum requirements for species survival, to implement preservation programs, and to monitor, record, and analyze the successes and failures of such programs. We must be willing to readjust our programs as more scientific information becomes available and to manage populations using genetic and ecological principles.

This chapter will first define rarity and endangerment. Certain population genetic variables which affect viability of habitats and biota and are readily altered by human induced stresses, will be reviewed briefly. These include decreased population sizes and numbers, which enhance genetic drift; increased inbreeding and its associated inbreeding depression; and decreased gene flow which reduces neighborhood size, and insularizes metapopulation complexes. Specific intrinsic characteristics as well as extrinsic interactions which may be crucial to a species survival will then be briefly mentioned. All of these factors taken together impose characteristic levels of variability partitioned within and among populations of a taxon. How this is measured and the levels found in nature are the topics of the subsequent section. Decisions regarding numbers and locations of populations needed to maintain "certain desirable" levels of variation with a "reasonable" assurance of survival are addressed in the section on minimal population size and population structure. Finally, general recommendations for conservation in practice are proposed in the final section of this chapter.

Rarity and endangerment

The definitions of rarity and endangerment have been somewhat arbitrary. To standardize them, Rabinowitz (1981) outlined a scheme for describing plant species rarity based on three sets of population biological parameters; geographic range, habitat specificity, and local population size. A variety of circumstances can produce rarity (Stebbins 1980, and Table 1), including newly evolved endemics (*Stephanomeria malheurensis*; Gottlieb 1973) or relics (*Sequoiadendron giganteum*; Raven and Axelrod 1977). The protection of "ephermeral" species may not be adopted as our responsibility or capability, but formerly well-established species which have been drastically reduced by human activities should be given priority. Thus, a fourth, historical dimension integrating data on range expansion or contraction should be added to the typology of rarity criteria. Many vernal pool taxa in family Limnanthaceae and grass tribe Orcuttieae, and two members of the sunflower genus (*Helianthus*) have suffered in this manner. This group also represents a range of economically, scientifically, and aesthetically important species. The genus *Helianthus* has a well established crop species, sunflower (*H. annuus*). The genus *Limnanthes* is a potential source of new industrial oil crop (Gentry and Miller 1965; Pierce and Jain 1977) and the tribe Orcuttieae possesses "the most unusual and rarest of the California grasses"

Table 1. Rare plant taxa and reasons for rarity.

Taxa	Reasons for rarity	Source*
Lupinus subcarnosus	Low variation, restricted to edaphic conditions, central Texas	Babbel and Selader 1974
Clarkia lingulata	Recent origin	Gottlieb 1974
Clarkia franciscana	Low variation, serpentine restricted, S.F.	Lewis and Raven 1958
Oenothera argillicola	Restricted to Appalchian shale barrens, relic	Levy and Levin 1975
Oe. gradiflora	Restricted to Appalachian shale barrens, relic	Levy and Levin 1975
Oe. organensis	Low variation, and seed dispersal, deep canyon habitat	Levin *et al.* 1979
Stephanomeria malheurensis	Recent origin, low variation	Gottlieb 1973
Plantago cordata	Low reproductive effort, low dispersal, habitat destruction	Meagher *et al.* 1978
Perennial grass sp.	Sparse species, low density, wide spread high dispersal, poor competitors	Rabinowitz 1978
Lephechinia spp.	Recent origin	Hart 1985
Pinus torreyana	Low variation, shrinking substrates, relics	Ledig and Conkle 1983
Sedum spp.	Rock outcrops, specific substrates, relics	Denton 1983
Sullivantia spp.	Low variation, specialized habitats	Soltis 1982
Astragalus spp.	Pollinator availability	Karron 1985
Lycopersicon parviflorum	Recent origin, self-pollinator	Rick *et al.* 1976

* For source references, see Hamrick *et al.* (1979), and Karron (1987).

(Crampton 1959). Among the 27 taxa listed, nine appear endangered (three rare rankings) and eight appear threatened (two rare rankings).

The current and historical status of many of these taxa has been roughly estimated from demographic studies and collection reports (Table 2). Regional (major watersheds) extirpation has been common for many of these taxa and seven deserve immediate attention. However, some rare taxa are persistent. The only two populations of the very restricted taxon *L. floccosa* ssp. *pumila* were collected on Table Rock in Oregon in 1886 by J. T. Howell

Table 2. Status of Limnanthes[a], Orcuttia and Tuctoria[b].

Taxon	Geographic Range (km)	No. of regions		No. of populations	
		Past	Present	Past	Present
Limnanthes					
macounii	10	2	2	28	25
bakeri	10	1	1	5	4
vinculans	15	2	1	22	19
sulphurea	95	3	2	7	4
gracilis	40	3	3	12[c]	8
pumila	1	1	1	2	2
grandiflora	5	1	1	2	2
californica	25	3	2	14	5
parishii	15	4	2	22	18
Orcuttia/Tuctoria					
tenuis	200	3	3	13	5
greenei	270	8	3	19	14
pilosa	350	4	3	15	10
inaequalis	60	6	3	16	13
viscida	5	1	1	3	3
californica	220	4	2	13	3
mucronata	1	1	1	1	1
fragilis	20	- - - - - - - - - - - -	no data	- - - - - - - - - - -	

* Data for *Limnanthes* from Mason 1952, Arroyo 1973, Ornduff 1968, Kesseli and Jain 1984, McNeill and Jain 1983, and personal observations by Kesseli and Jain.
[b] Data from Griggs and Jain (1983) and personal observations by T. Griggs.
[c] Orcuttieae from Reeder 1982, Griggs and Jain 1983, Crampton 1959, personal obs.

and are still present today. The stable persistence of some taxa is caused in part by the fortuitous location outside the region of human exploitation. Taxa residing near expanding urban or farming areas (e.g. San Francisco Bay, foothills around the Central Valley) are particularly susceptible to extirpation and one estimate showed that only 12% of the vernal pools once in existence in the Central Valley are still present (Holland 1976).

Effective population size and genetic drift

In order to reduce the stochastic process attributable to finite population sizes to its simplest form and to facilitate mathematical investigations of its effects, Wright (1968 and earlier) developed the concept of effective population size (N_e). Effective population size defines the number of breeding individuals in an idealized population with discreet generations, constant size, and no violations of the Hardy-Weinberg Law other than finite size. For finite populations which do not fit the idealized structure (most populations in nature), the actual number of breeding individuals can be converted into the effective number. Four examples are given here to show how various factors in nature might change N (for an excellent review, see Crow 1986, and Hedrick, Chapter 3).

In dioecious populations with unequal sex ratios the effective population size is less than the actual size. For example, small population with 90 females and 10 males has an effective population size of 36, as given by:

$$N_e = \frac{4N_m N_f}{N_m + N_f},$$

where N_m and N_f are the number of males and females, respectively. That is, the above population with 100 individuals will be subjected to the same stochastic consequences as an idealized population of 36 individuals. This seems reasonable since the 10 males of this population are contributing half of the total genetic information to the next generation.

The conditions for an ideal population do not stipulate that each parent contributes a constant number of progeny to the next generation, but that the contribution be a random variable with mean k = 2, and variance $s^2 = k$. The statistic s_k^2 is the variance for the average number of surviving progeny per parent (k). K, $N_e < N$, as is the case in most natural populations. For $s^2 \ll k$, we expect $N < N_e \leqslant 2N$, a possible outcome of manipulated populations. The expression,

$$N_e = \frac{4N - 2}{s_k^2 + 2},$$

is used to estimate the effective number caused by the skewed contributions of progeny to succeeding generations (Crow 1986).

If we assume no selection, i.e. only random survival, between seed production and adult survival the next generation, then the mean (k^-) and variance (s_k^2) in seed number can be converted to the expected mean (k = 2) and variance (s_k^2) of surviving progeny,

$$s_k^2 = s(1 - s)\,k' + R^2 s_k^2$$

were R = (k/k′ (Crow and Morton, 1955).

Table 3 gives estimate of N_e for the California endemic *Limnanthes douglasii* var. *rosea* in a 15 m^2 vernal pool. Quadrats were placed at intervals on transects through the 15 m^2 population. Plants were counted and scored for seed set in three consecutive years. Population sizes (N) were estimated to be as high as 4500 but effective population sizes (N_e) were often less than half the actual number.

The assumption of random mating is nearly always violated in plant populations. Outcrossing rates are often not 100%, and neighborhood structure with limited gene flow often yields consanguineous matings. This non-random mating features again reduces N_e so that.

$N_e = N/1 + F$, $F = (1 - \alpha/1 + \alpha)$, where F, Wright's inbreeding coeffi-

Table 3. Estimation of effective population size in a population of *L.d.* var. *rosea*.

Year	Plants/m^2	N	seeds/plant	$s^2_{k'}$	s^2	Ne′/N	N_e
1983	173	2595	9.2 ± 1.30	60.84	4.42	0.62	1490
1984	103	1545	25.4 ± 4.42	957.28	7.82	0.41	586
1985	312	4680	5.3 ± 0.77	37.95	6.73	0.46	1994

$\bar{N}_e$ (t = 3 generations) = 1042.

s^2_k = variance of seed set,
$s^2_{k'}$ = variance adjusted for $\bar{k}$ = 2 (see text),
N = estimated number of individual plants in the population.
$N'_e = (4N - 2)/(2 + s^2_k)$, and
$N_e = N_e/1 + F$; F is Wright's inbreeding coefficient, calculated using 0.85 as an estimate of outcrossing rate.

cient is a function of outcrossing rate α ($0 \leqslant \alpha \leqslant 1.0$). Outcrossing rates estimated for populations of *L.d.* var *rosea* average about 0.85 (Kesseli and Jain 1985). If we assume this value per generation, N_e is further reduced. Populations which periodically fluctuate for the number of breeding individuals have an effective number strongly influenced by the generations with the fewest. The harmonic mean of the numbers in each generation approximates the value of N_e. For the three generations of the *L.d.* var. *rosea* data (Table 3), $N_e = 1042$ from the equation:

$$N_e = t \Big/ \sum_{i=1}^{t} 1/N_i\,,$$

where t is the number of generations and N_i is the effective population in the ith generation. While there had been about 8820 individuals in this population over the three years, with respect to genetic drift the population behaved like one of 1042 individuals for one generation. Note that *Limnanthes* populations have probably witnessed much greater extremes than recorded in these three years (i.e., the California drought years of 1977 and 1978), thus our estimates of N_e are still probably high.

With effective population size defined, the outcome of violating the infinite population size requirements of the Hardy-Weinberg law can be investigated. In an infinitely large and expanding population, all gametes (and alleles) can be transmitted to succeeding generations. This is not possible, however, in finite populations and consequently only a small sample of the gametes (and alleles) are transmitted. The properties of finite populations are three; increased homozygosity, random extinction of alleles, and increased variances of allele frequency among populations. These changes occur at a rate of $1/2N_e$. Thus, for example, if $N_e = 10$, then 5% of the genetic variation will be lost each generation. In the *L.d.* var. *rosea* example, the loss of genetic variation due to drift would be 0.5% per generation. While this is not substantial, population sizes for several endangered taxa of *Limnanthes* and *Tuctoria* are often very low (N < 500, Table 4). Many of these taxa have higher selfing rates and severe population size fluctuations, therefore N_e may be even more reduced than it was for the population of *L.d.* var. *rosea.* For example, two of the known populations of *L.f. californica* have had less than 100 plants in 1984. Data on population sizes in several *Limnanthes* taxa, including *L. floccosa* ssp. floccosa and *L.f. californica*, provided estimates of the inter-year coefficient of variation and life history components of population regulation (Jain 1990).

Inbreeding

Non-random mating in natural populations is usually in the form of inbreeding. In plants, inbreeding is attributable to the developmental biology or population structure of a taxon. The developmental properties of the flower, viz. the physical relationship of anthers and stigma (e.g., exerted and inserted styles of *Lycopersicum* Rick 1976), the timing of stigma receptivity and anther dehiscence (e.g., protandry in *Limnanthes*) and the physical interaction of pollen and style (e.g., self-incompatibility systems in *Helianthus*; Heiser 1949) strongly affect the rates of self-pollination. These characteristics are peculiar to different species and even different populations (Schoen 1983). Their overall effect can be measured by comparing maternal genotypes with those of their progeny and deriving the statistic of outcrossing rate (Ritland 1983). Kesseli and Jain (1985) reported variation in outcrossing

Table 4. Diversity statistics based on 19 isozyme loci.*

Taxon	N	H_T	H_S	D_{st}	R_{st}
L. macounii	8	0.004	0.004	0.001	0.17
L. bakeri	3	0.01	0.07	0.02	0.47
L. vinculans	4	0.12	0.11	0.01	0.11
L. striata	8	0.26	0.14	0.12	1.01
L.d. rosea	13	0.18	0.14	0.03	0.26
L.d. inland	14	0.22	0.17	0.05	0.29
L.d. coastal	9	0.11	0.10	0.02	0.18
L.d. "Oregon"	2	0.10	0.06	0.04	1.20
inland, coastal and Oregon	25	0.25	0.13	0.11	0.88
All taxa	61	0.30	0.13	0.17	1.32

* Definitions in the text after Nei (1975).
H_T = total gene diversity
H_S = intra-population diversity
D_{st} = inter-population diversity

rates for three populations of *Limnanthes douglasii.* For the entire genus *Limnanthes* outcrossing rates (α) span the full range of values (Fig. 1). In addition to the role of protandry, cleistogamy and various floral features, population substructure accounts for inbreeding in the highly allogamous taxa. Because of generally short dispersal distances and non-random distribution (clustered around the maternal plant) of seeds, often adjacent individuals have a common ancestry. Thus, near neighbor matings may be consanguineous matings which can produce cumulative effects of inbreeding.

In natural populations, geneological data for estimating which allow one to "identity by descent" of different alleles, are not readily available, but as noted earlier, the expected equilibrium inbreeding coefficient F_e can be derived from outcrossing rate estimates (t). In addition, the cumulative effects of inbreeding, due to self-fertilization and consanguineous matings, on heterozygosity can be calculated from the fixation index, $F = 1 - H_o/H_e$ where H_o and H_e are respectively the observed and expected heterozygosity levels, assuming Hardy-Weinberg equilibrium conditions (Brown 1979). Data from *Limnanthes* species show some self-fertilization even in highly protandrous taxa and substantial inbreeding due to consanguineous matings (Kesseli and Jain 1985). For six populations of *L. douglasii*, consanguineous matings contributed an average of 50% to the total value of F at the zygote stage of the life cycle.

Changes in levels of inbreeding effect both the distribution of variability in populations (Jain 1975; Brown 1979) and the genome organization within individuals (Hedrick *et al.* 1978; Brown 1979). These changes can be brought about by disruption in the pollination system (e.g., loss of pollinators) or drastic reductions in population structure (e.g., bottlenecks). Significant fitness loss due to inbreeding depression' has been extensively discussed and quantified for animals (Chapter 3 for references). Inbreeding depression in natural populations of plants is also widespread but variable (Charlesworth and Charlesworth 1987, for review of methodology and many examples).

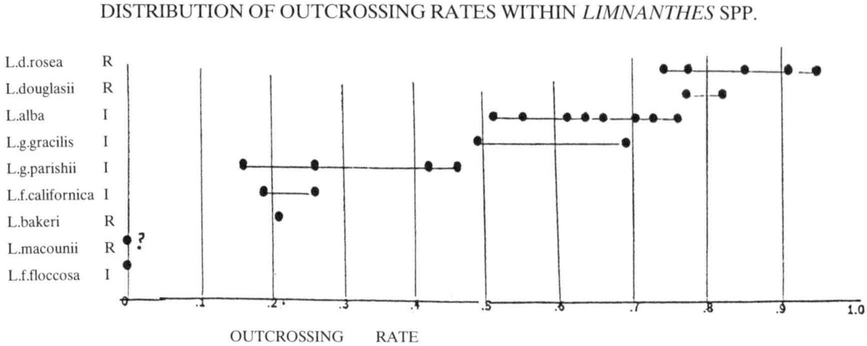

Fig. 1. Variation in mean outcrossing rate within and between *Limnanthes* taxa. Each dot represents a population estimate.

Gene flow

Gene flow is the process by which local populations exchange genetic variants. It is distinguished from migration (seasonal or cyclical movements of groups of animals) and dispersal (random and localized movement of individuals or propagules). Neither migration nor dispersal necessarily, produce the mixing of gene pools (Endler 1977). In plants, gene flow is mediated by pollen or seed dispersal and its consequences are higher amounts of variation within populations and smaller genetic differentiation among populations.

Direct measurements of gene flow are difficult. The tracking of dispersal units (pollen and pollinator or seed) is commonly used but does not measure the entry of the alien alleles into the target gene pools. These measures do, however, define life history characters and species interactions which do have significant influence on the rate of gene flow in natural populations (Levin and Kerster 1974; Beattie and Culver 1979). Genotypic or phenotypic changes from adult to offpsring provide rough estimates of gene flow, particularly if donor and target populations possess large genotypic differences. Selective forces (differential male reproductive capacities or seedling viabilities, etc.) might, however, confound the estimates. This method, however, has been used with some success in controlled spatial patterns designed to monitor intervarietal or crop-weed contamination (Bateman 1947) and in populations with step clines (Jain and Bradshaw 1966).

A more refined approach utilizes the dynamics of monitor unique alleles or genotypes. Schaal (1980) placed plants carrying a marker gene in the center of artificial populations. The progeny of the surrounding plants were then scored and gene flow distances were estimated. In natural populations this method can be used by transplanting plants with novel alleles or scoring and mapping native individuals with rare alleles. The lack of large numbers of naturally occurring rare alleles has been partially resolved by scoring multilocus genotypes, in certain paternity studies (Meagher 1986; Ellstrand and Marshall 1985).

A different approach was developed by Slatkin (1985) for estimating the number of migrants between populations based on the average frequency of so-called private (present in only one local population) alleles found among populations. Slatkin (1985) analyzed different data sets and found the average number of immigrants (Nm, where N = effective population size, m = mean migration rate among subpopulations) ranges from 0.16 to 42.0 per generation.

For several *Limnanthes* taxa within population gene flow due to pollen and seed was investigated. Using plots cleared of aboveground vegetation, seedling census gave dispersal distances of several meters in vernal stream population of *L. d.* var. *rosea* but less than two for vernal pool populations (Fig. 2). Gene flow via pollen is also generally restricted as measured by naturally occurring rare alleles. In a population of *L. d.* var. *rosea*, 85% of the plants received pollen from donors within 10 cm or less radius, as judged

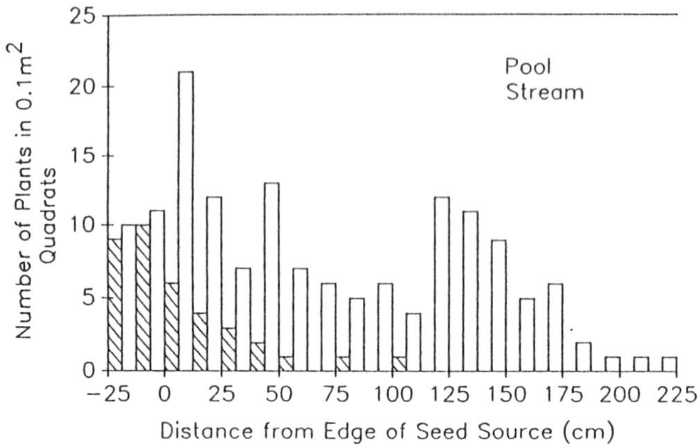

Fig. 2. Gene flow estimates based on the distribution of seed, estimated from plant counts, from a release print (or quadrat) into the adjacent vernal pools and along a stream.

by the rare allele data, whereas less than 5% plants would show pollen parents from as far as 80 cm (Fig. 3).

Long distance pollen dispersal between neighboring vernal pools does occur, however. R. Thorp (unpubl. data) marked pollen with florescent dye and found 97% of the dye was distributed by the pollinators within 5 meters although occasionally long distance dispersal up to 13 m did occur. Using Slatkins' (1985) model, I estimated the number of migrants among populations of three taxa (Kesseli 1984; Kesseli, unpubl. data). The Central Valley *populations of Limnanthes douglasii* var. *rosea*, which have no major ecogeographic barriers to gene flow, had a large Nm value (3.0) indicating a high

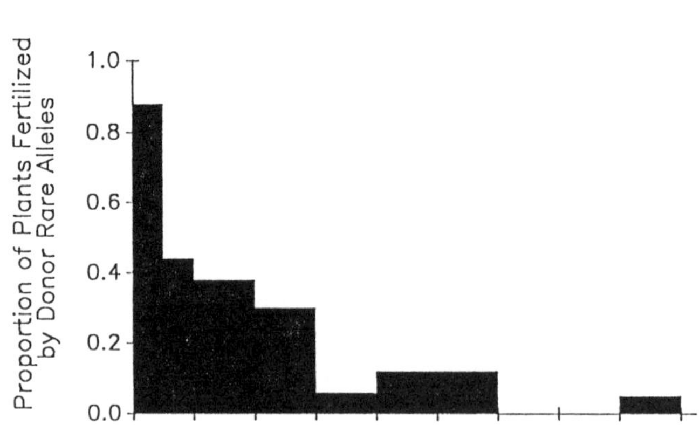

Fig. 3. Gene flow estimates based on pollen movement, estimated from the distribution of rare alleles among mapped individuals.

level of interpopulation gene flow. Two other taxa, *L. douglasii* var. *nivea* and *L. striata*, are located in the coastal mountain ranges and Sierra foothills, respectively, with a pattern of partial isolation among different watersheds. These taxa showed low interpopulation gene flow, Nm = 0.3 and 0.05, respectively.

The prospects that populations of endangered species can colonize new sites or replenish genetic variation depends on the rates of dispersal and gene flow. With an increasing loss of populations and destruction of habitat throughout the range of a species, rates of dispersal and gene flow are necessarily altered. Parameters of interpopulation dispersal, recruitment rates, patch turnover and gene flow in developing the design of reserves. Demographic variables are to be emphasized in developing genetic approaches to conservation (Jain 1991).

Population biology and ecology of target species

For conservation efforts to be successful, the peculiar life history characters and biotic and abiotic interactions of a species should be considered. For example, seed dormancy causes overlapping generations and produces population age structure similar to that in perennials. This feature buffers populations against periodic selective or stochastic stresses. It can retard the rate of allele frequency changes or change the eventual genetic outcome of selective interactions (Templeton and Levin 1979). Dormancy appears to be a major factor in the life histories of certain *Orcuttia* spp. and *Limnanthes* spp. Ritland and Jain (1984) modeled and discussed the evolutionary consequences of dormancy in a comparative study of taxa with short life cycle (*L.f.* ssp. *floccossa*) verus longer life cycle (*L. alba*).

Extrinsic factors like mutualistic biotic interactions must be known. The pollination vector of target species must be given adequate habitat in a nature reserve. In all *Limnanthes* spp. the major pollinators are the oligolectic solitary bees, *Andrena* spp. as well as the introduced *Apis mellifera*. The pollinator movements of *A. mellifera* are shorter than those of *Andrean* spp.; thus neighborhood size would be reduced with the loss of *Andrean* spp. and inbreeding may increase.

Probably more significant for the survival of a target species are the prevalence of antagonistic agents (Futuyma 1983). The California prairie, once dominated by perennial grasses like *Agrostis*, *Deschampsia*, and *Stipa* has been transformed in 200 years to an annual community dominated by *Avena* ssp. *Bromis mollis* and *Erodium* spp.. The weedy invaders coupled with intense cattle grazing destroyed the once highly productive system (Burcham 1957). These changes reduced the carrying capacity of the region and aided the elimination of many large mammals. The introduction of an alien pathogen (or removal of native ones) can also drastically alter selective regimes and eliminate target species (Futuyma 1983).

Levels of genetic variability

The above population genetics forces give species their unique characteristics and produce specific levels of variation distributed within and among populations. A major aspect of conservation biology is not to save only a few individuals of a species, but to preserve in "healthy" biological systems as much of a species variability spectrum as possible. Retention of this intraspecific variation is important on a practical level since species survival may depend on the presence of unique sets of genes. To assess this variability, what kinds of characters do we need to study and how do we identify and measure uniqueness and variation?

Two basic classes of characters are used, quantitative and discrete. Quantitative characters are measured ones such as height, number of flowers, physiological differences, and competitive ability, which are controlled by several to many genes and influenced by environmental conditions. To be useful as evolutionarily important characters they must be shown to be heritable. In the least common garden experiments can show this. For more sophisticated analyses, the variance of different quantitative characters can be partitioned into genetic and environmental components in experiments involving different lines with some degree of relatedness (parent-offspring regressions) or in artificial selection experiments. From these the statistic of narrow-sense heritability can be measured. By using many quantitative characters one can estimate the degree of differentiation among populations. The statistical framework of numerical taxonomy has been developed by Sneath and Sokal (1973).

Many of the evolutionary important characters of a taxon tend to be quantitative. Nearly all components of fitness show polygenic inheritance (Wright 1968; Lawrence 1984). Although they are critically important characters there are disadvantages to using them. One is the extra effort needed to confirm heritability. Another is the difficulty in standardizing the contribution of many different characters controlled by unequal numbers of genes and affected unequally by the environment.

Discrete characters are controlled by few genes (often one) and are minimally influenced by environmental factors. They can be scored at the whole plant level (seed coat or flower color, pubescence and even certain pathogen resistance genes), the cellular level (isozyme or biochemical markers), and the DNA level (base pair sequence data or location of restriction endonuclease cutting sites). These characters have the utility of having simple Mendelian inheritance and thus subjective interpretation is not necessary and standardization is possible. While measuring the variability at the DNA level is relatively slow (but quickening) and very expensive, isozymes and morphological marker data are easy and cheap to collect. These genetic differences can be readily transformed into identity statistics for population and taxon comparisons (Nei 1975, 1987).

A major disadvantage of these discrete characters is that they generally have low or unknown contributions to components of fitness. Even though much of the allelic variation may be selectively neutral, these markers can identify unique populations which often differ in whole suites of characters, some of which contribute to fitness. Our surveys of isozymes in *Limnanthes* spp. revealed several unique populations and regional clusters of populations, originally undetected in taxonomic descriptions, but later shown to be different for characters such as flowering time, flower shape, protandry, nutlet morphology (Fig. 4), and others (McNeill 1983; Kesseli and Jain 1984). Neutral markers are also useful in gene flow, drift, and breeding system studies.

Levels of variation detected by the different classes of characters tend to show concordance (Gottlieb 1981). In *Limnanthes*, the isozyme data generally agreed with subspecies and varietal classifications based on traditional morphological characters although there are notable exceptions. Our recent work with cultivated and wild species of lettuce (*Lactuca* spp.) again shows isozyme-morphological data concordance (Kesseli and Michelmore 1986) as well as general agreement of these with DNA (restriction fragment length polymorphism) data (Landry *et al.* 1987a, b).

Next, I shall note the utility of these single locus markers for partitioning variation into hierarchial levels of population structure. With allelic frequency data Nei's (1975) diversity statistics can readily partition total diversity (H_T) into intra- (H_s) and inter- (D_{ST}) populations components and the overall uniqueness of populations can be assessed. The range in diversity statistics for *Limnanthes* spp. (Table 4) reveal several generalizations about structure. The inbreeders (*L.f.* ssp. *floccosa*, *L. gracilis* vr. *parishii* and *L. bakeri*) while often retaining as much total variation as outcrossers partition

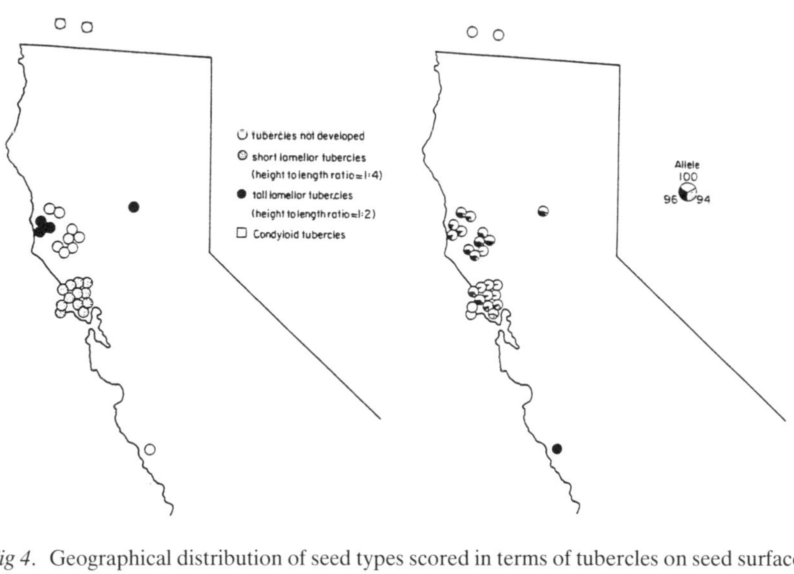

Fig 4. Geographical distribution of seed types scored in terms of tubercles on seed surface.

most of their variation among populations (also see Brown 1979; Gottlieb 1981; Hamrick 1983).

Taxa with large geographic ranges also possess more total diversity (H_T) and more alleles (Fig. 5). Broadscale comparisons of diverse taxonomic groups have also identified variability in life history relationships. Hamrick

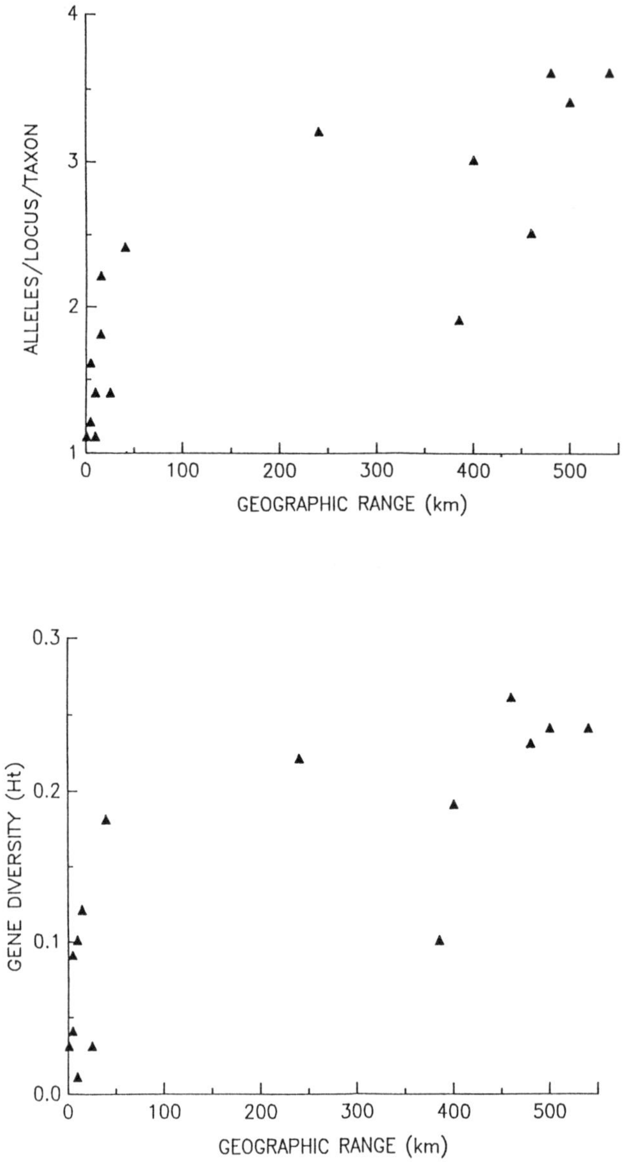

Fig. 5. Correlograms showing allelic diversity, gene diversity, and geographic range in *Limnanthes*.

et al. (1979) identified a positive association between longevity and heterozygosity and Brown (1979) noted that wind pollinated taxa have low interpopulation diversity (D_{ST}).

The explanations for the above generalizations are firmly based in population genetics theory. Interactions between breeding system, drift, gene flow, mutation and selection alter the structure of a taxon making conservation requirements diverse.

The conservation of crop genetic resources has features similar to that of conservation biology. Marshall and Brown (1975) make four points particularly relevant to the aims of conservation biology. First, considering the large amounts of variation present in many species and the potentially infinite number of genotypes present conservationists must focus on the variants which are locally common. Second, we must insure that sampling sites (or preserve establishment) represents as broad a range of environments as possible. Third, the knowledge of population structure, as affected by breeding systems and other forces, will alter our strategies. Fourth, capturing a species is very different from maintaining one. They also note that the number of alleles per locus is the simplest measure of diversity and that electrophoretic studies of enzyme loci are the easiest means to obtain allelic data.

A genetic typology, similar to the Rabinowitz (1981) species typology, can allow us to examine just how specific alleles are distributed among populations and to evaluate the importance of specific populations (Fig. 6). Note, the alleles are not evenly distributed. Some alleles (upper left box) would be captured regardless of which populations are chosen for preservation. Others are only rarely found (lower right box). The important category, however, is the geographically restricted but locally frequency class of alleles (lower left). Brown (1978) has estimated that about 20—30% of allozyme variation belongs to this category. We find levels somewhat higher in the inbreeder *L.f.* ssp. *floccossa* and lower in the outbreeder *L. alba* possibly caused by

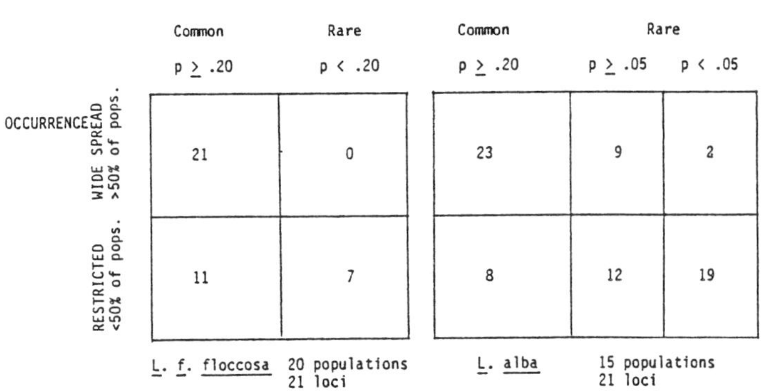

Fig. 6. Alleles at polymorphic loci in *Limnanthes* classified as widespread, restricted, common, and rare (see text).

gene flow differences. Morphological and physiological variants of this type in wild species are often important characters in the breeding programs of domesticated relatives (Rick 1976). In addition to these being important in artificial selection programs, they may indeed be critical adaptations of natural selection.

Since the numbers of populations which can be preserved are often limited, criteria for choosing populations could be useful. For example, 10 randomly chosen *L.f.* ssp. *floccossa* populations would capture less than 90% of the important locally abundant-geographically restricted alleles (Fig. 7). Using geographic distribution and ecological diversity data, we could choose as few as five populations which would possess 95% of these alleles.

Minimum viable population size and the effect of structure

Ewens *et al.* (1987), Shaffer (1984) and others have reviewed the inverse relationship of effective population size to the strength of stochastic processes of genetic drift and inbreeding. For example, Goodman (1987) showed that an effective population of a few hundred is generally sufficient to prevent demographic stochasticity from endangering a population over a long time scale. In a population of 50 to 100 plants, even with low growth rates of a few percent, populations would be expected to survive for hundreds to thousands of years.

Endangerment of small populations of an increasingly fragmented species range due to the loss of genetic variance or higher homozygosis due to genetic drift and inbreeding requires a knowledge of selective forces and population fitness measures. For neutral alleles, which could be selectively important in future populations, a very large Ne is necessary to maintain genetic variance. For *Limnanthes* species, expected values of H range from

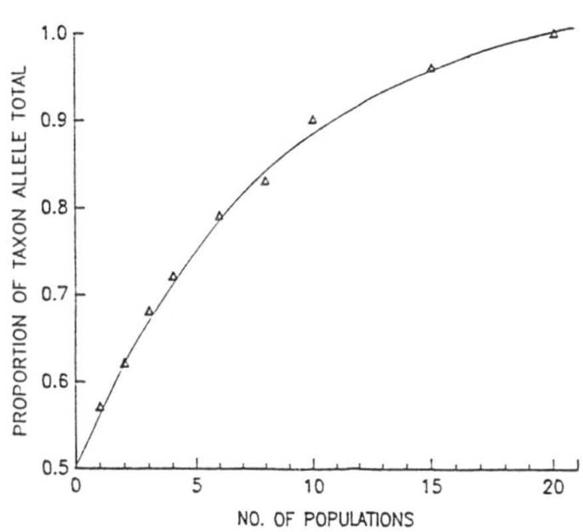

Fig. 7. Simulation of sampling process in *Limnanthes floccosa* ssp. *floccosa* for predicting allelic diversity in relation to the number of protected sites.

near 0.01 in *L. macounii* to 0.20 in *L. douglasii* (Kesseli and Jain 1984). With assumed mutation rates of 10^{-6} then N_e would range from about 2,500 to 62,500. Any populations reduction from these levels (and this has been observed for many of our taxa) could cause decreases in H and therefore fitness loss. Likewise, Nei *et al.* (1975) and Denniston (1978) demonstrated the effects of severe population reductions (bottlenecks). Heterozygosity is reduced, though no severely. Expected number of alleles per locus is substantially reduced. Assuming initial heterozygosity levels of 0.20 and 0.07, those of *L. d.* var *rosea* and *L. bakeri*, respectively, and population reductions to 1000 and 100 (found in several taxa) we can see that as much as one half the alleles but only 0.5% of the heterozygosity could be lost. Rare alleles are readily lost in bottlenecks yet contribute little to overall heterozygosity levels.

For quantitative characters, Franklin (1980) noted that the loss of genetic variance due to drift ($\sigma_g^2/1/2$ Ne) could be balanced by gain due to mutation for additive genetic variance. From data on quantitative characters in *Drosophila* (see Lande 1976), Franklin states that new variance produced by mutation (σ_m^2) is approximately one thousandth that of environmental variance (1×10^{-3} σ_e^2). He suggests that a mutation-drift balance could be reached if $N_e = 500$. Lande and Barrowclough (1987) showed that this is true if the heritability of a trait is 0.5, that is, the proportion of variance due to genetics equals the portion due to the environment ($\sigma_g^2 = \sigma_e^2$). He suggests that a mutation-drift balance could be reached if $N_e = 500$. Lande and Barrowclough (1987) showed that this is true if the heritability of a trait is 0.5, that is, the proportion of variance due to genetics equals the portion due to the environment ($\sigma_g^2 = \sigma_e^2$). This is roughly true for some quantitative characters. Thus, $\Delta\sigma_e^2 = \sigma_m^2 - \sigma_g^2$ (1/2 Ne). If the change in genetic variance $\Delta\sigma_g^2 = 0$, (i.e., equilibrium) and substituting the value for mutation rate, $\mu = 1 \times 10^{-3}$, then $N_e = 500$. In contrast to allelic variation at single loci, smaller numbers are needed to maintain additive genetic variance in a population.

The question of how to conserve a taxon shifts from how large should a population be, to how many should there be. The answer relies on two components of conservation: capturing and maintaining. As discussed earlier, if a species or taxon is a collection of populations with substantial variation apportioned among populations then the conservation of a species necessitates capturing not one but several populations. For the comparable problem of genetic resource collection strategies, Marshall and Brown (1978) suggest that sampling fewer individuals at many sites in geographically diverse areas will maximize the level of variation obtained. Yonezawa (1985) using a cost-benefit analysis of the problem also concludes that increasing the number of populations visited, not the sample size per site, is the most efficient collection strategy. Genetic conservation in *Limnanthes* will require protection of viable individual populations and a broad representation of populations which comprise a taxon. Clearly the specter of global extinction for a species is higher if these partially isolated reserves become too few and distant. In studies by Ehrlich (1983) local population extinction did not strongly threaten species survival because of recolonization by adjacent populations. Thus, the key parameter for species persistence, along with the asynchronous responses of populations to environmental stochasticity, is dispersal among populations. This extinction-dispersal interaction is the basis of the island biogeography work of MacArthur and Wilson (1967) and is broadly applicable to reserve design, and also the primary issue in the so-called SLOSS (one large or many small resources) debate (Chapter 10). Our studies on *Limnanthes* ssp. show the correlated response to environmental factors both within populations and among different but adjacent populations. Among quadrats placed on transects though populations of *L. d.* var. *rosea*, fluctuation in plant density and seed set are correlated.

Adjacent populations also show this effect. In a 40 km^2 region of the Sacramento Valley in which the endangered endemic *Limnanthes floccosa* ssp. *californica* resides, two populations showed similar changes in seed production over years. While some environmental changes will affect only one or two populations others may cause global responses. These populations of *L.f.* ssp. *california* are not independent in their response and are therefore extremely susceptible to extinction from random environmental pressures.

Conclusions

New technology and theory are not needed for the development of good conservation programs, yet much work is still needed on several fronts. For plants, the advantage of generally high densities and therefore potentially large population sizes as well as ease of propagation and storage of plant material should be exploited when mapping out future strategies. Disadvantages, including the difficulties of defining taxonomic or evolutionary units (which makes assessment and retention of variability problematic), as well as the general limited public awareness of most rare plant species, warrant intensified work by conservationists.

What approaches are needed for developing the theory and practice of biological conservation? First, the targeting of endangered taxa and habitats have been undertaken by organizations such as the Natural Diversity Data Base (NDDB) in many states. For rare plant species, they often glean from herbarium reports and scientific records the approximate number of populations. In California, rare plant taxa have been categorized. These organizations often do not have the resources to follow-up on initial reports and accurately verify citings. Many of the reported occurrences may be duplicated sites with poor location descriptions, misidentifications, or recently extirpated. Cooperation between local botanists, members of the academic community and state or local agencies can clarify the endangerment of a taxon. Initially, 30 occurrences were reported for *L.f.* ssp. *californica* by

NDDB. Of these, 11 were apparently duplicate citings, nine were misclassified, and three are extirpated. Thus, the taxon is in a far more critical state than previously suspected. Recent explorations of appropriate habitat have uncovered new sites in a local region (J. Dole, pers. comm.).

Given the severity and rapid pace of the extinction phenomenon, the second stage of conservation strategies warrants a more active approach by population biologists to the study of specific taxa and habitats. Based on these studies the prioritization of individual sites must be made so that educated hard decisions can be made if needed. Criteria for selecting populations should include: the likelihood that sites can be adequately protected which may include consideration of buffer zones surrounding reserves; the size and capacity of sites to maintain target species and associated essential species such as pollinators; the genetic uniqueness and inter- and intra-population diversity levels; and independence of populations from environmental stochastic events.

With this data and periodical monitoring of the above criteria, positive managerial decisions can be made. Two contrasting levels of management can be implemented, either of which is appropriate under certain circumstances. Choosing between a laissez-faire versus an interventionist approach can only be made with strong biological data. Certainly, the laisez-fire approach is the simplest but population biologists and managers should not hesitate to experiment and manipulate the demography and genetics of populations when backed by data supporting the need of such steps. Low level management may be simply collecting and preserving germplasm stocks to insure survival. More advanced intervention may include recolonization efforts, human-aided gene flow, or habitat alterations. Clearly, for conservation efforts to be successful academics and professionals must take a more active role in studying rare species and all must take an active role in increasing public awareness in the conservation of natural resources.

Literature cited

Allendorf, F. W. 1983. Isolation, gene flow, and genetic differentiation among populations. *In* C. M. Schonewald-Cox, S. M. Chambers, M. MacBryede and W. L. Thomas (eds.) Genetics and Conservation, Benjamin/Cummings, pp. 51–65.

Arroyo, M. K. 1973. A taximetric study of infraspecific variation in autogamous *Limnanthes floccosa* Limnanthaceae. Brittonia 25: 117–191.

Beattie, A. J. and D. V. Culver. 1979. Neighborhood size in *Viola.* Evolution 33: 1226–1229.

Boecklen, W. J. 1986. Optimal design of nature reserves: consequences of genetic drift. Biol. Conservation 38: 323–338.

Brown, A. H. D. 1978. Isozymes, plant population genetic structure and genetic conservation. Theor. Appl. Genet. 52: 145–157.

Brown, A. H. D. 1979. Enzyme polymorphisms in plant populations. Theor. Pop. Biol. 15: 1–42.

Burcham, L. T. 1957. California Range Land. Dept. of Natural Resources. Sacramento, CA.

Charlesworth, D. and B. Charlesworth. 1987. Inbreeding depression and its evolutionary consequences. Ann. Rev. Ecol. Syst. 18: 237—268.
Clarke, B. 1975. The contribution of ecological genetics to evolutionary theory: Detecting the direct effects of natural selection on particular polymorphic loci. Genetics 79S: 101—113.
Crow, J. F. 1986. Basic concepts in population, quantitative and evolutionary genetics. Freeman, New York.
Crampton, B. 1959. The grass general *Orcuttia* and *Neostapfia*: a study in habitat and morphological specialization. Madrono 15: 97—128.
Crow, J. F. and N. E. Morton. 1955. Measurement of gene frequency drift in small populations. Evolution 9: 202—214.
Denniston, C. 1978. Small population size and genetic diversity. Implications for endangered species. *In* S. A. Temple (ed.) Endangered Birds: Management Techniques for Preserving Threatened Species. pp. 281—290. Univ. of Wisconsin Press, Madison, WI.
Ehrlich, P. R. 1983. Genetics and the extinction of butterfly populations. *In* C. M. Schonewald-Cox, pp. 152—164.
Ellstrand, N. C. and D. L. Marshall. 1985. Interpopulation gene flow by pollen in wild radish *Raphanus sativa.* Am. Nat. 126: 606—616.
Endler, J. A. 1977. Geographic variation, speciation and clines. Princeton Univ. Press. Princeton, N.J.
Ehrlich, P. R. 1980. The strategy of conservation, 1980—2000. *In* M. Soule and B. A. Wilcox (eds.) Conservation Biology. p. 329—344. Sinauer Assoc., Sunderland, MA.
Frankel, O. H. and M. E. Soule. 1981. Conservation and evolution. Cambridge Univ. Press, Cambridge.
Franklin, I. 1980. Evolutionary changes in small populations. *In* M. Soule and B. Wilcox (eds.) Conservation Biology. pp. 135—149. Sinauer Assoc., Sunderland, MA.
Futuyma, D. J. 1983. Interspecific interactions and the maintenance of genetic diversity. *In* C. M. Schonewald-Cox, S. M. Chambers, M. MacBryde, and L. Thomas (eds.) Genetics and Conservation — A Reference for Managing Wild Animal and Plant Populations. pp. 364—373. Benjamin/Cummings Publ., Menlo Park, CA.
Gentry, H. S. and R. W. Miller. 1965. The search for new industrial crops. IV. Prospectus of *Limnanthes.* Econ. Bot. 19: 25—32.
Global 2000 Report to the President: Entering the twenty-first century, Vol. 2. Government Printing Office, pp. 328—331.
Goodman, D. 1987. Considerations of stochastic demography in the design and management of biological reserves. Nat. Res. Modeling 1: 205—274.
Gottlieb, L. D. 1973. Genetic differentiation, sympatric speciation and the origin of a diploid species of *Stephanomeria.* Am. J. Bot. 60: 545—553.
Gottlieb, L. D. 1981. Electrophoretic evidence and plant populations. Progress in Phytochemistry 7: 1—46.
Griggs, T. and S. Jain. 1983. Conservation of vernal pool plants in California. II. Population biology of a rare and unique grass genus *Orcuttia.* Biol. Conserv. 27:
Hamrick, J. L. 1983. The distribution of genetic variation within and among natural plant populations. *In* C. M. Schonewald-Cox, S. M. Chambers, B. MacBryde, and L. Thomas (eds.) Genetics and Conservation — A Reference for Managing Wild Animal and Plant Populations. pp. 335—348. Benjamin/Cummings Publ., Menlo Park, CA.
Hamrick, J. L., Y. B. Linhart, and J. B. Mitton. 1979. Relationships between life history characteristics and electrophoretically detectable genetic variation in plants. Ann. Rev. Ecol. Syst. 10: 173—200.
Hedrick, P., S. K. Jain and L. Holden. 1978. Multilocus systems in evolution. Evol. Biol. 11: 104—184.
Heiser, C. B. 1949. Study in the evolution of sunflower species *Helianthus annuus* and *H. bolanderi.* Univ. Calif. Publ. Bot. 23: 157—208.
Holland, R. F. 1976. The vegetation of vernal pools: A Survey. pp. 11—15, In: S. Jain (ed.) Vernal Pools: Their Ecology and Conservation. Inst. Ecol. Publ. 9, Univ. Calif., Davis.

Jain, S. K. 1975. Genetic resources. *In* O. H. Frankel and J. G. Hawkes (eds.) Crop Genetic Resources for Today and Tomorrow. Cambridge.

Jain, S. K. 1990. Variation and selection in plant populations, pp. 199—229. *In*: K. Wohrmann and S. Jain (eds.) Population Biology. Springer-Verlag Berlin.

Jain, S. K. 1992. Epilogue. *In*: P. Fiedler and S. Jain (eds.) Conservation Biology: Theory and Practice. Chapman and Hall, London.

Karr, J. K. 1982. Avian extinction on Barro Colorado Island, Panama: a reassessment. Am. Nat. 119: 228—239.

Karron, J. D. 1987. A comparison of levels of genetic polymorphism and self—compatibility in geographically restricted and widespread plant congeners. Evol. Ecol. 1: 47—58.

Kesseli, R. V. 1984. Biosystematics, evolution and breeding system analysis in *Limnanthes*. Ph.D. Diss., Univ. of California, Davis.

Kesseli, R. V. and R. W. Michelmore. 1986. Genetic analysis and phylogenies detected from isozyme markers among species of *Lactuca*. J. Hered. 77: 324—331.

Kesseli, R. V. and S. K. Jain. 1984a. New variation and biosystematic patterns detected by allozyme and morphological comparisons in *Limnanthes* sect. Reflexae (Limnanthaceae). Plant Syst. and Evol. 147: 133—165.

Kesseli, R. V. and S. K. Jain. 1984b. An ecological genetic study of gynodioecy in *Limnanthes douglasii*. Amer. J. Bot. 71: 775—786.

Kesseli, R. V. and S. K. Jain. 1985. Breeding systems and populations structure in *Limnanthes*. Theor Appl. Genet. 71: 292—299.

Kimura, M. and J. Crow. 1964. The number of alleles that can be maintained in a finite population. Genetics 49: 725—738.

Lande, R. 1976. The maintenance of genetic variability by mutation in a polygenic character with linked loci. Genet. Research 26: 221—235.

Lande, R. and G. F. Barrowclough. 1987. Effective population size, genetic variation, and their use in population management. pp. 87—123. *In*: M. E. Soule (ed) Viable Populations for Conservation, Cambridge.

Landry, B. S., R. Kesseli, H. Leung, and R. W. Michelmore. 1987a. Comparison of restriction endonucleases and sources of probes for their efficiency in detecting restriction fragment length polymorphisms in lettuce (*Lactura sativa* L.). Theor. Appl. Genet. 74: 646—653.

Landry, B. S., R. V. Kesseli, B. Farrara and R. W. Michelmore. 1987b. A genetic map of lettuce (*Lactuca sativa* L.) with restriction fragment length polymorphisms, isozymes, disease resistance genes and morphological markers. Genetics 116: 331—337.

Lawrence, M. J. 1984. The genetical analysis of ecological traits. *In* B. Shorrocks (ed.) Evolutionary Ecology. pp. 27—64. Blackwell Scientific Publ., Oxford.

Levin, D. A. and H. W. Kerster. 1974. Gene flow in seed plants. Evolutionary Biology 7: 139—220.

MacArthur, R. H. and E. O. Wilson. 1967. The theory of island biogeography. Princeton Univ. Press, Princeton, N.J.

Marshall, D. R. and A. H. D. Brown. 1975. Optimum sampling strategies in genetic conservation. *In* O. H. Frankel and J. G. Hawkes (eds.) Crop Genetic Resources for Today and Tomorrow. pp. 53—80. Cambridge Univ. Press, Cambridge.

McNeill, C. I. 1983. Genetic differentiation, species relationships and the evolution of mating systems in the plant genus *Limnanthes* (section Inflexae). Ph.D. Diss., Univ. of California, Davis.

Megher, T. R. 1986. Analysis of paternity within a natural popu lation of *Chamaelirium luteum*. I. Identification of most likely male parents. Am. Nat. 128: 199—215.

Meagher, T. R., J. Antonovics and R. Primack. 1978. Experimental ecological genetics in *Plantago*. III. Genetic variation and demography in relation to survival of *Plantago cordata*, a rare species. Biol. Conserv. 14: 243—257.

Myers, N. 1983. A wealth of wild species. Storehouse for Human Welfare. Westmier Press. Boulder, CO.

Nei, M. 1975. Molecular population genetics and evolution. American Elsevier, New York.

Nei, M. 1987. Molecular evolutionary genetics. Columbia Univ. Press, N.Y.
Nei, M. T. Maruyama and R. Chakraborty. 1975. The bottleneck effect and genetic variability in populations. Evolution 29: 1—10.
Olivieri, A. M. and S. K. Jain. 1977. Variation in the *Helianthus exilis-bolanderi* complex: a reexamination. Madrono 24: 117—189.
Pierce, R. O. and S. K. Jain. 1977. Variation in some plant and seed oil characteristics of Meadowfoam. Crop Sci. 17: 521—526.
Rabinowitz, D. 1981. Seven forms of rarity. *In* H. Synge (ed.) The Biological Aspects of Rare Plant Conservation. pp. 205—217. John Wiley and Sons, Ltd. N.Y.
Raven, P. H. and D. I. Axelrod. 1977. Origin and relationships of the California Flora. Univ. of California Press, Berkeley.
Reeder, J. R. 1982. Systematics of the tribe Orcuttieae (Gramineae) and the description of a new segregate genus *Tuctoria.* Amer. J. Bot. 69: 1082—1095.
Rick, C. M. 1976. Natural variability in wild species of *Lycopersicom* and its bearing on tomato breeding. Genet. Agr. 30: 249—257.
Ritland, K. and S. K. Jain. 1984. The comparative life histories of two annual *Limnanthes* species in a temporally variable environment. Amer. Nat. 124: 656—679.
Schaal, B. 1980. Measurement of gene flow in *Lupinus texensis.* Nature 284: 450—451.
Schoen, D. J. 1983. Relative fitness of selfed and outcrossed progeny in *Gilia achilleifolia.* Evolution 36: 361—370.
Slatkin, M. 1985. Rare alleles as indicators of gene flow. Evolution 39: 53—65.
Sneath, P. and R. Sokal. 1973. Numerical taxonomy. W. H. Freeman, San Francisco.
Soule, M. 1980. Thresholds for survival: maintaining fitness and evolutionary potential. *In* M. Soule and B. Wilcox (eds.) Conservation Biology. pp. 151—170. Sinauer Assoc., Sunderland, MA.
Sousa, W. P. 1984. The role of disturbance in natural communities. Ann. Rev. Ecol. Syst. 15: 353—391.
Stebbins, G. L. 1980. Rarity of plant species: a synthetic viewpoint. Rhodora 82: 77—86.
Templeton, A. and D. Levin. 1979. Evolutionary consequences of seed pools. Amer. Nat. 144: 232—249.
Wright, S. 1931. Evolution in Mendelian populations. Genetics 16: 97—159.
Wright, S. 1968. Evolution and the genetics of populations, Vol. 1. Genetic and biometric foundations. Univ. of Chicago Press, Chicago.
Yonezawa, K. 1985. A definition of the optimal allocation of effort in conservation of plant genetic resources with application to sample size determination for field collection. Euphytica 34: 345—354.

5. Genetics of weed invasions

SPENCER C. H. BARRETT

Abstract

Understanding the genetics of colonization is important not only to evolutionary studies and population biology, but also to agriculture and conservation. Basic features of weediness and successful worldwide colonists are briefly reviewed with emphasis on genetic processes. Based on a detailed review of breeding system and genetic variation patterns in *Echinochloa* (barnyard grass) and *Eichhornia* (water hyacinth), two major aquatic weeds, strong evidence is found for the effects of founder events as well as strong selective pressures during colonization. Information on the genetic structure, reproductive biology, and evolutionary dynamics in the past and present would be helpful in devising effective methods of biological control. For example, there is indirect evidence for greater success in the biological control of asexually reproducing weeds. This is clearly a largely unexplored area in population biology.

Introduction

While weeds have a long history of association with man and his crops, the scientific study of their biology is little more than a century old. The numerical abundance of weeds in areas of human settlement, their detrimental effects on crop yields and the rapidity with which populations appear and disappear have stimulated naturalists, botanists and agricultural scientists to investigate the success of weed invasions in terms of ecological and historical factors, the morphological and physiological pre-adaptations of weeds, and more recently in the context of chemical control methods (Gray 1879; Zinger 1909; Harper 1960; Salisbury 1961; Crafts and Robbins 1962; Baker and Stebbins 1965; King 1966; Muzik 1970; Baker 1974; Holm *et al.* 1977). The changing approaches to the study of weed biology reflects an increased knowledge of their distribution and ecology as well as the growth of ideas in various biological sub-disciplines. In the past decade the development of population biology as a vigorous field of inquiry has resulted in the frequent use of weed groups as experimental systems for studies on a variety of topics including demography (Mack and Pyke 1983; Burdon *et al.* 1983), life history theory (Law *et al.* 1977), the significance of genetic polymorphism

S. K. Jain and L. W. Botsford (eds), Applied Population Biology, 91–119.

and phenotypic plasticity to colonizing ability (Moran *et al.* 1981; Martins and Jain 1978) and mating system evolution (Clegg *et al.* 1985; Barrett and Shore 1987).

Studies of the population biology of weeds frequently involves a synthesis of information on the ecology, physiology, and genetics of populations as attempts are made to understand the development of colonizing ability in the context of both community interactions and evolution change. While population studies of weeds have enriched our understanding of many aspects of plant systematics, ecology and evolution, in general they have had relatively little impact on applied research on weeds where the primary objective is usually the eradication of populations. This situation, which has been discussed in detail by Sagar (1968), is partly reinforced by the institutional divisions that occur between biology and agriculture in most universities, particularly in N. America. Researchers in the two fields often work in isolation, affiliated with different departments and with only limited opportunities for interaction.

A considerable amount of the botanical work conducted on weeds that is relevant to applied research has been concerned with identification problems in taxonomically difficult weed groups (Warburg 1960; McNeill 1976; Barrett and Strother 1978), studies of the factors that influence the dormancy and germination characteristics of populations (Roberts 1964; Popay and Roberts 1970; Barrett and Wilson 1983) or the magnitude of weed competition on crop yields (Bleasdale 1966; Firbank and Watkinson 1986). Until recently there has been relatively little work on the genetics of weed populations, aside from early investigations of race formation which were largely undertaken within a biosystematic framework.

In this chapter I review the information currently available about the genetic characteristics of weed populations and in particular focus on some of the ecological and genetic consequences of long distance colonizing episodes that are a feature of many cosmopolitan weed species. I hope to show that basic information on the levels and organization of genetic diversity in weed populations is of value to applied biologists concerned with weed control. Since plant reproductive systems are of prime importance in determining the amounts of genetic variation in populations, I shall also discuss how studies of the reproductive biology of weed populations can have implications for the type of control practices employed to eradicate weed populations. Before I review these examples, which come from our own studies of *Eichhornia* (Water Hyacinth) and *Echinochloa* (Barnyard grass). I briefly discuss some of the prevailing ideas about the ecological and genetic attributes of successful weed species and what is known about the patterns of genetic variation in populations.

General features of weeds

Weediness has evolved many times among angiosperm families and weed species can be found occupying a variety of different types of disturbed environments. As a consequence, a broad range of colonizing strategies are employed by weed species. Since weed invasions occur on very different scales, from range extensions and local habitat shifts to intercontinental migration, it has been suggested by Brown and Marshall (1981) that the selection pressures acting on weed populations will vary with species and event such that each case is unique. Despite this obvious complexity, many attempts either through floristic surveys (Stebbins 1965; Price and Jain 1981; Newsome and Noble 1986) or by comparative studies of related taxa (Baker 1965; Jain 1969; Barrett and Wilson 1981) have been made to identify shared attributes that are commonly found in successful weed species (Baker 1974; Brown and Marshall 1981; Barrett and Richardson 1986).

Some general features of weeds that are relevant to their genetics and evolution are summarized in Table 1. Of course many cases can be found where a particular weed species possesses traits not normally associated with high colonizing ability (e.g. Heiser 1965; Barrett 1978). Where this occurs it is often valuable to examine whether possession of such traits constrains weediness in any way, and if not, what other attributes contribute towards colonizing success. What usually emerges is that a suite of co-adapted traits are involved in determining weediness and it is the coordinated response and joint evolution of these traits that determine the particular fates of individual biological invasions. Quantitative genetic analysis of these character syndromes in conjunction with demographic and life history studies of natural and artificially established colonies may offer the most promise in untangling those features of weeds that are critical to colonizing success (also see Vrijenhock 1990, for an excellent review of asexuality in colonizing species).

Patterns of genetic variation in weeds

The earliest studies of variation in weed species were largely directed towards describing the patterns of ecotypical differentiation among populations in relation to easily measured environmental factors. In widespread species, marked inter-population genetic differentiation can be regarded as the rule and individual populations usually display distinctive morphological and ecological properties. The amount of variation is associated with adaptation to the local environment as opposed to genetic drift has rarely been investigated although reciprocal sowing and transplant studies can be useful in this regard. Applied biologists should be aware that race formation in weeds is both ubiquitous and rapid although the high phenotypic plasticity of many weed species can complicate attempts to assess the nature of genetic differentiation based strictly on field observations. By growing samples collected from different sites under uniform garden or glasshouse conditions, the magnitude of population differentiation can be assessed directly. When studies of this type are undertaken it is usually apparent that a considerable amount of both morphological and ecological variety is subsumed within the

Table 1. Some ecological and genetical features of weeds (modified from Baker 1965 and Brown and Marshall 1981).

1. Well developed dormancy mechanisms with germination requirements fulfilled in many environments.
2. Rapid growth, development and reproductive maturation.
3. High phenotypic plasticity often with the ability to reproduce as long as growing conditions permit.
4. Unspecialized pollination mechanisms involving self-pollination, wind-pollination or generalist insects.
5. Uniparental reproductive systems involving selfing, apomixis, or clonal propagation.
6. High reproductive capacities through seed or vegetative propagules.
7. Mechanisms for the short- and long-distance dispersal of weed.
8. Polyploidy with attendant fixed heterozygosity.
9. Genetically depauperate populations with strong multilocus associations that originate from a limited number of colonizing genotypes.
10. Substantial interpopulation differentiation owing to founder effects, drift and variable environments.

latin binomial. As a general rule statements about the ecological characteristics of weed species A or B based on studies of single populations should be treated with caution! Further discussion of the factors influencing race formation in weeds can be found in Baker (1974) and Barrett (1982, 1983).

Detailed studies of the patterns of genetic variation *within* weed populations are more recent and are largely in their infancy. Work of this type is important from both the pure and applied standpoints since the amount, kinds and organization of genetic diversity in populations will largely determine their capacity to respond to the local selection pressures imposed by the physical and biotic environment. Understanding the selection response of weed populations to specific agricultural practices might be particularly useful since it is possible that the crop environment could be manipulated to take advantage of the nature of genetic variation for important life history traits.

For example, the recent trend to shorten the growing period of many rice varieties in California may eventually lead to a loss of the rice mimic *Echinochloa phyllopogon* from fields as a result of the limited amounts of genetic variation in populations of the weed for maturation time (Barrett 1983). Seeds of the rice mimic are frequently harvested with those of cultivated rice at harvest and redistributed in contaminated rice seed the following year. Unfortunately there has been relatively little work on the quantitative genetics of agricultural weeds nor on the evolutionary responses of weed populations to agronomic practices. In some ways this is surprising since in comparison with plant populations from uncultivated land, agricultural weeds inhabit a relatively uniform environment in which the selective pressures are often easily identified and closely controlled. Furthermore weeds are abundant, possess rapid life cycles and are easily grown and measured. As such they provide an excellent source of experimental material for the evolutionary geneticist.

The highly developed dispersal mechanisms and vagile life styles of most weeds have important consequences for the genetic structure of populations. In many cases the invasion of new territory is associated with frequent genetic bottlenecks, small population sizes, inbreeding and strong directional selection (Barrett and Husband 1989). These influences can result in rapid selection responses and an erosion of genetic variation in populations. Studies of electrophoretic variation in weed populations have revealed a large number of cases in which populations contain very low levels of genetic polymorphism or are completely uniform at isozyme loci. A variety of other factors can counteract these influences; these include the multiple introduction of genotypes from geographically separated and hence genetically differentiated source populations, outbreeding, introgressive hybridization with related taxa, and the rapid expansion of populations after founding. Table 2 lists some of the major factors influencing the levels of genetic variation in weed populations and the topic is covered well in several recent reviews (e.g., Brown and Marshall 1981; Clegg and Brown 1983; Jain 1983; Barrett and Richardson 1986; and Barrett and Shore 1988).

Reproductive systems

The methods by which a species reproduces can play a major role in determining the levels and pattern of genetic variation in populations. In perennial weeds the emphasis on clonal propagation and the frequency of sexual reproduction and seedling recruitment can have a profound influence on the genetic diversity of populations. In species where reproduction is principally

Table 2. Some factors that influence the level of genetic variation in weed populations (After Barrett 1981).

Factor	Influence
Founder effect	Number and types of immigrants
Genetic drift	Frequency of genetic bottlenecks during colonizing episodes
Population age and size	Stability of habitat and life history features of species
Degree of sexuality	Importance of cloning and frequency of sexual reproduction
Matin system	Level of inbreeding
Hybridization	Opportunities for gene exchange with related taxa
Environmental heterogeneity	Spatial and temporal variation of habitat

by seed, the mating system, the frequency of cross- and self-fertilization, controls the character of genetic transmission and the levels of heterozygosity in a population. Surveys of genetic variation among species with contrasting mating systems provide evidence of the importance of this component of the genetic system in regulating the level of genetic diversity within populations. In general outbreeding species display higher levels of genetic polymorphism, allelic variation and heterozygosity than predominantly selfing species (Brown 1979; Hamrick *et al.* 1979). However, since the mating system can be influenced by a range of environmental and demographic factors (Loveless and Hamrick 1984) and in addition can itself exhibit genetic variation, a wide frequency of outcrossing levels can be displayed by different populations of the same species (Schemske and Lande 1985). Accordingly, levels of genetic variation can also vary dramatically among different populations of a single species (Rick *et al.* 1979; Schoen 1982; Glover and Barrett 1987).

Most weed species are capable of some form of uniparental reproduction either through selfing, apomixis or clonal propagation (Baker 1965; Brown and Burdon 1987) and shifts in mating systems towards increased selfing are often associated with the colonization of unoccupied territory (Baker 1955; Lloyd 1980). Nevertheless, this pattern is not always found and in several studies (reviewed by Brown and Marshall 1981) increased outcrossing rates and heterozygosity have been found in colonizing populations in comparison with populations from presumed source regions. The diversity of reproductive systems that can be found both within and between plant species as well as the importance of local ecological factors in influencing the nature of reproduction should caution us against extrapolation of a species' reproductive behavior from one area to another based on studies from only a limited portion of its range. The following case studies of two weed groups will clearly illustrate this point.

Case studies of weed groups

The use of the comparative approach in studies of related taxa of colonizing species, originally pioneered during the 1960's by J. L. Harper (reviewed in Harper 1977) and Baker (1965), has provided valuable information on those ecological attributes that are likely to influence colonizing success. In recent years comparative studies have been extended to include experimental demographic studies of natural populations of a variety of weed species with contrasting life history characteristics and genetic systems (e.g. Sarukhan 1974; Solbrig and Simpson 1974; Jain and Martins 1979; Warwick 1980; Law 1981). These detailed studies have largely concerned ecological processes operating in one or a limited number of sites in close proximity to one another. Comparative ecological studies of populations from different portions of the range of widely-distributed weed species have been less frequently conducted because of the logistical difficulties inherent in such work. In contrast genetic studies of geographically separate populations of cosmopolitan species are more easily undertaken since much of the experimental work is conducted at one location on plant material collected from different parts of the range. Valuable information on both the regional and geographical patterns of genetic diversity in wide ranging weed species has been obtained by using this approach (e.g. Clegg and Allard 1972; Kahler *et al.* 1980; Giles 1983; Bosbach and Hurka 1981; Lyman and Ellstrand 1984; Warwick *et al.* 1987). The next step in studies of the genetic structure of weed populations will be investigations of the temporal patterns of micro-evolutionary change within populations. To date there are few chronological studies of the flux of genotypes or phenotypes in weed populations through time (Burdon *et al.* 1983), despite the relevance of such an approach to the process of natural selection.

Our own work on the ecological genetics of weeds began with a careful consideration of which weed groups could be most profitably studied from both the biological and applied viewpoints. It was decided to investigate two groups of weeds of wide distribution and major economic importance but which displayed contrasting colonizing strategies. The two chosen were the free-floating aquatic weed *Eichhornia crassipes* (Water Hyacinth) and members of the *Echinochloa crus-galli* complex (Barnyard Grass), weeds of a broad range of crops. Perusal of the literature in the mid-seventies when the work began indicated that remarkably little was known of the ecological genetics of either species despite the fact that both are among the world's most noxious weeds (Table 3) and are excellent experimental organisms, being easily grown and widely available. The remainder of this chapter summarizes work on these two groups of weeds and their close relatives.

The water hyacinth genus (Eichhornia)

Eichhornia is perhaps the classic example of where a single species (*E. crassipes*) in a genus has become a serious weed whereas the remaining taxa shown only a limited tendency to become aggressive. All members of *Eichhornia* have widespread distributions in their native range, a feature of many aquatic plants. However, none have shown the dramatic world-wide spread of *E. crassipes*, despite the fact that several of the species (*E. azurea*, *E. paniculata*) have been used as pond ornamentals in various parts of the world and presumably have had opportunities to escape.

Comparisons of the genetic systems of *Eichhornia* species are largely uninformative in accounting for the difference in behavior between *E. crassipes* and its relatives (Table 4). Although predominant self-fertilization has evolved on a number of occasions in the genus, in association with the breakdown of trimorphic incompatibility (Barrett 1985a), the selfing species have shown no tendency to become aggressive weeds, despite the increased colonizing potential selfing might be expected to convey. While *E. crassipes*

Table 3. A ranked list of the world's worst 10 weeds of crops (after Brown and Marshall 1981 based on information in Holm *et al.* 1977).

Species	Distribution		Habit[a]	Pollination	Ployploid	Reproductive capacity
	Countries	Crops				
1. *Cyperus rotundus*	92	52	Per	Cross	*	tubers: 40 tons/ha
2. *Cynodon dactylon*	80	40	Per	Cross	*	stolons
3. *Echinochloa crus-galli*	61	36	Ann	Self	*	40,000 seeds/plant
4. *Echinochloa colonum*	60	35	Ann	Self	*	1,000 seeds/plant
5. *Eleusine indica*	60	46	Ann	Self?	*	135,000 seeds/plant
6. *Sorghum halepense*	53	30	Per	Cross	*	rhizomes: 33 tons/ha
7. *Imperata cylindrica*	73	35	Per	Cross?	?	rhizomes: 4.5×10^6 shoots/ha
8. *Eichhornia crassipes crassipes*	>65	waterways	Per	Cross[b]	*	stolons: 0.04 ha/plant/season
9. *Portulaca oleracea*	81	45	Ann	Self	*	10,000 seeds/plant
10. *Chenopodium album*	47	40	Ann	Self	*	0.5×10^6 seeds/plant

[a] Per = perennial; Ann = annuasl.
[b] But see Table 4.

Table 4. Genetic systems and weediness in *eichhornia* species (modified from Barrett and Richardson 1986).

Taxon	Ploidy	Major breeding system*	Asexual reproduction	Distribution	Weed Status
E. paniculata	2×	tristylous	none	N. Brazil, Jamaica & Cuba	occasional in rice fields
E. paradoxa	2×	s. homostylous	none	neotropics; rare, disjunct	no reports
E. meyeri	2×	s. homostylous	none	Paraguay & Argentina, rare	no reports
E. azurea	4×	tristylous	moderate	neotropics; widespread	local problem
E. crassipes	4×	tristylous	highly developed	worldwide	noxious aquatic weed
E. heterosperma	4×	s. homostylous	moderate	netoropics; local	no reports
E. diversifolia	4×	s. homostylous	moderate	neotropics; widespread	occasional in rice fields
E. natans	—	s. homostylous	moderate	Africa; widespread	occasional in rice fields

* Tristylous species are largely outcrossing; semi-homostylous species are predominantly selfing. In each of the three tristylous species semi-homostylous races occur. The clonal structure of *E. crassipes* populations results in considerable inbreeding as a result of geitonogamy. For further details see Barrett 1977, 1979, 1985a.

is polyploid, so are several other species in the genus that exhibit no weedy tendencies.

Water Hyacinth's invasive powers are principally the result of its free-floating life form and prolific powers of asexual reproduction by stolons. No other member of the genus possesses the clonal architecture and free-floating habit which are adaptations to the periodic water-level fluctuations and temporary habitats that are a feature of Water Hyacinth's native Amazonian habitats. These observations of the genus *Eichhornia* support Stebbins' (1965) view that "no particular type of mating system or chromosomal condition is either necessary or generally favorable for pre-adapting a group of species to evolve in the direction of weediness. A hereditary predisposition toward weediness or the ability to colonize areas disturbed by human interference must reside in certain morphological and physiological characteristics of the plants concerned". In *E. crassipes* these are easily identified.

A. *Geographical distribution of style morphs*

In many parts of the adventive range of Water Hyacinth, it is represented by a limited number of asexually reproducing clones and extensive areas of genetic uniformity prevail as a result of the rampant clonal multiplication of water-dispersed vegetative fragments. Water Hyacinth exhibits the rare genetic polymorphism tristyly (Barrett 1977) and observations of the geographical patterns of style morph distribution are a guide to the history of colonization, the occurrence of sexual reproduction, and the extent of genetic uniformity in an area. In the native Amazon Basin, Mato Grosso and parts of Argentina, all three style morphs occur although individual populations are often monomorphic. In the adventive range only two morphs are found, and the mid-styled morph predominates in most regions. This uneven distribution of style morphs is the result of founder effects and prolific clonal propagation (Fig. 1). Absence of the short-styled morph of *E. crassipes* from the introduced range is probably the result of chance. Short-styled genotypes were apparently not among the initial introductions to North and Central America and the Old World. As a result of the genetic basis of tristyly (two loci *S*, *M*, with two alleles at each locus and *S* epistatic to *M*), the short-styled morph cannot arise by segregation from crosses or selfs of long- and mid-styled morphs. Thus a separate introduction of this morph would be necessary for colonization of the alien range. This has not occurred.

The unusual pattern of style morph distribution in *E. crassipes* has led to several misconceptions about the reproductive biology of the species. Studies restricted to the alien range led to the conclusion that the short-styled morph was missing from the species and that the true breeding system was in fact

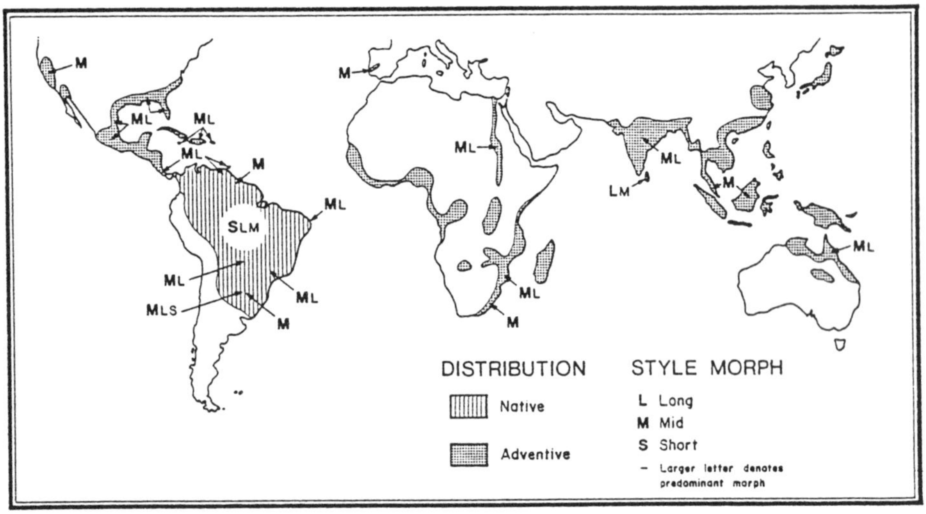

Fig. 1. The geogrpahical distribution of style morphs in *Ecihhornia crassipes.* After Barrett (1977) and Barrett and Forno (1982).

distyly (Faegri and van der Pijl 1971; Mulcahy 1975). In addition, because the species is heterostylous it has been repeatedly stated in the literature that individual clones are self-incompatible and that the absence of sexual reproduction in populations was because of their monomorphic structure (e.g. Baker 1965; Richards 1986). The apparent sterility of clones has led to the assumption that the destruction of vegetative colonies by herbicides or desiccation would largely eradicate the weed problem. However, while most heterostylous taxa are strongly self-incompatible, *E. crassipes* is an exception, with high levels of self-fertility a feature of most clones (Fig. 2). Our own field studies indicate that populations composed of a single style morph usually produce abundant seed, although seed germination and seedling establishment are usually restricted in many artificial water bodies (such as canals or reservoirs) owing to the absence of suitable safe sites for germination and establishment (Barrett 1985b).

Current management techniques employed for aquatic weed control in various parts of the world (e.g. "drawdowns" in Southeastern U.S.A.) involve the removal of water from infested lakes and reservoirs during certain periods of the year. This can result in the drying out and death of vegetative fragments. However, since large quantities of seed are also produced in most populations, these practices can provide an opportunity for seed germination and the establishment of seedlings, particularly if rains occur following the drying-out period. Interestingly, these weed management practices mimic the natural water level fluctuations that are a feature of Amazonian habitats to which the Water Hyacinth's sexual reproductive cycle is adapted. Weed control practices that enhance sexual reproduction can favor the build-up of genetic diversity in Water Hyacinth populations.

An indication of the successful occurrence of sexual reproduction in Water Hyacinth can be seen from observations of the patterns of style morph distribution in particular regions. Where water level fluctuations characterize an area, either as a result of seasonal climates (e.g. parts of India, Queensland)

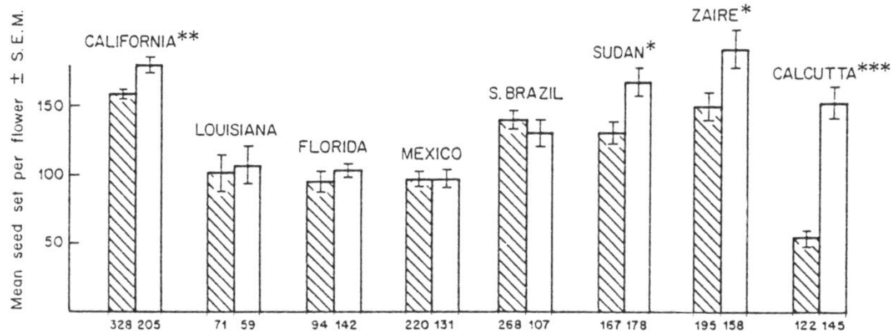

Fig. 2. Seed roduction following controlled self- and cross-pollinations of *Eichhornia crasspies* clones. All cross-polllinations involved a single clone from Costa Rica. Sample sizes are the number of flowers pollinated; *P < 0.025, **P < 0101, *** P< 0.001. After Barrett (1981a).

or management techniques (S.E. U.S.A.) the long-styled morph is frequently found at low frequency in mid-styled populations. This morph arises through segregation as a result of self- and geitnogamous pollinations of mid-styled clones heterozygous at the *M* locus (*ssMm*). In California, where only the mid-styled morph occurs, clones are heterozygous at the *M* locus, and seed is produced in most populations. But the absence of suitable safe sites for seedling establishment prevent segregation of long-styled plants (Barrett 1980b).

Electrophoretic surveys of clonal diversity in different parts of the native and adventive range of Water Hyacinth would be valuable to establish whether the patterns evident for style length loci reflect the overall levels of genetic diversity in populations. We have used this approach in the related diploid *Eichhornia paniculata* which exhibits a remarkably similar pattern of style morph distribution associated with its migration from lowland South America to the Caribbean region.

In contrast to *E. crassipes*, reproduction and dispersal in *E. paniculata* is primarily by seed and its Caribbean distribution (Cuba and Jamaica) pre-dates human introduction suggesting natural spread to the region. Surveys of style morph distribution in populations of *E. paniculata* from N.E. Brazil indicate that the majority are tristylous although dimorphic (L, M) and monomorphic (M) populations occur. However, on Jamaica the short-styled morph is absent and monomorphic populations (M) predominate with dimorphic populations (L, M) occurring rarely (Barrett 1985a). Thus in both *Eichhornia* species the short-styled morph is largely restricted in its distribution to N. Brazil while throughout the remainder of the range the mid-styled morph predominates with the long-styled morph occurring less often. In both species these changes in population structure from trimorphism to monomorphism are associated with colonization of new territory and the breakdown of tristyly to homostyly (Barrett 1979, 1985b).

B. *Genetic variation in continental and island populations*

Investigations of the mating systems and genetic structure of *E. paniculata* populations from N.E. Brazil and Jamaica indicate the importance of founder effects and increased levels of self-fertilization on genetic diversity. A survey of 21 allozyme loci in 11 populations revealed that Jamaican populations were genetically depauperate in comparison with those from Brazil (Table 5). Outcrossing rates (t) were significantly correlated with the number of polymorphic loci, and mean observed heterozygosity of populations (Barrett and Husband 1990). This demonstrates the important influence of the mating system on levels of genetic variation in populations (Fig. 3).

The low levels of allozyme polymorphism on Jamaica probably result from a restricted number of long-distance dispersal events to the island (Husband and Barrett 1991). Since self-pollinating variants of the mid-styled

Table 5. Summary of genetic differences between Brazilian and Jamaican populations of eichornia paniculata based on a survey of 21 isozyme loci (after Glover and Barrett 1987).

Parameter	Symbol	Brazil	Jamaica
% loci polymorphic per pop.	P	23.8	7.6
Observed % loci heterozygous	H_O	7.8	2.0
Total gene diversity	H_t	0.15	0.06
Gene diversity within pops.	H_s	0.091	0.027
Gene diversity between pops.	G_{st}	0.40	0.57
Inter- to intra-pop. ratio	R_{st}	0.81	1.7
Av. genetic distance	$\bar{d}$0.085	0.049	

Brazil N = 6 populations, Jamaica N = 5 populations

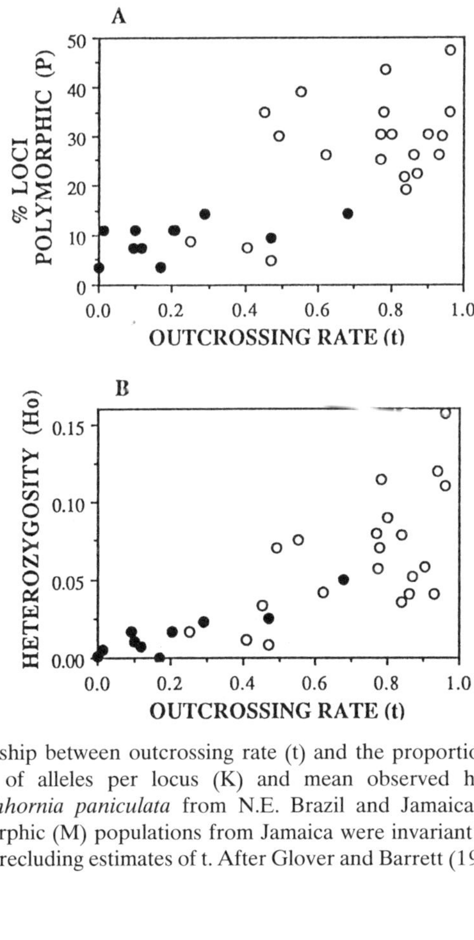

Fig. 3. The relationship between outcrossing rate (t) and the proportion of loci polymorphic (P), mean number of alleles per locus (K) and mean observed heterozygosity (Ho) in populations of *Eichhornia paniculata* from N.E. Brazil and Jamaica. Two additional self-pollinating monomorphic (M) populations from Jamaica were invariant at the 21 isozyme loci that were surveyed precluding estimates of t. After Glover and Barrett (1987).

morph predominate on Jamaica it seems likely that they were favored following establishment after long-distance dispersal. The primarily selfing mode of these variants in conjunction with genetic drift may have resulted in further losses of genetic variation during colonization. Differences in the mating system of *E. paniculata* populations in the two regions probably also account for differences in the partitioning of genetic variation within and among populations. Although populations within both regions are highly differentiated from one another, in N.E. Brazil populations contain more variation than is distributed among them, while in Jamaica the reverse pattern exists (Table 5).

Although the data for both style morph and allozyme loci indicate that Jamaican populations are relatively uniform with respect to genetic variation, it does not follow that they are entirely devoid of variation at other gene loci. A study of quantitative variation in several life history traits conducted in five Jamaican populations demonstrated significant between family variation for most traits within each population (S.C.H. Barrett, unpubl. data). Two of the populations included in the study were fixed for selfing variants of the mid-styled morph and contained no allozyme variation at the 21 loci surveyed. This disparity is in accord with several other studies that have compared variation in isozymes and quantitative traits and have found considerable variation in life history traits in species uniform with respect to electrophoretically detectable enzyme variation (Moran *et al.* 1981; Giles 1983). Further discussion of the complex relationships between isozyme polymorphisms and variation in quantitative traits can be found in Price *et al.* (1984), Lewontin (1984) and Brown and Burdon (1987). The latter authors suggest that because of the cohesive nature of the genome in selfers and apomicts, allozymes are likely to serve as useful neutral markers to index variation, without necessarily being causative agents.

Barnyard grasses (Echinochloa spp.)

Among the world's most noxious weeds of agriculture are members of the cosmopolitan genus *Echinochloa* (Barnyard Grass). The genus contains approximately 50 species of annual or perennial C_4 grasses that inhabit marshes, lake margins, riverbanks and other seasonally inundated habitats. Where man provides disturbed open environments with an assured moisture supply, such as with irrigated arable land, Barnyard Grasses often rapidly colonize and become serious weed problems. The most widespread and economically important member of the genus is the annual *E. crus-galli* which ranges from 50 °N to 40 °S latitude and is reported as a weed in 36 different crops in 61 countries (Holm *et al.* 1977; Maun and Barrett 1986). Other weeds of agriculture include *E. colona*, *E. crus-pavonis*, *E. oryzoides*, *E. phyllopogon* and *E. microstachya.* Superficial similarities between taxa in gross morphology, combined with their highly plastic nature, often results in difficulties of identification for weed control specialists. This problem is particularly striking where cultivated rice is grown and it is commonplace to find 3—6 species of *Echinochloa* as weeds of rice in a particular region (Michael 1983).

Our research on this group has involved comparative studies of the ecological genetics of *Echinochloa* spp. that have invaded the rice agroecosystems of California and New South Wales, Australia. Historical links between the two regions suggest that several barnyard grasses now present in Australian rice fields were introduced to the continent with imported rice seed stocks from California (McIntyre and Barrett 1986). Studies of the patterns of genetic variation in populations from the two regions have been used to substantiate this proposal as well as to provide insights into the processes of genetic differentiation following intercontinental dispersal (Barrett and Husband 1989).

The correct identification of taxa is a critical first step in any comparative study of the population biology of a group of closely related weeds. Thus at the outset we focussed our attention on unravelling the complex patterns of phenotypic variation in the two regions to determine how many taxonomic (and biological) species were present in each. The multiple introduction of genetic material in alien taxa, predominant self-fertilization, a high degree of phenotypic plasticity and considerable genetically-based morphological polymorphism all contribute to the identification problems in *Echinochloa.* In California all variation has been treated as *E. crus-galli* when in reality several distinct species with contrasting ecologies and which pose different threats to yield reduction occur. Cytological and electrophoretic studies indicated that in California the variation can be organized into 4 separate intersterile assemblages each of which deserves specific rank. In Australia three of these taxa (*E. crus-galli*, *E. oryzoides*, *E. microstachya*) occur. The distribution and history of introduction to the two regions of the Barnyard Grasses studied are presented in (Table 6). Two important points are worth noting. First in both areas *E. crus-galli* is by far the most abundant and widespread species although the introduction of permanent deep water culture in Californian rice fields has reduced population densities dramatically so that it no longer poses such a serious threat to crop yields (Barrett and Wilson 1983). In both countries the arrival of *E. crus-galli* predates the commencement of commercial rice production, which took place during the period 1912—15 in California (Barrett and Seaman 1980) and in the year 1926 in New South Wales (McIntyre and Barrett 1986).

The second point is that the earliest herbarium records of *E. oryzoides* and *E. phyllopogon* in California and *E. oryzoides* and *E. microstachya* in N. S. Wales are from rice experiment stations at Biggs and Leeton, respectively. This indicates that their introductions to both regions were most likely associated with the importation of contaminated rice seed stocks. Two lessons can be drawn from these observations. Clearly rigorous seed certification schemes need to be maintained to prevent the accidental introduction of noxious weeds from one region to another. In addition, the fact that alien taxa are often collected for the first time in a country in the vicinity of research stations suggests that the early identification and elimination of potentially weedy populations should be encouraged soon after a species is introduced by chance or design.

The four species of Barnyard Grass in Table 6 can be divided into two broad ecological groups on the basis of their geographical distributions, habitat preferences and colonizing ability. *Echinochloa crus-galli* and *E. microstachya* are small seeded, highly fecund generalists capable of colonizing a broad spectrum of wet disturbed land both agricultural and ruderal (Barrett and Wilson 1981). The two species are very similar in overall appearance to the extent that flora writers rarely distinguish the two species and specimens of *E. microstachya* are usually filed as *E. crus-galli* in most herbaria. In contrast, *E. oryzoides* and *E. phyllopogon* are satellite weeds of rice with large seeds and lower fecundity (Barrett 1983). Despite their abundance in rice fields, populations rarely occur outside the rice agroecosystem and thus they can be viewed as obligate rice weed specialists. Figure 4 illustrates the differences in distribution of *E. crus-galli* and *E. oryzoides* in California.

A. *Genetic variation in generalist and specialist weeds*

The contrasts in distribution and ecological amplitude between the two *Echinochloa* groups raise the issue as to whether different patterns of genetic variation are associated with the generalist and specialist weed strategies. To examine this question a comparative study of the levels of genetic variation for isozyme loci and quantitative traits was undertaken in California an

Table 6. Distribution and history of introduction of barnyard grasses to California (U.S.A.) and New South Wales (Australia).

Taxon origin	Ploidy	California	New South Wales
E. crus-galli Eurasia	(6×)	Widespread weed (introduced mid-19th Century)	Widespread weed (introduced late-19th Century)
E. oryzoides Asia	(6×)	Restricted to Californian rice fields earliest records 1912–15	Restricted to N.S.W. rice fields (earliest record 1938)
E. phyllopogon	(4×)	Restricted to Californian rice fields earliest records 1912–15	Absent
E. microstachya N. America	(4×)	Widespread in riverbanks, ditches, wasteground	Largely restricted to rice-growing areas (earliest record 1938)

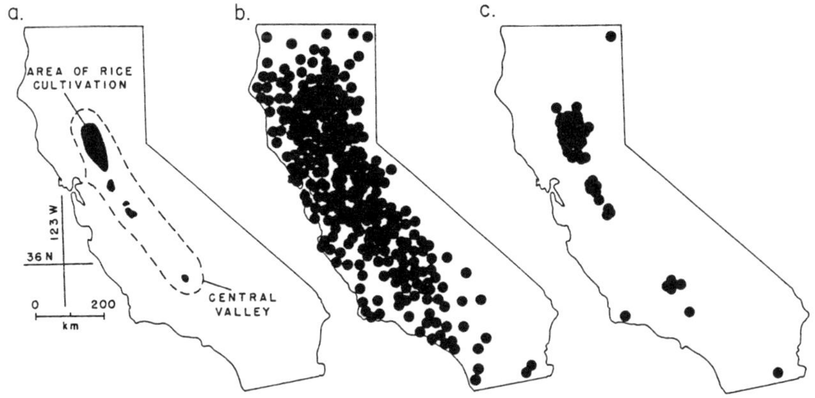

Fig. 4. The distribution of the generalist weed *Echinochloa crus-galli* and the specilaist rice weed *Echinochloa oryzoides* in relation to the areas of rice cultivation in California.

Australian populations of the two hexaploid species *E. crus-galli* and *E. oryzoides* (S. C. H. Barrett and A. H. D. Brown unpubl. data). In the Central Valley of California at each of 10 rice field sites, open pollinated families were collected of the two species. Populations of *E. oryzoides* were sampled from the flooded interior of rice fields whereas those of *E. crus-galli* were obtained from the mosaic of wet disturbed habitats around the edges of rice fields. Starch gel electrophoresis was then employed to assess the magnitude of enzyme polymorphism within and among populations of the two species.

Results of the study indicated that both species were relatively low in genetic diversity with virtually no heterozygosity at polymorphic loci and a high degree of genetic differentiation among populations. This finding is expected in highly self-fertilizing species and has often been found in isozyme surveys of self-pollinating weeds (Brown and Marshall 1981; Brown and Burdon 1983, 1987). At virtually all sites sampled, populations of the generalist *E. crus-galli* were more variable at isozyme loci both in terms of the proportion of genes that displayed polymorphism and the number of alleles present within populations (Table 7).

Of particular interest were the differences in genetic polymorphism for loci controlling the enzyme alcohol dehydrogenase (*Adh*) in the two Barnyard Grass species. In *E. crus-galli*, a total of 12 homozygous multilocus genotypes were evident as a result of polymorphism at 1-3 loci. All populations in California contained at least one of the genotypes, with some composed of up to six, and most were characterized by different combinations of *Adh* genotypes. In contrast all populations of *E. oryzoides* except one, contained the same multilocus genotype. The exceptional population was fixed for a variant allele at a single locus. In a parallel survey of the other rice field specialist, *E. phyllopogon*, no variation was detected at *Adh* genes.

The almost complete absence of *Adh* variation in populations of the two

Table 7. Comparison of isozyme variation in Californian ricefields populations of *echinochla crus-galli* and *echinochloa oryzoides*. S. C. H. Barrett and A. H. D. Brown, unpublished data.

	echinochla crus-galli (n = 31 loci)			*echinochloa oryzoides* (n = 32 loci)		
Site	No. of polymorphic loci	No. of alleles at polymorphic loci	Diversity	No. of polymorphic loci	No. of alleles at polymorphic loci	Diversity
A	8	26	1.899	2	4	0.569
B	6	14	2.788	1	2	0.391
D	8	18	2.543	3	6	0.375
E	7	15	1.992	2	4	0.356
F	3	7	1.647	2	4	0.615
G	7	16	1.854	2	4	0.605
H	3	7	0.816	3	6	0.996
I	9	8	2.327	2	4	0.250
J	3	6	0.800	2	4	0.836
K	3	14	2.077	1	2	0.231
L	6	12	0.897	2	4	0.977
Mean	5.7	13.0	1.785	2.0	4.0	0.564

specialist rice weeds may reflect selection of an "optimum" *Adh* genotype adapted to the relatively uniform conditions of flooded rice fields. In contrast, polymorphism at *Adh* genes in the generalist *E. crus-galli* may be maintained by the heterogeneous nature of the disturbed wetland habitats it occupies. Unfortunately, although there is considerable evidence that *Adh* enzymes function to enable plants to tolerate the anaerobic conditions associated with flooded environments the low levels of recombination in selfing species and strong linkage disequilibrium complicate attempts to determine the adaptive significance of variation at individual isozyme loci (Lewontin 1974). Surveys of *Adh* variation in other parts of the range of these Barnyard Grass species would, however, be valuable to establish whether the patterns observed in California occur elsewhere.

Studies of quantitative variation in life history traits in California populations of the two Barnyard Grass species were largely in accord with data obtained from isozymes. Populations of *E. crus-galli* contained significantly more between family variation than *E. oryzoides* for most traits that were examined. Particularly striking were the differences between the two species with respect to the time taken to flower from germination. Populations of *E. oryzoides* displayed strong synchrony in flowering time and a relatively small amount of inter-population genetic differentiation in comparison with *E. crus-galli.*

Historical factors in the form of founder effects and frequent genetic bottlenecks have undoubtedly played a major role in determining both the levels and pattern of genetic variation in populations of the two *Echinochloa* species in California. Since agricultural weeds often initially enter a new region as crop seed contaminants, multiple introductions of genetic material would be expected in a generalist weed such as *E. crus-galli* which is associated with a range of different crop plants. In addition, its longer residency in the state would also provide more opportunities for this process to occur. In contrast the low overall levels of genetic variation in *E. oryzoides* may simply reflect a limited number of introductions of seed in contaminated rice stocks. In both species occasional outcrossing among different biotypes may enable some mixing of genetic variation although frequent colonizing episodes and the transient nature of many agricultural habitats are likely to retard the build up of genetic diversity in populations and the development of locally adapted races.

B. *Genetic bottlenecks associated with continental invasions*

The success of commercial rice production in California's Sacramento Valley provided the impetus to develop rice growing in the Murrumbidgee Irrigation Areas of N.S.W. (Australia) after observations of the similarities in soils and climate between the two regions. The first successful rice crop in Australia was sown in 1922 using imported Californian rice varieties (McIntyre and Barrett 1986). As mentioned earlier, among the earliest records in Australia of the rice weed specialist *E. oryzoides* and the N. American *E. microstachya* are from in and around rice fields at Leeton Rice Experiment Station (Table 6). This points to California as the most likely source area for the Australian invasion of the two Barnyard Grass species. Historical hypotheses such as this which involve tracing the origin of weed invasions can be assessed by the use of isozymes as genetic markers, as long as sufficient polymorphism occurs in the species under study. Isozymes can also be used to examine whether intercontinental dispersal events such as those proposed for the two Barnyard Grass species are associated with significant bottlenecks causing a reduction in genetic variation in the introduced area.

An isozyme survey of 10 rice field populations of *E. oryzoides* from N.S.W. indicated that all populations were genetically uniform at the 32 loci that were screened. The only variation observed involved a single population that was fixed for a variant allele at a locus coding for the enzyme *Pgm.* The Australian sample represents a highly limited extract of the genetic variation present in California, a pattern consistent with the historical information. A comparison of quantitative variation in California and Australian populations further indicates that populations from the two regions are for the most part indistinguishable from one another (Fig. 5). This suggests that relatively little genetic differentiation has occurred in response to local conditions following the species introduction to Australia. Two factors are likely to be important in this regard. First, the ecological conditions present in rice fields in the two

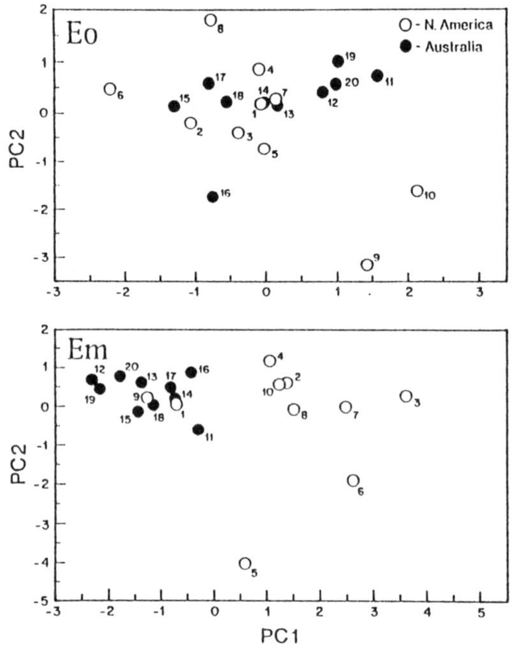

Fig. 5. Principal components analysis of quantitative variation in twelve life history traits in populations of *Echinochloa oryzoides* (Eo) and *Echinochloa microstachya* (Em) from N. America (open circles, 1—10) and Australia (closed circles, 11—20). The twenty populations of each species were grown under uniform glasshouse conditions. Note the absence of differentiation between populations from the two regions in *E. oryzoides* and in *E. microstachya* the occurrence of two N. American populations (1, 9) that cluster with the Austrlaian sample. The two populatiosna re from N. California (S. C. H. Barrett and A. H. D. Brown unpubl. data).

regions are quite similar and, secondly, the selection responses of Australian populations may be limited by a restricted amount of genetic variation present in the founding stocks.

The invasion of Australian rice fields by *E. microstachya* is of particular interest because it has no previous history as an agricultural weed and is limited in distribution for the most part to its native North America. The species is widespread in its native range as a ruderal weed of roadside ditches and open, naturally disturbed sites. It occurs commonly in N. California and at the present time is only occasionally found as a weed of rice. In common with *E. crus-galli*, the deep water conditions in California rice fields prevent successful establishment of *E. microstachya* and populations are largely restricted to levees and rice field edges. Prior to the introduction of permanent flooding in Californian rice fields, the species may have been a more troublesome weed problem but because of its similarity to *E. crus-galli* it was never recorded as such.

An isozyme survey of 20 N. American populations of *E. microstachya* revealed a high degree of inter-population genetic differentiation with each population largely composed of a unique multilocus isozyme genotype. However, the parallel survey of Australian populations collected from in and around N.S.W. rice fields revealed a markedly different pattern and one very similar to that observed in *E. oryzoides.* Of the 20 populations screened, 18 were genetically uniform and composed of the same multilocus genotype. Two variant gene loci (*Hk*, *Lap*) were evident in the remaining populations. Once again the Australian sample provides evidence for a marked genetic bottleneck associated with intercontinental migration. Of particular interest was the finding that the predominant Australian genotype could be identified from among the N. American sample of populations. The genotype occurred in a population from N. California close to Biggs Rice Experiment Station, the likely exit point for cultivated rice varieties shipped to Australia in the 1920's.

Comparisons of quantitative variation among populations from the two regions gave results that were in accord with those from the isozyme survey. A high degree of genetic differentiation was evident among the N. American sample, a result anticipated on the basis of the species wide distribution and concurrent likelihood of regional differentiation. In contrast a restricted range of variation was present among the Australian populations, a pattern indicative of a genetic bottleneck (Fig. 5). Interestingly, two populations of *E. microstachya* from N. California clustered with the Australian sample as a result of their similar life history attributes. Once again this points to N. California as the likely source region for the Australian invasion.

While the degree of genetic differentiation among Australian populations was restricted in comparison with the N. American sample, it should be noted that some local differentiation among populations was evident. In particular, there was small but significant genetic differentiation both within and between populations for the time taken to reach anthesis. This variation could have arisen since introduction to Australia as a result of the recombination and selection of polygenic mutation. Considerable amounts of quantitative genetic variation can be maintained by this process even in highly selfing species (Lande 1977). The variation may also have originated through recombination following hybridization between different founding stocks. Field studies would be required before we can conclude whether this variation reflects local adaptation to Australian conditions.

Relevance of genetic studies to weed control

The preceding case studies illustrate how in two widely distributed weed groups, founder effects and genetic bottlenecks can have an important influence on the levels of genetic diversity in a region. These processes are likely to be more pronounced in weeds than in many animal pests where it has been argued that introduced populations maintain most of the genetic variation found in native populations (Myers and Sabath 1980). If this generalization holds up to further empirical study, it will likely reflect the prevalence of uniparental methods of reproduction in many weedy species. The occurrence of high levels of selfing, apomixis and clonal propagation can result in dramatic contrasts in population structure and levels of genetic diversity in populations occurring in different parts of the geographical range. In particular long distance migration on a continental scale is likely to be frequently associated with reduced levels of genetic diversity in newly occupied territory.

How can studies of the geographical patterns of genetic variation in weed species contribute to the development of effective weed control methods? Before we can examine this question it is important to distinguish the three main methods by which weed populations can be eradicated; i.e. biological, cultural and chemical control. Population studies are likely to have different implications for each of these methods, and to involve both direct and indirect benefits.

The conscious release of insect pests or fungal pathogens that destroy a specific weed species may be used to dramatically reduce, although not completely eradicate, populations. Of critical importance to the success of biological control programs is the likelihood of weed populations evolving resistance to pest or pathogen attack. There is some evidence that an association exists between the degree of control that has been achieved in biological control programs and the reproductive system of the target species. Asexually reproducing species appear to be more effectively controlled than those that reproduce by sexual means (Burdon and Marshall 1981). These authors suggest that successful biological control is favored by limited amounts of genetic variation in weed populations and that the reproductive system of a weed is one of the prime determinants of population genetic structure. Of course, from the viewpoint of biological control, the only genetic variation of relevance in the target weed is that which determines the differences in resistance or tolerance to control agents. Unfortunately at this time little is known about the patterns of host resistance in natural plant populations although a start has been made in examining this variation (Burdon 1985).

It is possible that if isozyme surveys of weed populations provide some index of the levels of genetic diversity in populations, they may be useful for providing a guide to the potential likelihood of the evolution of host resistance. A comparison of the patterns of genetic diversity revealed by isozyme studies in Australian populations of *Chondrilla juncea* (Burdon *et al.* 1980) and *Echium plantagineum* (Brown and Burdon 1983, 1987) is instructive in this regard. While the apomict *C. juncea* appears to be represented by only three genotypes in Australia, populations of the outbreeding *E. plantagineum* contain very high levels of genetic variation with virtually every member of the population possessing a unique genotype. The high level of polymorphism in *E. plantagineum* seems to have risen from multiple introduction of diverse genetic material for ornamental use followed by subsequent hybridization among biotypes. These effects complicate any attempt to locate source areas within the native range for collection of potential control agents. In contrast, in *C. juncea* source regions for the three apomictic triploid clones present in Australia may be more easily located and genotype specific strains of potential biological control agents identified. In *C. juncea* there is already evidence of the presence of a genotype specific strain of the rust *Puccinia chondrillina* and this has been successfully employed for biological control in Australia (Cullen and Groves 1977; Burdon *et al.* 1981).

Information on genetic diversity may be particularly valuable for the control of clonally propagating aquatic weeds such as Water Hyacinth where only a small portion of the genetic diversity present in the native region has been introduced to the adventive range. In addition, the use of isozyme for pinpointing likely source areas of weed invasions may aid in increasing the efficiency of biological control programs by reducing the effort spent in searching for race or genotypic specific control agents in regions distant from source areas. It is noteworthy that while it has long been stressed in the biological control literature that comprehensive genetic sampling of potential control agents is necessary, the sampling of the range of genetic variation in the target weed has received much less attention. Adequate sampling of the weed is as important as that of the control agent if the aim is to ensure the best possible match between the two (Marshall *et al.*1980).

Cultural control methods encompass a broad range of agricultural practices including rotation and cultivation methods, water management, crop spacing and density, the timing of fallow periods, etc. Obviously a sound knowledge of the life history characteristics and population dynamics of problem weeds is critical in formulating cultural control methods (Crawley 1987). Mortimer (1983) has discussed at some length the importance of these approaches for understanding the mechanisms of population regulation in weeds (also see Weiner 1990). It is more difficult, however, to see how information on the genetics of weed populations can aid directly in the development of more effective cultural control methods. Nevertheless, it would seem important that weed control specialists are familiar with the range of genetic variation present in weed species and in particular are aware of the high degree of genetic differentiation that exists among weed populations for many ecologically important traits such as dormancy (Naylor and Jana 1976), development rate (Kadereit and Briggs 1985), competitive ability (Solbrig and Simpson 1977) and fecundity (Barrett and Wilson 1981). As yet we are some way from being able to predict how the introduction of particular cultural control methods will influence the genetic structure and evolution of weed populations or vice versa (Barrett 1988). However, it is clear that a prerequisite for such an approach is a sound knowledge of the quantitative genetics of life history traits in weed populations. So far there are relatively few studies on this topic, particularly of agricultural weeds.

The most effective and widespread method of weed control is through the use of herbicides. Despite their obvious potency as selective agents, relatively little is known about the influence of herbicides on the genetic characteristics of weed populations. Genetically based shifts in ecological traits such as germination behavior, growth rate or flowering time might be anticipated as a result of the evolution of various "avoidance strategies" or, alternatively, weed populations may simply respond through phenotypic plasticity (Putwain *et al.* 1982). Studies on the levels of genetic variation for herbicide tolerance in weed populations (Price *et al.* 1983; Thai *et al.* 1985) are important since information on the potential likelihood for the evolution of tolerance is of value in the planning and assessment of herbicide programs as well as the implementation of crop rotation schemes. Despite earlier predictions to the contrary (Harper 1956; Gressel and Segel 1978), an increasing number of cases of the evolution of herbicide tolerance is coming to light, particularly in response to the use of the S-triazine herbicides (LeBaron and Gressel 1982). Integrated control strategies for aquatic weeds, reviewed by Ashton and Mitchell (1989), require further research on the ecological genetic and demographic aspects. Many theoretical and empirical advances and research needs are elegantly reviewed in Roughgarden *et al.* (1989); clearly, weed control is a closely interwoven subset of problems in the management of populations or ecosystems.

Conclusions

The input of principles and methods developed in population biology to weed control practices is likely to be gradual at first since population biology is a relatively young discipline and the interactions between fields are at present limited. Nevertheless, if weed biology is going to move from its present, largely descriptive state to a situation where it is possible to predict the influence of weeds on crop yields, and to manipulate both the ecological and genetic characteristics of weed populations through management techniques, then the approaches and methods used in population biology must be integrated into the framework of weed science (Kluge *et al.* 1986). Collaborative research studies between agriculturalists and population biologists as well as interdisciplinary programs are required to nurture the growing field of applied population biology.

Literature cited

Ashton, P. and D. S. Mitchell. 1989. Aquatic plants: Patterns and modes of invasion, attributes of invading species and assessment of control programmes. In: Drake, J. A. *et al.* (eds.) Biological Invasions: A Global Perspective. John Wiley, Chester.

Baker, H. G. 1955. Self-compatibility and establishment after "long-distance" dispersal. Evolution 9: 347—349.

——. 1965. Characteristics and modes of origins weeds, pp. 141—172. In H. G. Baker and G. L. Stebbins (eds.). The Genetics of Colonizing Species. Academic Press, London.

——. 1974. The evolution of weeds. Ann. Rev. Ecol. Syst. 5: 1—24.

Baker, H. G., and G. L. Stebbins (eds.) 1965. The Genetics of Colonizing Species. Academic Press, London.

Barrett, S. C. H. 1977. Tristyly in *Eichhornia crassipes* (Mart.) Solms (Water Hyacinth). Biotropica 9: 230—238.

——. 1978. Heterostyly in a tropical weed: the reproductive biology of the *Turnera ulmifolia* complex (Turneraceae). Can. J. Bot. 56: 1713—1725.

——. 1979. The evolutionary breakdown of tristyly in *Eichhornia crassipes* (Mart.) Solms. Evolution 33: 499—510.

——. 1980a. Sexual reproduction in *Eichhornia crassipes* (Water Hyacinth). I. Fertility of clones from diverse regions. J. Appl. Ecol. 17: 101—112.

——. 1980b. Sexual reproduction in *Eichhornia crassipes* (Water Hyacinth). II. Seed production in natural populations. J. Appl. Ecol. 17: 113—124.

——. 1982. Genetic variation in weeds, pp. 73—98. In R. Charudattan and H. L. Walker (eds.), Biological Control of Weeds with Plant Pathogens. John Wiley & Sons, New York.

——. 1983. Crop mimicry in weeds. Econ. Bot. 37: 255—282.

——. 1985a. Floral trimorphism and monomorphism in continental and island populations of *Eichhornia paniculata* (Spreng.) Solms (Pontederiaceae). Biol. J. Linn. Soc. 25: 41—60.

——. 1985b. Ecological genetics of breakdown in tristyly, pp. 267—275. In: J. Haeck and J. W. Woldendorp (eds.), Structure and Functioning of Plant Populations 2: Phenotypic and Genotypic Variation in Plant Populations. North Holland Publishing Company, Amsterdam.

——. 1988. Genetics and evolution of agricultural weeds, *In*: M. Altieri and M. Liebman (eds.), Weed Management in Agroecosystems: Ecological Approaches. CRC Press Inc., Boca Raton, Florida.

Barrett, S. C. H., and I. W. Forno. 1982. Style morph distribution in new world populations of *Eichhornia crassipes* (Water Hyacinth). Aquatic Bot. 13: 299—306.

Barrett, S. C. H. and B. C. Husband. 1989. The genetics of plant migration and colonization. *In*: A. H. D. Brown, M. T. Clegg, A. L. Kahler, and B. S. Weir (eds.), Plant Population Genetics, Breeding, and Genetic Resources, pp. 254—277. Sinauer, Sunderland.

Barrett, S. C. H. and B. C. Husband. 1990. Variation in outcrossing rates in *Eichhornia paniculata*: The role of demographic and reproductive factors. Plant Species Biol.

Barrett, S. C. H., and B. J. Richardson. 1986. Genetic attributes of invading species, pp. 21—30. *In*: R. H. Groves and J. J. Burdon (eds.) Ecology of Biological Invasions: An Australian Perspective. Australian Acad. of Sci., Canberra.

Barrett, S. C. H. and D. E. Seaman. 1980. The weed flora of Californian rice fields. Aquatic Bot. 9: 351—376.

Barrett, S. C. H., and J. S. Shore. 1987. Variation and evolution of breeding systems in the *Turnera ulmifolia* complex (Turneraceae). Evolution 41: 340—354.

Barrett, S. C. H., and J. S. Shore. 1988. Isozyme variation in colonizing species. *In*: D. Soltis and P. Soltis (eds.). Isozymes in Plant Biology.

Barrett, S. C. H., and J. L. Strother. 1978. Taxonomy and natural history of Bacopa (Scrophulariaceae) in California. Syst. Bot. 3: 408—419.

Barrett, S. C. H., and B. F. Wilson. 1981. Colonizing ability in the *Echinochloa crus-galli* complex (barnyard grass). I. Variation in life history. Can. J. Bot. 59: 1844—1860.

—— and ——. 1983. Colonizing ability in the *Echinochloa crus-galli* complex (barnyard grass). II. Seed biology. Can. J. Bot. 61: 556—562.

Bleasdale, J. K. A. 1966. Plant growth and crop yield. Ann. Appl. Biol. 57: 173—182.

Bosbach, K., and H. Hurka. 1981. Biosystematic studies of *Capsella bursa-pastoris* (Brassicaceae): Enzyme polymorphism in natural populations. Plant Syst. Evol. 137: 73—94.

Brown, A. H. D. 1979. Enzyme polymorphism in plant populations. Theor. Pop. Biol. 15: 1—42.

Brown, A. H. D., and J. J. Burdon. 1983. Multilocus diversity in an outbreeding weed, *Echium plantagineum* L. Austr. J. Biol. Sci. 36: 503—509.

——, and ——. 1987. Mating systems and colonizing success in plants, pp. 115—132. *In*: A. J. Gray, M. J. Crawley and P. J. Edwards (eds.), Colonization, Succession and Stability. British Ecological Soc. Symp. No. 26, Blackwell, Oxford.

Brown, A. H. D., and D. R. Marshall. 1981. Evolutionary changes accompanying colonization in plants, pp. 351—363. *In*: G. G. E. Scudder and J. L. Reveal (eds.), Evolution Today, Proceedings of the Second Internationl Congress of Systematic and Evolutionary Biology. Carnegie-Mellon Univ., Pittsburgh.

Burdon, J. J. 1985. Pathogens and the genetic structure of plant populations, pp. 313—326. *In* J. White (eds.), Studies on Plant Demography. Festsch. for John L. Harper. Academic Press, London.

Burdon, J. J., and D. R. Marshall. 1981. Biological control and the reproductive mode of weeds. J. Appl. Ecol. 18: 649—658.

Burdon, J. J., D. R. Marshall, and R. H. Groves. 1980. Isozyme variation in *Chondrilla juncea* L. in Australia. Aust. J. Bot. 28: 193—198.

Burdon, J. J., R. H. Groves and J. M. Cullen. 1981. The impact of biological control on the distribution and abundance of *Chondrilla juncea* in southeastern Australia. J. Appl. Ecol. 18: 957—966.

Burdon, J. J., D. R. Marshall and A. H. D. Brown. 1983. Demographic and genetic changes in populations of *Echium plantagineum*. J. Ecol. 71: 667—679.

Clegg, M. T., and R. W. Allard. 1972. Pattern of genetic differentiation in the slender wild oat species *Avena barbata*. Proc. Nat. Acad. Sci. U.S.A. 69: 1820—1824.

Clegg, M. T., and A. H. D. Brown. 1983. The founding of plant populations, pp. 216—228. *In*: C. M. Schonewald-Cox, S. M. Chambers, B. MacBryde, and W. L. Thomas (eds.), Genetics and Conservation. Benjamin/Cummings, California.

Clegg, M. T., D. J. Schoen and B. K. Epperson. 1985. The interactions between phenotypic diversity and mating patterns in plant populations, pp. 287—298. *In*: J. Haeck and J. W. Woldendorp (eds.), Structure and Functioning of Plant Populations. 2. Phenotypic and Genotypic Variation in Plant Populations. North-Holland Publishing Company, Amsterdam.

Crafts, A. S. and W. W. Robbins. 1962. Weed Control, McGraw-Hill, New York.

Crawley, M. J. 1986. The population biology of invaders. Phil. Trans. R. Soc. Lond. B. 314: 711—731.

Cullen, J. M. and R. H Groves. 1977. The population biology of *Chondrilla juncea* L. in Australia. Proc. Ecol. Soc. Austr. 10: 121—134.

Faegri, K. and L. van der Pijl. 1971. The Principles of Pollination Ecology (2nd ed.) Pergamon, Oxford.

Firbank, L. G., and A. R. Watkinson. 1986. Modelling the population dynamics of an arable weed and its effect upon crop yield. J. Appl. Ecol. 23: 147—159.

Giles, B. E. 1983. A comparison between quantitative and biochemical variation in the wild barley *Hordeum murinum*. Evolution 38: 34—41.

Glover, D. E., and S. C. H. Barrett. 1987. Genetic variation in continental and island populations of *Eichhornia paniculata* (Pontederiaceae). Heredity 59: 7—17.

Gray, A. 1879. The pertinacity and predominance of weeds. Amer. J. Sci. 118: 161—167.

Gressel, J., and L. A. Segel. 1978. The paucity of plants evolving genetic resistance to herbicides: possible reasons and implications. J. Theor. Biol. 75: 349—371.

Hamrick, J. L., Y. B. Linhart, and J. B. Mitton. 1979. Relationships between life history characteristics and electrophoretically detectable genetic variation in plants. Ann. Rev. Ecol. Syst. 10: 173—200.

Harper, J. L. 1956. The evolution of weeds in relation to resistance to herbicides. Proc. 3rd British Weed Control Conf. 1: 179—186.

——. 1960. Biology of Weeds. British Ecol. Soc. Symp. No. 1, Blackwell, Oxford.

——. 1977. Population Biology of Plants. Academic Press, London.

Heiser, C. B. Jr. 1965. Sunflowers, weeds and cultivated plants, pp. 391—401. *In*: H. G. Baker and G. L. Stebbins (eds.). The Genetics of Colonizing Species. Academic Press, London.
Holm, L. G., D. L. Plucknett, J. V. Pancho, and J. P. Herberber. 1977. The World's Worst Weeds: Distribution and Biology. University of Hawaii Press, Honolulu.
Husband, B. C. and S. C. H. Barrett. 1991. Colonization history and population genetic structure of *Eichlornia paniculata* in Jamaica. Heredity (in press).
Jain, S. K. 1969. Comparative ecogenetics of two *Avena* species occurring in Central California. Evol. Biol. 3: 73—118.
——. 1983. Genetic characteristics of populations, pp. 240—256. *In*: H. A. Mooney and M. Gordon. (eds.), Ecological Studies: Analysis and Synthesis. Springer-Verlag, Berlin.
Jain, S. K., and P. S. Martins. 1979. Ecological genetics of the colonizing ability of rose clover (*Trifolium hirtum* All.). Amer. J. Bot. 66: 361—366.
Kadereit, J. W., and D. Briggs. 1985. Speed of development of radiate and non-radiate plants of *Senecio vulgaris* L. from habitats subject to different degrees of weeding pressure. New Phytol. 99: 155—169.
Kahler, A. L., and R. W. Allard, M. Krzakowa, C. F. Wehrhahn, and E. Nevo. 1980. Associations between isozyme phenotypes and environment in the slender wild oat (*Avena barbata*) in Israel. Theor. Appl. Genet. 56: 31—47.
King, L. J. 1966. Weeds of the World: Biological Control. Hill, London.
Kluge, R. L., H. G. Zimmerman, C. J. Culliers, and G. B. Harding. 1986. Integrated control of invasive alien weeds. *In*: I. A. W. Macdonald, F. J. Kruger, and A. A. Ferrar (eds.) The Ecology and Management of Biological Invasions in Southern Africa. pp. 295—303. Oxford, Cape Town.
Lande, R. 1977. The influence of the mating system on the maintenance of genetic variability in polygenic characteristics. Genetics 86: 485—498.
Law, R. 1981. The dynamics of a colonizing population of *Poa annua.* Ecology 62: 1267—1277.
Law, R., A. D. Bradshaw, and P. D. Putwain. 1977. Life history variation in *Poa annua.* Evolution 31: 233—246.
LeBaron, H. M., and J. Gressel (eds.). 1982. Herbicide Resistance in Plants. John Wiley, New York.
Lewontin, R. C. 1974. The Genetic Basis of Evolutionary Change. Columbia Univ. Press, New York.
——. 1984. Detecting population differences in quantitative characters as opposed to gene frequencies. Amer. Nat. 123: 115—124.
Lloyd, D. G. 1980. Demographic factors and mating patterns in Angiosperms, pp. 67—88. *In*: O. T. Solbrig (ed.), Demography and Evolution in Plant Populations. Blackwell, Oxford.
Loveless, M. D., and J. L. Hamrick. 1984. Ecological determinants of genetic structure in plant populations. Ann. Rev. Ecol. Syst. 15: 65—95.
Lyman, J. C., and N. C. Ellstrand. 1984. Clonal diversity in *Taraxacum officinale* (Compositae), an apomict. Heredity 50: 1—10.
Mack, R. N. and D. A. Pyke. 1983. The demography of *Bromus tectorum*: variation in time and space. J. Ecol. 71: 69—93.
Marshall, D. R., J. J. Burdon, and A. H. D. Brown. 1980. Optimal sampling strategies in the biological control of weeds. Proc. V Intern. Symp. Biol. Control of Weeds. Brisbane, pp. 103—111.
Martins, P. S. and S. K. Jain. 1978. The role of genetic variation in the colonizing ability of rose clover (*Trifolium hirtum* All.). Amer. Nat. 114: 591—595.
Maun, M. A. and S. C. H. Barrett. 1986. The biology of Canadian weeds. 74. *Echinochloa crus-galli* (L.) Beauv. Can. J. Plant Sci. 65: 739—759.
McIntyre, S., and S. C. H. Barrett. 1986. A comparison of weed communities of rice in Australia and California. Proc. Ecol. Soc. Austr. 14: 237—250.
McNeill, J. 1976. The taxonomy and evolution of weeds. Weeds Res. 16: 399—413.
Michael, P. W. 1983. Taxonomy and distribution of *Echinochloa* species with special refer-

ence to their occurrence as weeds of rice, pp. 391—306. *In* Weed Control in Rice. Proc. Conf. Int. Rice Res. Inst. Los Banos, Philippines.
Moran, G. F., D. R. Marshall, and W. J. Muller. 1981. Phenotypic variation in the colonizing species *Xanthium strumarium* L. (Noogoora Burr). Aust. J. Biol. Sci. 34: 639—648.
Mortimer, A. M. 1983. On weed demography, pp. 3—41. *In* W. W. Fletcher (ed.), Recent Advances in Weed Research. Commonwealth Agric. Bur., Farnum Royal.
Mulcahy, D. L. 1975. The reproductive biology of *Eichhornia crassipes* (Pontederiaceae). Bull. Torr. Bot. Club 102: 18—21.
Musik, T. J. 1970. Weed Biology and Control. McGraw-Hill, New York.
Myers, J. H., and M. D. Sabath. 1980. Genetic and phenotypic variability, and the success of establishment of insect introductions for the biological control of weeds, pp. 91—102. *In*: Proc. V Intern. Symp. Biol. Control of Weeds, Brisbane, Australia, CSIRO, Melbourne.
Naylor, J. M., and S. Jana. 1976. Genetic adaptation for seed dormancy in *Avena fatua*. Can. J. Bot. 54: 306—312.
Newsome, A. E., and I. R. Noble. 1986. Ecological and Physiological Characters of Invading Species, pp. 1—20. *In* R. H. Groves and J. J. Burdon (eds.). Ecology of Biological Invasions: An Australian Perspective Austr. Acad. of Sci., Canberra.
Popay, A. I., and E. H. Roberts. 1970. Factors involved in the dormancy and germination of *Capsella bursa-pastoris* (L.) Medik and *Senecio vulgaris* L.J. Ecol. 58: 103—122.
Price, S. C., J. E. Hill, and R. W. Allard. 1983. Genetic variability for herbicide reaction in plant populations. Weed Sci. 31: 652—657.
Price, S. C., K. M. Shumaker, A. L. Kahler, R. W. Allard, and J. E. Hill. 1984. Estimates of population differentiation obtained from enzyme polymorphisms and quantitative characteristics. J. Heredity 75: 141—142.
Price, S. C., and S. K. Jain. 1981. Are inbreeders better colonizers? Oecologia 49: 283—286.
Putwain, P. D., K. R. Scott, and R. J. Holliday. 1982. The nature of resistance to triazine herbicides; case histories of phenology and population studies, pp. 115—132. *In*: H. M. LeBaron and J. Gressel (eds.), Herbicide Resistance in Plants. John Wiley, New York.
Richards, A. J. 1986. Plant Breeding Systems. George Allen & Unwin, London.
Rick, C. M., J. F. Fobes, and S. D. Tanksley. 1979. Evolution of mating systems in *Lycopersicum hirsutum* as deduced from genetic variation in electrophoretic and morphological characters. Pl. Syst. Evol. 132: 279—298.
Roberts, H. A. 1964. Emergence and longevity in cultivated soil of seeds of some annual weeds. Weed Res. 4: 296—307.
Roughgarden, J., R. M. May, and S. A. Levin (eds.). 1989. Perspectives in Ecological Theory. Princeton Univ. Press, Princeton.
Sagar, G. R. 1968. Weed biology — a future? Neth. J. Agric. Sci. 16: 155—164.
Salisbury, E. J. 1961. Weeds and Aliens. Collins, London.
Sarukhan, J. 1974. Studies on plant demography in *Ranunculus repens* L., *R. bulbosus* L. and *R. acris* L. II. Reproductive strategies and seed population dynamics. J. Ecol. 62: 151—177.
Schemske, D. W., and R. Lande. 1985. The evolution of self-fertilization and inbreeding depression in plants. II. Empirical observations. Evolution 39: 41—52.
Schoen, D. J. 1982. Genetic variation and the breeding system of *Gilia achilleifolia*. Evolution 36: 361—370.
Solbrig, O. T., and B. B. Simpson. 1974. Components of regulation of a population of dandelions in Michigan. J. Ecol. 62: 473—486.
——, and ——. 1977. A garden experiment on competition between biotypes of the common dandelion *Taraxacum officinale*. J. Ecol. 65: 427—430.
Stebbins, G. L. 1965. Colonizing species of native Californian flora, pp. 173—192. *In* H. G. Baker and G. L. Stebbins (eds.). The Genetics of Colonizing Species. Academic Press, London.
Thai, K. M., S. Jana, and J. M. Naylor. 1985. Variability for response to herbicides in wild oat *Avena fatua* populations. Weed Sci. 33: 829—835.

Vrijenhock, R. C. 1990. Genetic diversity and the ecology of asexual populations, pp, 175—198. *In*: K. Wohrmann and S. Jain (eds.) Population Biology. Springer-Verlag, Berlin.
Warburg, E. F. 1960. Some taxonomic problems in weedy species, pp. 43—47. *In* J. L. Harper (eds.), The Biology of Weeds. British Ecol. Soc. Symp. No. 1, Blackwell, Oxford.
Warwick, S. I. 1980. The genecology of lawn weeds. VII. The response of different growth forms of *Plantago major* L. and *Poa annua* L. to simulated trampling. New Phytol. 85: 461—469.
Warwick, S. I., B. K. Thompson and L. D. Black. 1987. Genetic variation in Canadian and European populations of colonizing weed species *Apera spica-venti.* New Phytol. 106: 301—318.
Weiner, J. 1990. Plant population ecology in agriculture, pp. 235—262. *In*: C. R. Carroll, J. H. Vandermeet, and P. M. Rosset (eds.) Agroecology. McGraw-Hill, New York.
Zinger, H. B. 1909. On species of *Camelina* and *Spergularia* occurring as weeds in sowings of flax and their origin. Trudy Bot. Muz. Imp. Akad. Nauk 6: 1—303.

6. Population management in new plant breeding approaches

SUBODH K. JAIN

Abstract

This chapter briefly reviews some examples of the application of population and quantitative genetics to the development of new breeding methods. Plant breeding methods are all centered on the basic concepts of exploiting natural or artificially induced genetic variation in crop plants through selection, agronomic trials, and repeated cycles of this controlled evolutionary scheme. Methods vary in terms of the breeding system of the crop species (e.g. outbreeding, inbreeding, asexual), breeding objectives and the plant breeder's primary training or research area. Over the past three decades a significant shift to population genetic and evolutionary thinking has modified many breeding approaches in terms of methods for synthesizing initial gene pools, recurrent selection procedures for improving source populations, and the desired genetic heterogeneity in the end product cultivars. This shift has been induced by several factors: greater recognition of the role of variation in populations due to increase in evolutionary training, greater availability of genetic resources, ecological and environmental issues, and changes in agriculture toward long-term biocontrol and multi-species strategies. Theoretical ideas have also played a role in understanding these new methods although most basic principles are too general or elementary to be cited as the direct basis for developing any particular breeding methods. For example, recurrent selection involves some general background in polygenic inheritance models and in the basic theory of selection response. However, even with much empirical work from numerous crop breeding projects, only general guidance, not specific rules, can be obtained from them. In managing genetically heterogeneous host populations for developing durable disease resistance, indeed several interesting discoveries have been made on the genetic basis of resistance, cost of virulence, pathogenicity in mixed populations, epidemiology of different fungi, and so on.

In the management of forestry and rangeland communities, selection criteria include numerous population and community ecological variables related to persistence (recruitment), niche evolution, coexistence, community response to invasive species, and herbivory. Likewise, the development of new crops from wild biota often requires a basic understanding of the variation patterns, adaptations, reproductive biology, and role of natural enemies. Assembly of their genetic resources and initial domestication steps require perhaps a few major genes controlling such traits as seed retention at maturity, earliness, or apomixis; a great deal of population management using quantitative genetics then yields improvements in yield or crop adaptability to a region and an agroecosystem. There is clearly a need to involve some population biology in the training of plant breeders.

S. K. Jain and L. W. Botsford (eds), Applied Population Biology, 121–147.

Introduction

Plant breeding, during the early stages of domestication and prior to the foundation of genetics, always relied upon observation of the economic performance of populations and selection of desired individuals. Mendelian genetics provided the basic logic for manipulating inherited traits. It should be recognized emphatically that population and quantitative genetic ideas were inspired by breeding work from the start; populations and quantitative genetic variation are the bases of scientific breeding work. However, breeding methods are often defined also in terms of major gene manipulation, more so now with the molecular genetic advances in biotechnology. Allard (1960) and Simmonds (1962), on the other hand, ushered in a distinctly "population school" for understanding genetic variation in breeding materials as well as some new kinds of end-product, commercial varieties. Populations are not simply a mixture (beanbag) of genotypes (phenotypically distinct and independent) from which a breeder picks out one or a few most productive lines. We must recognize the interactive and recombinational dynamics of gene pools, the competitive and facilitative genotype interactions, responses to selection, continuing needs for the creation and maintenance of variation, and finally, the evolutionary advantages of genetically variable populations as varieties *per se*. This is what we signify as the "new population management approaches" reviewed in this chapter. Examples of global or national breeding programs with this new thinking (a "revolution" or slowly emerging new biology) are few, but, I believe, on the rise, and demand more attention from the students of population biology and evolution than ever before.

Recurrent selection (population improvement)

The possible choices of intrapopulation methods of recurrent selection depend on the resourcefulness of the breeder. Refinements of techniques for increasing effectiveness of the methods are continually being made as information is accumulated and methods are applied — to a greater array of crop species (Fig. 1, for use of a method in *Brassica*). Intrapopulation recurrent selection is amenable to most crop species and traits within crop species (Table 1). They have been applied to the autogamous (inbreeding) and allogamous (outbreeding) crop species, to the crop species that emphasize forage and grain production, and to the traits that vary in the complexity of their inheritance. The frequency of their use among crop species has varied primarily as to the level of difficulty in producing progenies, intermating of selected progenies, and competitiveness with other ongoing breeding programs. For obligate self-fertilizing species (e.g., oats [*Avena sativa* L.], barley [*Hordeum vulgare* L.], wheat [*Triticum vulgare* L.], and soybean [*Glycine max* (L.) Merrill]), development of inbred progenies is relatively easy, but intermating of selected progenies is relatively difficult. In contrast, the use of

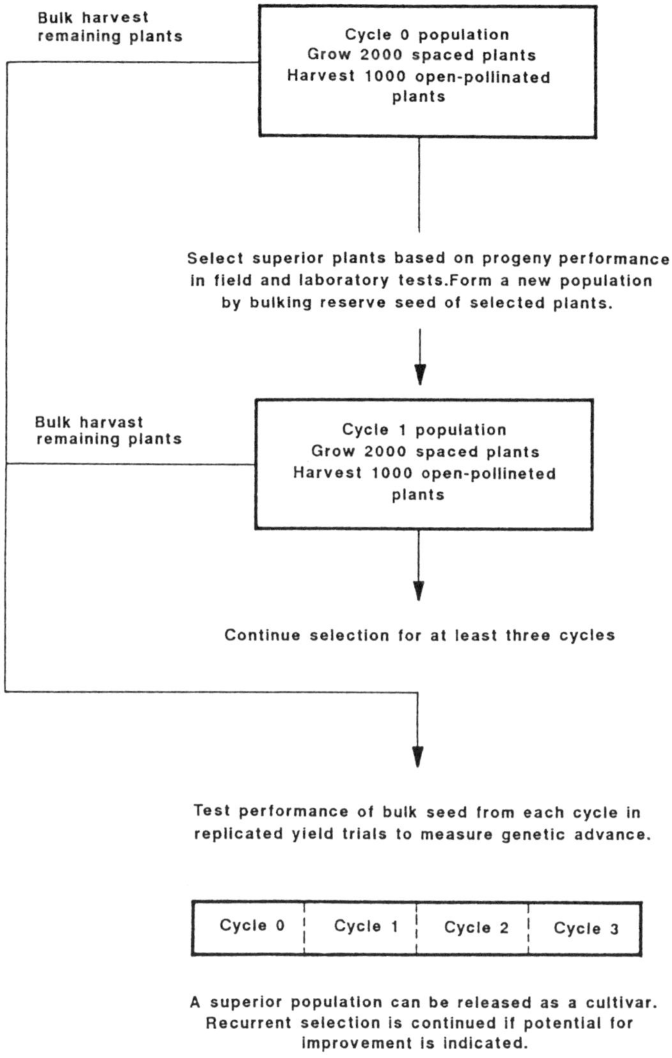

Fig. 1. Recurrent selection breeding scheme for *B. campestris*, as currently practiced at the Saskatoon Research Station. The initial composite could be an F_2 or any segregating population.

half-sib recurrent selection has been extensively used in the improvement of maize, but the production of half-sib progenies and their intermating has to be done either manually or in special isolation blocks that require detasseling (Table 2). Most of the methods, however, have been modified for use in most crop species. Doggett and Eberhart (Eberhart *et al.* 1967), Brim and Stuber, and others adapted half-sib recurrent selection for use in sorghum, soybeans, and wheat by incorporating a genetic male-sterile system in their populations

Table 1. Some recurrent selection programs in autogamous species (after Hallauer 1987).

Crop	Initial material	Number of cycles	Selected character	Gain/cycle (%)	Source
Pearl millet	158 lines and ms forms	3	Yield	2.5	Doggett (1972)
Tobacco	8-line hybrid	5	Height, low	−4.9	Matzinger *et al.* (1977)
			Leaf number	7.0	
Tobacco	8 lines	3	Leaf weight	7.2	Gupton (1981)
		5	Alkaloid content	5.2	
Cotton	2 lines	4	Yield	9.9	Miller and Rawlings (1967)
Soybean	10 lines	3	Yield	5.4	Kenworthy and Brim (1979)
Soybean	40 lines	3	Yield, early	4.3	Sumarno and Fehr (1982)
			Yield, midseason	0.0	
			Yield, late	1.2	
Soybean	3 lines	4	Oil content	5.5	Burton *et al.* (1983)
Barley	2 lines	3	Content, high	7.4	Byrne and Rasmusson (1974)
			Content, low	−12.2	
Barley	7 lines	1	Yield	0.0	Patel *et al.* (1985)
Barley	7 lines	1	Yield	39.8	Delogu *et al.* (1988)
Wheat	12 lines	2	Protein	7.8	McNeal *et al.* (1978)
Wheat	10 lines	4	Kernel weight	5.8	Busch and Rasmusson (1974)
Wheat	2 lines	3	Content, high	12.4	Byrne and Rasmusson (1974)
			Content, low		
Oats	12 lines	3	Yield	3.8	Payne *et al.* (1986)
Oats	4 lines	1	Yield	2.5	Vanselow (in press)

Table 2. Mean performance for six traits using three methods of recurrent selection in maize (after Hallauer 1987).

Method of selection	Populations per se	Trait: Grain Yield q ha^{-1}	Grain Moisture %	Days to silk no.	Ear height cm	Lodging % Root	Lodging % Stalk
Mass	BSKCO	47.3	22.6	81.5	148	37.2	22.3
	BSK(M)C14	51.2	25.0	85.5	172	40.3	22.8
	b*	0.24**	0.14**	0.22**	1.75**	0.22**	0.09
S_1	BSKCO	41.3	22.6	81.8	130	37.2	18.4
	BSK(S)C8	60.9	23.1	76.5	112	12.9	31.6
	b	3.98**	0.00	−0.34**	−2.89**	−2.97**	2.04**
Half-sib	BSKCO	41.3	22.6	81.8	130	37.2	18.4
	BSK(HI)C8	60.1	24.6	76.6	114	31.5	20.0
	b	3.85**	0.20**	−0.36**	−1.82*	−0.60	0.56

*b = regression coefficient measuring gain per cycle.
* = P = .05; ** = P = .01.

to facilitate production of half-sib progenies and recombination in crop species that are primarily self-fertilizing (Hallauer 1987; Jensen 1988). Several parameters directly affect the expected gains from recurrent selection (Table 3). Quantitative genetic theory provides guidelines for designing different methods for different crops.

Application of recurrent selection methods in different crop species has

Table 3. Parameters affecting expected genetic gain for different methods of recurrent selection (after Hallauer 1987).

Method	h^2 (additive and dominance variance components)		Number per cycle: Years	Seasons
Mass				
No male control	1/2	$\sigma_A^2 + \sigma_D^2$	1	1
Male control	1	$\sigma_A^2 + \sigma_D^2$	1	1
Ear-to-row				
Modified	1/2	$(1/4)\sigma_A^2$	1	1
Modified-Modified	1	$(1/4)\sigma_A^2$	2	2
Half-sib				
Remnant half-sib	1	$(1/4)\sigma_A^2$	2	4
S_1	2	$(1/4)\sigma_A^2$	2	4
Full-sib	1	$(1/2)\sigma_A^2 + (1/4)\sigma_D^2$	2	3–4
S_1	1	$\sigma_A^2 + (1/4)\sigma_D^2$	2	3–4
S_2	1	$(3/2)\sigma_A^2 + (3/16)\sigma_D^2$	3	4–5

increased in recent years. For forage crops, recurrent selection has been used frequently for the past 30 years, but reported use of recurrent selection in autogamous species has been most frequent in the past 5 to 10 years. This was primarily because of the problems of intermating and the types of cultivars used by the farmers. Use of recurrent selection methods in autogamous crop species has been for a broad array of traits, ranging from seed quality to yield and components of yield. The interest in recurrent selection methods applied to autogamous crop species changes as breeders of autogamous crop species become more familiar with the potential of recurrent selection. Mass selection and inbred progenies developed by single-seed descent seem to be the most popular recurrent selection methods currently used for the autogamous crop species. In often cross-pollinated crops such as sorghum, cotton, fava bean and amaranths which have populations with high outcrossing rates, choices of breeding methods are even more flexible than the extremes of predominant selfers or outcrosses. Amaranths have sufficiently high outcrossing rates, and now both genetic and gene-cytoplasmic male sterility are available (Jain *et al.* 1985); its genetics and breeding are reviewed later in this chapter.

In an elegant review Ramage (1987) summarized historical shifts in the breeding methods for barley as follows:

> Bulk population methods were developed to extract pure lines from land races and, later, from progenies of crosses. Proponents of bulk methods, who usually were not concerned with maintaining a particular background genotype, believed that it was advantageous to handle as many crosses as possible, thus increasing the chances of obtaining superior gene combinations. Population breeding methods, such as composite crosses, evolutionary plant breeding, diallel selective mating, and male sterile facilitated recurrent selection, developed from this philosophy. Most pedigree breeding programs are committed to maintaining and improving a particular background genotype, such as one for malting and brewing quality. Population and recurrent selection breeding methods are used in situations where yield is more important than background genotype.

Modifying reproductive systems

Breeders often wish to develop inbred lines from allogamous (outbreeding) populations in order to select for maximum hybrid vigor (Fig. 2). Corn and alfalfa provide classic examples of inbreeding depression studies in diploid and autopolyploid crops, respectively, which do differ in the rate of inbreeding progress (homozygous) and expected heterozygous of superior hybrids. In many self-incompatible fruit and vegetable crops, selection for self-fertility is an important breeding goal. Genetic and ecological studies of devices controlling outbreeding systems have always attracted a great many population biologists and biosystematists. Likewise, predominant inbreeders are

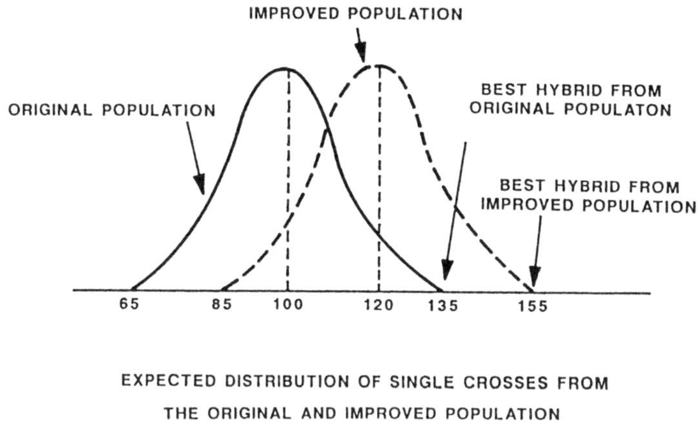

Fig. 2. A model based on recurrent selection for population improvement (Eberhart 1970).

often subjected to various forced open-pollinated systems, as noted earlier, for increasing recombination-derived genetic variation. Many species with mixed selfing and outbreeding systems may have large interpopulation variation in the reproductive patterns; breeders have often sought ways to modify self-fertility and seed yield. Research on self-fertility in rye and broad bean (*Vicia faba*) provides classic examples. Populations differ in their genetic and environmental components of seed yield, recovery from inbreeding depression, floral traits, and specific genes affecting selfing rates. Breeders design selection experiments under different pollination regimes.

Investigation of some ecotypes of *V. faba* showed the percentage of outcrossing to vary between 15 and 46%, with an increase in the crossing rate as plant densities increased. Analysis of plants derived from seeds segregating within pods indicated that when pollinating insects visit flowers, both cross-pollination and self-pollination occur in nearly equal proportions (Cubero and Moreno 1986). On the basis of the results obtained in a long-rum experiment on pollination in *Vicia faba* carried out in Southern Italy, a scheme of modified selection was outlined: this is based on a four-year cycle, two of self- and two of open-pollinations, with selection applied in the second and fourth year, respectively, for self-fertility and for combining ability. By using bees for intercrossing after some selfed pods have already been obtained, the cycle was reduced to three years. The effect of one self-pollination applied in different generations was studied by comparing homogenous populations of common origin; Table 4 shows that yield was reduced by 40% when selfing was applied in the last generation, and by 15% when it was applied in the last but-one generation, but no effect was obtained with selfing was applied at the last but-two generation.

Table 4. Influence of self- and open-pollination on the yield of dry seed (kg) in *Vicia faba* (Cubero and Moreno 1986).

Period / Population	1977–1979	1978–1980	1979–1981	1980–1982	1981–1983	Mean yield (%)
PPP	2.68	3.36	3.65	2.12	2.72	100
PPS	1.80	1.98	2.39	1.62	0.98	60
PSP	2.22	2.97	—	2.06	2.60	85
SPP	2.40	3.60	—	2.07	2.83	94
SSS	1.01	1.46	1.74	1.19	1.25	100
SSP	1.99	2.70	3.17	1.92	2.80	189
SPS	1.37	2.62	—	1.30	1.18	122
PSS	0.98	1.63	—	1.51	0.94	95

P = open-pollin; S = selfing; population with three generations of P or S in various sequences.

Population genetics and ecology in pasture crop improvement

Breeding of pasture crops often requires improvement in at least two distinct steps: first, collection of natural population samples in selected species (such as white clover, subclover or Hardinggrass) to develop by selection and testing a few cultivars, and, second, testing under multispecies sward conditions in which interspecies mixtures are evaluated in a community framework. Populations are often genetically heterogeneous and evolve (coevolve among interacting species). Numerous recent studies have initiated testing of such evolving mixtures and modelling of both population genetic and ecological features. Significant improvements have been made in several species during the first step, and now even greater strides are expected from the new approaches at the multispecies level. Pasture species are not bred as intensively as the food or fiber crops since a great many population variables enter in to the pasture management, but clearly population biology research is rewarding in dealing with genetic variation, response to grazing, survival and recruitment under natural environments, and adaptation to stress. Several modified breeding methods have been developed to deal with competition, use of heterosis, stability of mixed species stands, and continuing genetic inputs using new landraces.

Research at ICARDA on pasture legumes (Cocks *et al.* 1985) has shown that (a) native medics are more resistant to frost than the Australian cultivars (Table 5), (b) the advanced selections of *M. rigidula* and *M. rotata* are very resistant to frost, (c) frost resistance in native populations of annual legumes is related to frost frequency in their native habitats, (d) herbage and seed yield of susceptible species is greatly reduced by frost, (e) frost resistance is affected by plant density, and much of the variation in frost resistance with different densities and species is associated with variation in leaf-area ratio.

Table 5. Statistical parameters for the relationship between genotype survival and minimum temperature of the site at which genotypes were collected: n, number of genotypes collected; b, slope of the regression line for relationship; and 100 r^2, percentage of variation accounted for (Cocks *et al.* 1982).

Species	n	b	100 r^2
M. rigidula	50	NS	0
M. polymorpha	102	0.523***	13.2
M. turbinata	23	0.607*	20.1
M. orbicularis	43	0.240*	11.6
M. minima	43	NS	1.2
M. scutellata	11	1.000***	56.4

* = P = .05; ** = P = .01; *** = P = .001.

Basic population dynamics of coexistence and coevolution in pastures are described in terms of species diversity as well as genetic variation within species. Management of rangelands includes introduction of species and of improved cultivars in few selected species. Population biologists would find excellent research opportunities here.

Quantitative genetics, selection methods, and varietal

Testing in amaranth

Quantitative genetic studies are important for obtaining estimates of heritability of various plant type and yield traits, the role of heterosis, and character correlations. We have conducted two types of studies that have produced preliminary estimates of heritability for some quantitative characters. Studies of the genetic variation within and between families in landrace populations produced broad sense heritability estimates. These estimates are based on the between-family components of genetic variance for characters such as days to flowering, plant height, leaf length, and single plant yield. A mass selection experiment was used to estimate realized heritabilities for plant height and yield in one population each of *A. cruentus* and *A. hypochondriacus* (Table 6). Four different selection experiments are designed to investigate days to flowering, leaf length, yield, harvest index, and other predictors of yield. Development of appropriate breeding methods depend on an increased knowledge of the breeding system. Several types of selection methods are compared in order to compare breeding in a few accessions as base varieties versus population improvement schemes which involved wider gene pools of intervarietal crosses. Clearly, these methods satisfy both the short- and long-term breeding objectives (Simmonds 1979). Mean yields of the top three germplasm accessions from 1981 to 1983 are listed in Table 7. An experiment to test the effect of visual selection for high yielding plants within the highest yielding accessions was initiated in 1982.

Table 6. Average percentage gains from different selection schemes in two populations (tested at Davis in 1984) (after Ayiecho 1985).

Selection criterion and method	Mean									
	Plant height	Plant weight	Head length	Head weight	Threshing % ratio	Seed yield: height	Days to flowering	Seed weight	Harvest index	Seed yield per plant
UC 87										
Harvest index M1 + M2	0.0	10.2	1.8	16.2	4.7	23.7	−2.3	0.0	10.2*	21.8
Harvest index S1 + S1 · M1	0.7	8.9	2.6	13.9	3.4	14.5	−0.9	0.11	7.2*	14.2
Seed yield:height M1 + M2	3.6	15.5	4.3	20.4	2.7	24.6*	−1.9	8.7	25.3	
Seed yield:height S1 + S1 · M1	0.0	5.2	1.8	10.8	3.1	18.9*	−1.0		8.8	13.8
UC 99										
Harvest index M1 + M2	8.3	19.3	4.8	33.9	35.1	57.8	−6.1	8.0	47.0*	60.0
Harvest index S1 + S1 · M1	8.3	22.8	5.5	28.5	33.3	44.7	−4.4	4.7	36.4*	50.7
Seed yield:height M1 + M2	5.3	13.7	3.9	24.7	34.7	44.5*	−5.0	44.4	45.6	
Seed yield:height S1 + S1 · M1	6.6	17.7	6.7	25.3	31.3	40.6*	−4.8	—	36.2	44.4

M1 = Cycle 1 of mass selection.
M2 = Cycle 2 of mass selection.
S1 = Cycle 1 of S1 selection.
* = Direct response.

The effect of visual selection in five populations was evaluated in a combined trial during the summer 1984; significant responses to selection for higher plant yield, summarized for three of the populations (Table 7), suggested significant scope for further improvement.

Other selection experiments are in progress; S_1 recurrent selection and mass selection were compared in one study (Ayiecho 1985); results of the first two cycles showed significant responses in harvest index as well as the ratio of seed yield to plant height. Successful long-term breeding methods in grain amaranth will almost certainly involve a recurrent selection scheme with a program of interpopulation hybridization. In addition, major gene loci controlling such traits as plant height, flowering response, and male sterility should be transferred into a variety of genetic backgrounds to evaluate their usefulness.

Domestication of Limnanthes (meadowfoam) as a new oil crop

Limnanthes (family Limnanthaceae) species are herbaceous winter annuals with seed oil consisting of a high proportion of long chain fatty acids (C_{20}, C_{22}) of many potential industrial uses. Domestication efforts have largely involved biosystematic studies on genetic resources, agronomic evaluation of yield, and cultural practices and population improvement using mass and family selection methods. Significant gains have been made in seed yield and its various components which clearly make it a promising new crop. Search for both naturally-occurring and induced mutations for modifying plant architecture, seed retention, and increased self-fertility will be needed to facilitate commercial production as well as further gains in yield and adaptability, utilizing a combination of Mendelian and biometrical/population-based approaches.

Nearly 120 populations of all known taxa have been observed in natural stands as well as field or greenhouse plantings. Four species (Table 8) appeared to be relatively more promising candidates for domestication (Jain 1989); Table 9 summarizes over 10 years of cumulative information on various key traits which vary among populations or taxa more frequently than within populations.

Multilocus enzyme assays of parent-progeny similarities provided estimates of outcrossing rates and levels of inbreeding. A large amount of variation in outcrossing rate was found both among plants within a population and among populations of several species. This variation was correlated with the levels of autofertility and several floral traits (viz. petal length, stamen position at dehiscence) only at the interspecies level but not between families or populations. Apparently, the outcrossing rate is under numerous non-genetic controls and only weakly heritable. Variation in floral biology and degree of selfing among five different races of inbreeder *L. floccosa* might also be under polygenic control with low heritability. A selection experiment

Table 7. Response to two cycles of visual selection* for high yield/plant in three populations (Jain, unpublished data).

Population	Base (S_0)				S_1 cycle				S_2 cycle			
	Pl. yd. $\bar{x}$	(g.) sd	Infl. $\bar{x}$	(cm.) sd	Pl. yd. $\bar{x}$	(g.) sd	Infl. $\bar{x}$	(cm.) sd	Pl. yd. $\bar{x}$	(g.) sd	Infl. $\bar{x}$	(cm.) sd
R 104 (Mexican *cruentus*)	18.0	9.1	35.4	6.1	20.2	11.3	35.0	5.9	25.0	11.2	37.4	6.0
UC 87 (Guatemalan *cruentus*)	14.7	5.0	—	—	17.4	6.8	—	—	20.9	11.2	—	—
UC 100 (Mexican *cruentus*)	15.1	11.6	35.8	7.2	23.0	12.0	37.5	5.4	—	—	—	—

* Selection intensity was nearly 7%, except for S_1 cycle in R104 with 1.6% selection intensity.

Table 8. Some characteristics of the more promising four species (Jain 1989).

Species/Taxon[b]	Some notable characters
L. alba	Margins of rice fields; requires less water; large number of varieties; outcrossing rates 40—90%; seed size (wt.) 4—8 mg; *SSE*[a] 1 to 4; growth habit semi-prostrate to erect.
L. floccosa	Five varieties, all large-seeded, with high C_{22} (8—15%); two vars. almost fully cleistogamous and *SSE* high; early-maturing; high dormancy.
L. gracilis	Mixed mating (10—30% outcrossing); growth habit more upright; few branches with synchronous flowering.
L. douglasii var *rosea*	Early high *RGR*; early maturing; high seed yield, low seed oil % (< 18%); low seed dormancy.
var *nivea*	Excellent crop establishment; high yield; seed oil content (20—30%); very asynchronous; high seed shattering; need better plant architecture.

[a] SSE = Seed set efficiency, based on number of nutlets per flower, on the main branch.
[b] *alba, floccosa, gracilis* (section Inflexae) can be intercrossed; so can *douglasii, bakeri, striata* (section Reflexae) be inter-crossed.

Table 9. Evaluation of genetic resources in *Limnanthes* (Jain 1989).

Trait	Many popns.	Few popns.	Inter-pop. hybrids	Inter-specific hybrids	N.A.
1. Non-shattering[b]	—	X	—	X	
2. Early flowering[b]	—	X	X	X	
3. Loss of dormancy	—	X	?	?	
4. Upright, taller habit	—	—	—	—	X
5. Fewer synchronous flowers		—	—	X	X
6. Natural selfing:					
a. Cleistogamy	—	X	?	?	
b. Lacking protandry	—	—	X	X	
c. High auto-fertility	X	—	X	X	
7. High relative growth rate (RGR)	—	X	X	X	
8. Determinate habit[a]	X	—	X	X	
9. Large seed size with high % oil[b]					
10. Male sterility[a]	—	X	X	X	

N.A. = Not available in any accession.
X = available in the named materials.
? = genetics still unconfirmed.
[a] Major genes were identified for only these traits. In addition, we have found major genes controlling flower color, anther color, gynodioecy, leaf and stem pubescence, varous plant part pigmentation patterns, earliness, and possibly nutlet morphology.
[b] High h^2 values have been found in some populations and crosses. Selection response in preliminary efforts appears encouraging.

was designed to increase autofertility in *L. alba* by mass selection, and to analyze morphological correlates at the end of selection experiment.

Overall improvements in yield, as noted above, have been rather impressive in both *L. alba* (300—500 kg/ha in 1977 to 1100—1300 kg/ha in 1985) and *L. douglasii* (600—900 kg/ha in 1977 to 1200—1400 kg/ha in 1985). Artificial selection for earliness, higher seed set, and better plant type has been effective but even greater changes in traits such as higher germinability, reduced seed shattering, larger seed size, and overall adaptability have possibly occurred under automatic (natural) selection (Jain 1989). In order to accelerate selection gains, we need better methods for single plant screening and controlled pollinations; we need a critical analysis of heterosis versus evolutionary adjustments to higher selfing, and we seek a better understanding of genetic and environmental correlations among different yield components.

Breeding for durable disease resistance

Burdon (1987) thoroughly discussed the issues of variation in natural populations and coevolution in terms of host-pathogen agroecosystems. Table 10 shows very large ranges of variation in pathogenicity and racial composition among populations, presumably related to current evolutionary transilience as well as chance.

As Harper (1977) pointed out, many of the ideas and techniques of plant pathology are also relevant to this area of population biology and should not be ignored. The continuing use of multilines and varietal mixtures for agricultural crops provides opportunities to explore competitive interactions in the presence of disease on a large scale, thus hopefully overcoming the interference problems that plague small plot experiments. Equally, each new biological control program incorporating fungal pathogens represents a 'on-off' field experiment in which many of the ideas discussed here can be examined.

Mundt (1990) noted that for the logistic model, initially both pure stand and mixtures show similar disease curve, but with time, carrying capacity of pure stand becomes a limiting factor (less tissue available for infection), whereas in a mixture, disease rate is larger beyond certain "threshold" time — that is, disease is delayed! (Fig. 3). He concluded, "Mixtures should be more effective against diseases that increase at moderate to slow rates over long periods of time (e.g., stripe rust of small grains, powdery mildew) than epidemics that increase rapidly over shorter periods of time (e.g., stem rust of small grains and downy mildew)". Also, the factors such as HR varieties or unfavorable weather slow down the epidemic increase which makes mixtures more effective.

In both cereals and vegetables (Table 11), mixtures showed reduced disease and insect damage levels compared with the pure (single cultivar)

Table 10. Representative examples of the range and mean level of resistance to fungal pathogens found in a variety of host-pathogen combinations (after Burdon 1985).

Host-pathogen combination	Number of populations	Mean sample size per population	Range of resistance[a] (%)	Mean level of resistance	Number of races[b]	comments[c]	Reference
Avena sp./	31	20	0.0—?	12.4	2	SR	Dinoor 1970
Puccinia coronata	(i)[d] 11	44	0.0—95.0	27.3	4	SR	Burdon *et al.* 1983
	(ii)[d] 10	46	0.0—5.0	1.0	4	SR	Burdon *et al.* 1983
Helianthus annuus/ *Puccinia helianthi*	142	c20	0.0—c80.0	12.8	1		Zimmer & Rehder 1976
Helianthus petiolaris/ *Puccinia helianthi*	57	c20	0.0—100.0	85.4	1		Zimmer & Rehder 1976
Hordeum spontaneum/ *Erysiphe graminis*	15	17	0.0—100.0	50.8	composite	SR	Moseman *et al.* 1983
	77	35	0.0—45	?	composite	AR	Fischbeck *et al.*
1976							
Pinus taeda/ *Cronartium fusiforme*	3	11	12.1—21.2	16.1	composite		Barber 1986
Triticum dicoccoides/ *Erysiphe graminis*	10	23	16.7—100.0	55.3	composite	AR	Moseman *et al.* 1984
Phlox pilosa detonsa/ *Erysiphe cichoraceaum*	10[e]	19	57.2—96.3	77.1	1		Jarosz 1984
Phlox drummondi mcallisteri/ *Ersiphe cichoracearum*	4[e]	26	43.3—74.6	62.3	1		Jarosz 1984 M.Levy's Lab
Phlox pulcherrima/ *Erysiphe cichoracearum*	5[e]	13	63.4—83.3	74.0	1		Jarosz 1984

[a] Resistance measured in a variety of ways; see original references for explicit details.
[b] Number of pathogen races inoculated individually.
[c] SR = seedling resistance/ AR = adult plant resistance.
[d] (i) and (ii) = environments favorable and unfavorable for the pathogen, respectively (see text).
[e] Including only those populations with a sample size > 10.

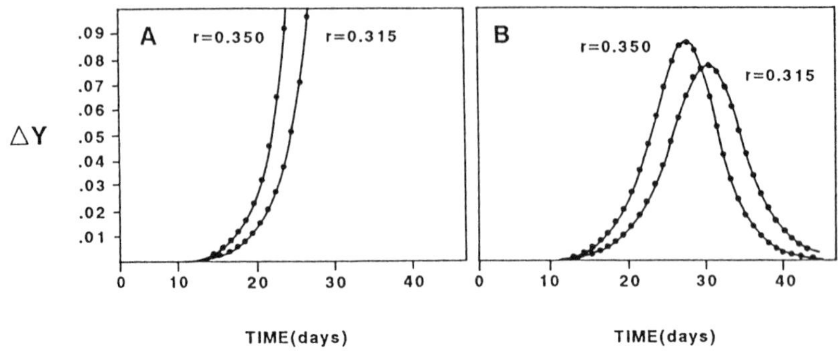

Fig. 3. Predicted daily incremental increase (ΔY) for wheat stem rust on susceptible plants for a pure stand of susceptible plants (circles) and a 1:1 mixture of resistant:susceptible plants (x's), assuming (A) logarithmic and (B) logistic increase of disease (after Mundt 1990).

Table 11. Comparison of infection and infestation rates of susceptible plants in pure and mixed cultures (after Trenbtah 1977; Chang and Oka 1984).

Crop	Partner, mixed with	Measurement	Pure	Mixed	% Difference*
Fungal diseases:					
Wheat	Resistant var. wheat	% rust cover	90	0–60	−33 to −100
Oat	Resistant var. oat	Infection rate Stem rust	0.338	0.272	−20
Celery	Tolerant var.	Infection rate *Cercospora apii*	0.128	0.042	−67
Cress	Grass	Spread (cm/day) *Pythium* sp.	1.14	0.72	−37
Insects:					
Corn	Groundnut	Corn borer tunnels	115	89	−23
Carrot	Shallots	% affected by carrot root fly	44	23	−48
Collard	Tomato	Feeding holes/cm^2 by flea beetles	5.4	1.4	−74

* Percentage derivation of "mixed" for "pure".

plantings. Explanations vary from epidemiological to selective interaction models but a great deal of research is needed to make such population responses more predictable on the field scale.

Wolfe and Barrett (1985) reported a barley study which included six two-way, and four three-way combinations of Athene, Igri, Gerbel, and Sonja. The summarized scores in Table 12 show that, despite the similarity in mildew resistance between the varieties, there was a general reduction in infection in the mixtures compared with the means of pure stands. As is commonly observed, the greatest overall reduction in infection occurs in mixtures containing the most susceptible component, in this instance, Gerbel. Among the three-component mixtures, Athene-Gerbel-Igri and Athene-Igri-Sonja reduced the mildew infection relatively more than did Athene-Gerbel-Sonja and Gerbel-Igri-Sonja; Athene-Igri-Sonja reduced mildew incidence. The authors recommended integrated control measures (chemicals, resistance, epidemiology) by which pathogen could be entirely eliminated if the composition of both winter and spring barley mixtures were monitored so as to include improved varieties and fungicides.

Disease and pest resistance in an agroeconomic context

Host plant (crop) variation at all levels, within and between individuals or within and among populations, provides several modes of defense against pests and diseases. Breeding for disease resistance seeks out genetically controlled vertical (major genes, race specific) or horizontal resistance (polygenic, non-race specific). Populations of cultivars may consist of one or more sources of such resistance, its durability in relation to the "arms race" with coevolving pathogens or herbivores depending on numerous evolutionary genetic as well as ecological conditions.

Denno and McClure (1983) review in detail how host plant variation in morphology, allelochemistry and nutrition poses major problems for herbivores and affects their population regulation; understanding this variation genetics and ecology may promote better pest management programs (e.g., spatial and temporal deployment of resistant crop varieties). Kareiva (1983) strongly advocated the need for large scale field studies on host-plant interactions. In a thorough review, Gould (1983) showed evidence for widespread genetic variation in both insect pests and host plants, both monogenic and

Table 12. Average percentage mildew infection in plots of pure stands and two-component and three-component mixtures in all possible combinations involving Athene, Gerbel, Igri, and Sonja (Wolfe and Barrett 1985).

	Pure stands and mixtures involving				Overall
	Athene	Gerbel	Igri	Sonja	average
Pure stands	4.4	14.1	7.4	7.9	100
2-components	6.4	7.6	6.8	7.0	84
3-components	5.4	6.9	5.9	6.2	71

polygenic, so that a plant breeder, working with an evolutionary ecologist, should determine stability of resistance in crop varieties. Cost of pest resistance, again, is a major issue. Gould concluded, "The rate and direction of change in herbivore adaptation to plant genotypes depends on interactions between the ecology and the genetics of the specific herbivore and plant populations". We may briefly note that the Integrated Pest Control (IPM) has been hailed as one of the most powerful payoffs from applied ecology (Krogan 1986), which still lacks a strong theoretical and evolutionary genetic background.

In a review of herbivorous insect dynamics, Power and Kareiva (1990) discussed the role of crop genetic variability in the development of insect and disease resistant cultivars or mixtures. The "fly paper" effect suggested for reduced disease incidence in mixtures depends on the passive dispersal of spores which probably fits relatively few insects. Several examples (Table 13), however, suggest potential role of mixtures for aphids and leafhoppers in which population reduction may be due to higher emigration rates. This is an area greatly in need of detailed field studies, especially with the rapid literature growth in agroecology on pest's population biology, the factors that influence abundance, and interactions with the crop genotypic arrays.

Several chapters in Wolfe and Caten (1987) give excellent examples to show how genetic variation in crop fields affects the occurrence of plant disease epidemics; population genetics and specific host-pathogen coevolutionary factors affect the disease control strategies. Genetic vulnerability of

Table 13. Response of herbivorous insects to plant genetic diversity[a] (Power and Kareiva 1990).

Insect	Stage	Response[b]	Reference
Potato leafhopper	Numphs	−	Cantelo and Sanford (1984)
Bean leafhopper	Adults	0	Power *et al.* (1987)
Corn leafhopper	Adults	−	Power (1988)
Cabbage aphid	Adults and nymphs	−	Altieri and Schmidt (1987)
English grain aphid	Adults and nymphs	−	Power, unpubl. data
Oat-bird cherry aphid	Adults and nymphs	−	Power, unpubl. data
Cereal leaf beetle	Eggs	−	Casagrande and Haynes (1976)
Cereal leaf beetle	Adults	±	Casagrande and Haynes (1976)
Cereal leaf beetle	Adult feeding damage	±	Casagrande and Haynes (1976)
Mexican bean beetle	Adults	0	Cantelo and Sanford (1984)
Imported cabbageworm	Eggs	+	Cantelo and Sanford (1984)
Imported cabbageworm	Larvae and pupae	+	Cantelo and Sanford (1984)

[a] Field experiments where insects were free to move among treatments (pure stands or mixed stands).

[b] Insect response to increased genetic diversity: − disease, + increase, ± variable response, 0 no response.

monocultures (large scale single species cultivation) is particularly increased by the lack of genetic variation in highly bred pure lines or hybrid cultivars. Pathogenicity is under genetic control with the gene-for-gene host-pathogen interactions in many cases; races of pathogens evolve as we introduce new disease resistance genes. Two kinds of resistance (vertical and horizontal) are defined in terms of race specificity and number of genes, such that breeding for more durable, polygenic, horizontal resistance is now sought in the new breeding procedures. Research is needed on population genetic factors of evolution (mutation, migration, directional and stabilizing selection; polymorphic race composition of pathogens, sexual recombination and ecological/epidemiological variables). Natural populations provide information on ideas for balanced plant protection, spatio-temporal environments, and sources of useful germplasm. It is important to note that the breeders' ideas in this important area have changed dramatically.

Breeding in long-lived crop species

Burdon (1989) illustrated the role of population genetic methodology in the breeding progress of *Pinus radiata* (Fig. 4). Seed orchards produce seed from different open-pollinated selected lines by which large gains have been made in growth rate, stem straightness, and branching pattern. More and more attention is now paid to recurrent selection modes (larger population, cycles of recombination), overlapping generations, and use of wider based genetic stocks; there is now better know-how of adaptational diversity in natural populations of different provenances, and durable disease resistance based on genetic diversity appears to be a plausible goal. Burdon (1989) reviews many quantitative genetic issues.

Development of perennial sorghums (*Sorghum bicolor*), weedy, perennial (*S. halepense*), produces rhizomes, fleshy underground stems capable of winter survival. At the Land Institute in Kansas, Kulakow and Ennis (1988) crossed the perennial weedy species to the artificially produced tetraploid form domesticated annual *S. bicolor* in order to develop perennial, high-yielding forms. Due to serious potential risks of weediness, the selected lines will be closely monitored to avoid such risks. Patterns of morphological variation in F_1 and F_2 suggests promise: large leaf number, low tillering, good seed characteristics of cultivated parent might be recombined with perennial reproduction of weedy parent. Table 14 shows F_2 data in terms of variation for 15 characters among and within 10 groups of crosses; except for rhizome number and percent seed set, most key traits showed significant variation among crosses as well. Rhizome production appeared to be simply inherited as a nongenic, dominant class. However, it would certainly become a major problem in field studies to deal with many correlated responses and constraints in life history shifts — perenniality often requires major shifts in resource allocation as well as demographic features. Authors also noted the

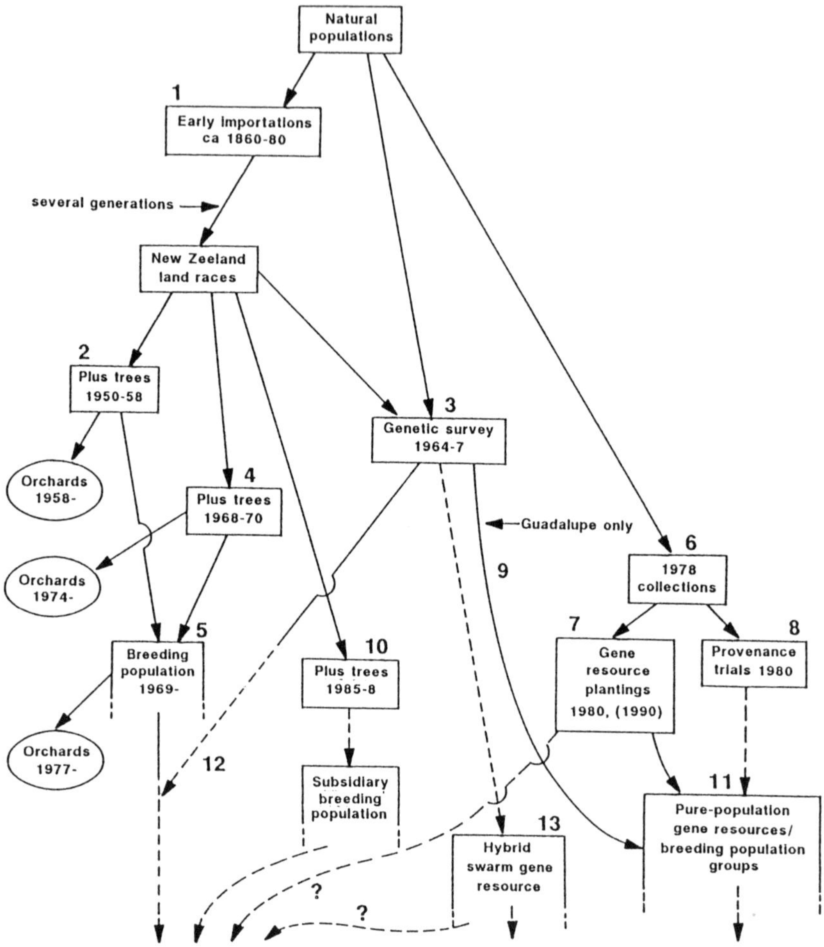

Fig. 4. Flow chart showing deployment and proposed deployment of genetic material for overall breeding population in New Zealand's *Pinus radiata* breeding program. Bold arrows show major movement of material; light arrows, minor movements; dashed — (Burdon 1988).

need for several cycles of recurrent selection to break the association between rhizome presence and seed shattering.

Genetic resources: sampling, ecogeography, and conservation

Plant breeders derive genetic materials from landraces, wild, and weedy crop relatives which have been collected and studied worldwide for variation. Population genetic models have been applied to such assays in order to infer past evolution, dispersals, adaptive or neutral variation, and centers of

Table 14. Analysis of variance among and within crosses for 15 characters scored in the F_2 generation from 10 groups of crosses involving different *S. halepense* parents (after Kulakow and Ennis 1988).

Character	Source	df	Mean square
Leaf length	among	9	127.80
	within	230	91.40
Leaf width	among	9	7.60[a]
	within	230	1.10
Leaf number[b]	among	9	0.11[a]
	within	229	0.04
Plant height	among	9	3673.90[a]
	within	229	669.20
Internode length	among	9	72.40[a]
	within	229	5.50
Tiller number[c]	among	9	1.94
	within	230	0.29
Reproductive culms	among	9	5.60[a]
	within	230	1.40
Days to flowering+	among	9	0.12[a]
	within	229	0.04
Panicle exertion	among	9	147.10[a]
	within	229	49.00
Panicle length	among	9	86.50[a]
	within	229	30.30
Panicle width	among	9	120.40[a]
	within	229	38.30
Percentage seed set[d]	among	9	0.16
	within	195	0.10
Seed weight[c]	among	9	4.82[a]
	within	172	2.02
Rhizome number[b]	among	9	0.77
	within	230	0.50
Rhizome node number[b]	among	9	2.52[a]
	within	230	0.99

[a] $p < 0.01$.
[b] analysis on log transformed data.
[c] analysis on square root transformed data.
[d] analysis on arcsin on square root transformed data.

diversity. Genetic conservation, that is conservation of genetic resources, involves *in situ* and *ex situ* (use of gene banks, botanical gardens, etc.) projects. Sampling genetic variation so as to maximize the retention of natural diversity is based on models of ecogeographic patterns — clinal, regional/ecotypic, racial, etc. Marshall (1989) reviewed various key issues in the genetic resource conservation area. Brown (1978 and later) reviewed strategies for sampling genetic variation in landrace collections. Qualset (1975) discovered an altitudinal cline for barley yellow dwarf resistance in

Ethiopia. Many rice disease resistant types are so rare that there appears to be no pattern of coevolved indemism (Harlan 1975). Nevo (1983) described variation in the wild summer wheat (*Triticum dicoccoides*) in Israel and near East center of diversity, notable for wild adaptive diversity against multiple and variable pathogens, pests, and diverse ecological stresses. He found this wild wheat to have "unique population structure" with subdivisions in an "archipelago" setting.

Comparisons of wild ancestors and derived cultivated landrace populations have been reported in many crop genera. Molecular variation seems to have been reduced under domestication (e.g., radish, Table 15). However, Jain *et al.* (1986) showed in grain amaranth that morphological and isozyme variation patterns do not show concordance. In many field crops and colonizing pasture species (e.g., rose clover, see Molina and Jain 1992), genetic variation for only a few polymorphic linked gene systems may be critical for evolutionary changes and agricultural success.

Cytoplasmic male sterility has been found in wide crosses between the crop species and their wild relatives. This may be associated with increased disease susceptibility in some cases (e.g., the well-known case of Southern leaf blight and T cytoplasm in maize, certain races of leaf rust in wheat). Mapping and genetic analyses of mt DNA are of special interest, as are the surveys of variation in natural or domesticated populations of numerous plant species. Several examples (amaranth, Peters and Jain 1988; meadowfoam, Kesseli and Jain 1985; rose clover, Molina and Jain, in press) have been recently discovered and analyzed for specific roles in modifying the reproductive systems. Plant breeders have shown great interest in these evolutionary studies for an understanding of the fitness components of male sterility primarily introduced to enhance interbreeding/hybrid seed production in their breeding nurseries.

Attempts to modify chemical composition of seeds (e.g., lowered toxic glycosinolates in rapeseed, low alkaloid in lupin, higher lysine in maize and

Table 15. Distribution of allozyme variation in *Raphanus sativus* (after Ellstrand and Marshall 1985).

Locus	Cultivars				Wild populations				
	H_T	H_S	D_{ST}	G_{ST}	H_T	H_T	H_S	D_{ST}	G_{ST}
IDH	0.44	0.36	0.08	0.17	0.51	0.42	0.09	0.18	
PGM-1	0.61	0.42	0.19	0.31	0.48	0.46	0.02	0.04	
PGM-2	0.58	0.34	0.24	0.41	0.65	0.57	0.09	0.13	
PGD	0.46	0.27	0.19	0.41	0.45	0.44	0.01	0.02	
EST	0.33	0.22	0.11	0.33	0.57	0.54	0.07	0.06	
LAP	0.54	0.40	0.14	0.26	0.71	0.66	0.05	0.07	
TPI	0.00	0.00	0.00	—	0.47	0.16	0.31	0.65	
FOR	0.00	0.00	0.00	—	0.52	0.47	0.05	0.09	

sorghum, coumarin-free cultivares of *Melilotus* spp., and high oleic lines of safflower) using mutants and/or "recurrent selection" have provided important goals in breeding safe and nutritious food sources. Genetic engineering is particularly applicable in these projects. However, the pleiotropic effects on yield or overall agronomic performance in the background of various populations used as cultivars are largely unknown. In this and several other kinds of breeding goals, enhanced risks of crop-weed gene transfers warrant a rigorous scrutiny of associated risks. Hence, a panel recommended high priority on research on documenting the distribution and potential interbreeding of wild relatives of U.S. agronomic crops. Kogan (1986) provided a nice review (see her Table 5.1) of certain landmark development in the theory and practice of herbivore/plant resistance development. Evidently plant defense strategy are multifaceted (physical, e.g., hairiness, biochemical; e.g., coumarin or alkaloids; phenological, e.g., asynchrony of life cycles; or physiological, e.g., shifts in resource allocation). Resistance breeding often focusses on one trait or one kind of defense. New breeding approaches should take cognizance of the population resources and history of changes under domestication, as well as a proper accounting for tradeoffs/character complexes. These ideas seem to trace back through the years to publications in the 60s but perhaps now the agroecology movement would truly accomplish those goals.

A NAS/NRC publication on Alternative Agriculture (p. 179) provided examples in which the cost of resistance breeding (Hessian fly and stem sawfly on wheat, spotted alfalfa aphid, European corn borer) is less than $10 million but annual benefits over hundreds of millions of dollars. IPM is now in use for 12 major U.S. crops over 14—80% acreage; breeding plays a key role and so would new genetic engineering methodologies. Another closely related issue, not covered in this book, is the management of pest evolution such as to deter pesticide resistance (NAS 1986); this issue needs the help of evolutionary biologists as well in order to attempt to integrate chemical and nonchemical control methods, genetic and nongenetic, epidemiological and biochemical-physical. Via's (1986) treatment of quantitative genetic models for pesticide resistance evolution is particularly useful. The evolutionary biology of crops, weeds and pests (herbivores) as well as their coevolutionary outcomes, given genetics, ecology and effects of human intervention are all important, yet must also be viewed considering the effects of fungicides, herbicides and pesticides to which pathogens, weeds and pests evolve resistance. Breeders' work on crop resistance management has to be an integral part of these coevolutionary networks and the response to chemicals. The National Academy of Sciences (US) (1986) publication is an excellent source for looking at these dimensions together and comparatively. Genetic variation, the nature of inheritance, selection pressures and time scales of evolutionary respons — these are all central issues. Wolfe and Barrett (1986) showed how mean pathogenicity of fungi interacts with resistance to fungicides and varies with crop varieties carrying certain disease resistance genes.

Hence, mixtures of varieties and mixes of fungicides, with properly monitored applications, clearly warrant research.

Research on grasslands and forage crops often involves population studies on intra- and interspecies interactions and manipulation of several improved cultivars. A model of pasture community evolution (Fig. 5) shows how an understanding of evolutionary and genetic processes would be required for designing management of forage production and use. The new agroecological literature has brought out sharply the comparisons between natural and agricultural ecosystems (or wild biota and domesticates). Breeders will begin to consider these features in developing cultivars for sustainable agriculture: low input of energy and chemicals. Greater persistence and competing ability, adaptability to move heterogeneous environments and coevolutionary

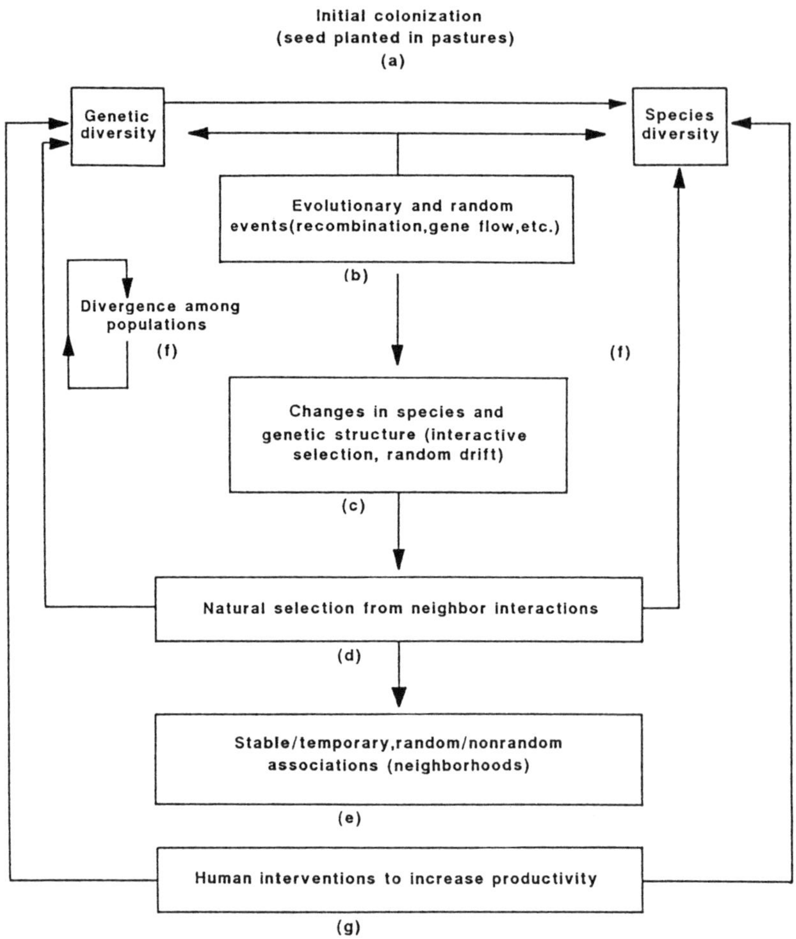

Fig. 5. Forces affecting pasture community evolution (adapted from Aarssen and Turkington 1985).

properties of multispecies neighborhoods. The Land Institute, at Salina, Kansas, for example, has started pioneering breeding work along "nature as model" ideas (Annual Reports 1987—91). Population genetic and ecological principles will clearly come into focus in this work.

Selected readings

Aarsen, L. W. and R. Turkington. 1985. Biotic specialization between neighboring genotypes in *Lolium perenne* and *Trifolium repens* in a permanent pasture. J. Ecol. 73: 605—614.

Allard, R. W. 1960. Principles of Plant Breeding. Wiley, New York.

Allard, R. W. 1988. Genetic changes associated with the evolution of adaptedness in cultivated plants and their wild progenitors. J. Heredit 79: 225—238.

Allard, R. W. and P. E. Hansche. 1964. Some parameters of population variability. Adv. Agron. 16: 281—325.

Andow, D. A. and P. M. Rosset. 1990. Integrated pest management. *In*: Agriculturl Ecology (C. R. Carroll, J. H. Vandermeer, and P. Rosset eds) pp. 413—440. McGraw-Hill, New York.

Baker, L. H. and R. N. Curnow. 1969. Choice of population size and use of variation between replicate populations in plant breeding selection programs. Crop Sci. 9: 555—559.

Barker, J. S. F. and R. H. Thomas. 1988. A quantitative genetic perspective on adaptive evolution. *In*: V. Loeschcke (ed). Genetic Constraints in Adaptive Evolution. Springer-Verlag, pp. 3—23.

Barrett, J. A. 1980. Pathogen evolution in multilines and varietal mixtures. J. Plant Dis. 87: 383—396.

Bond, D. A. 1987. Recent developments in breeding field beans (*Vicia faba* L.). Plant Breeding 99: 1—26.

Burdon, J. J. 1980. Intra-specific diversity in a natural population of *Trifolium repens.* J. Ecology 68: 717—735.

Burdon, J. J. 1987. Diseases and Plant Population Biology. Cambridge Univ. Press.

Burdon, R. D. 1989. Recruitment for breeding populations: Objectives, genetics, and complementation. *In*: B. S. Weir, E. J. Eisen, M. M. Goodman and G. Namkoong, eds., Quantitative Genetics, pp. 555—572. Sinauer, Sunderland.

Cervantes, S. T., M. Goodman, E. D. Casas and J. O. Rawlings. 1978. Use of genetic effects and genotype by environmental interactions for the classification of Mexican races of maize. Genetics 90: 339—348.

Chevalt, C. 1989. Control of genetic drift in selected populations. *In*: B. S. Weir, E. J. Eisen, M. M. Goodman and G. Namkoong, eds., Quantitative Genetics, pp. 379—394. Sinauer, Sunderland.

Cocks, P. S., M. J. Mathison and E. J. Crawford. From wild plants to pasture cultivars: annual medics and subterranean clover in Southern Australia.

Cox, T. S., L. R. House and K. J. Frey. 1985. Trait associations in introgressed populations of sorghum. Zeitsel. fur Pflanzenz. 94: 265—277.

Cubero, J. I. and M. T. Moreno. 1986. Breeding for self-fertility. *In*: P. D. Hebbelthwrite, T. C. K. Dawkins, M. C. Heath and G. Lockwood (eds.), *Vicia faba.* Agronomy, Physiology and Breeding.

Donnelly, E. D. 1979. Selection for chasmogamy in *Sericea lespedeza.* Crop Sci. 19(4): 528—531.

Eberhart, S. A., M. N. Harrison and F. Ogada. 1967. A comprehensive breeding system. Zuchter 37: 169—174.

Evans, D. R., J. Hill, T. A. Williams and I. Rhodes. 1985. Effects of conexistence on the performance of white clover-perennial ryegrass mixtures. Oecologia 66: 536—539.

Francis, C. A. (ed.). 1986. Multiple cropping: Practices and potentials. MacMillan, New York.
Francis, C. A. 1986. Variety development for multiple cropping systems. CRC Critical Rev. In Plant Sci. 3: 133—168.
Gregorius, H. R. 1985. (ed) Population Genetics in Forestry. Springer-Verlag, Berlin.
Hallauer, A. R. 1987. Compendium of recurrent selection methods and their application. CRC Critical Reviews in Plant Sci. 3(1):
Harper, J. L. 1977. Population Biology of Plants. Academic Press, New York.
Hill, J. and P. T. Michaelson-Yeates. 1987. Effects of competition upon the productivity of white clover-perennial ryegrss mixture. Genetic analysis. Plant Breeding 99: 239—250.
Hill, W. G. and P. D. Keightley. 1989. Interrelations of mutation, population size, artificial and natural selection. In Quant. Genetics (B. S. Weir *et al.*) Sinauer.
Humphreys, L. R. 1981. Environmental adaptation of tropical pasture plants. MacMilland, London.
ICARDA. 1985. Pasture, forage and livestock. Intern. Center for Agric. Res. in the Dry Areas. Annual Report.
Jain, S. K. 1989. Domestication of *Limnanthes* (meadowfoam) as a new oil crop. Intern. Atomic Energy Agency, Vienna.
Jain, S. K. 1990. Variation and selection in plant populations. *In*: K. Wöhrman and S. Jain (eds.) Population Biology, Chapter 6, Springer-Verlag, Heidelberg. pp. 199—230.
Jain, S. K. P. Kulakow and I. Peters. 1986. Genetics and breeding of grain amaranths. Proc. 3rd Amaranch Conf., Rodale Press.
Jensen, N. F. 1988. Plant Breeding Methodology. Wiley, New York.
Kogan, M. (ed.) 1986. Ecological Theory and Integrated Pest Management Practice. Wiley, New York.
Kulakow, P. and J. Ennis. 1988. Variation in the F_1 and F_2 generations of crosses between tetraploid *Sorghum bicolor* and *S. halepense.* The Land Inst. Report 5: 23—29.
Lange, W., A. C. Zeven and N. G. Hogenboom. 1984. (eds.). Efficiency in Plant Breeding. PUDFOC, Wageningen.
Mareck, J. H. and C. O. Gardner. 1979. Response to mass selection in maize and stability of resulting populations. Crop Sci. 19: 779—783.
Marshall, D. R. 1989. Crop genetic resources: Current and emerging issues. *In:* A. H. D. Brown, M. T. Clegg, A. Kahler and B. S. Weir (eds.), Plant Populatin Genetics, Breeding and Genetic Resources. Sinauer, Sunderland.
Matzinger, D. F. and E. A. Wernsman. 1968. Four cycles of mass selection in a synthetic variety of an autogamous species *Nicotiana tabacum* L. Crop Sci. 8: 239—243.
Miller, P. A. and J. P. Rawlings. 1971. Breakup of linkage blocks in cotton, *Gossypium hirsutum.* Crop Sci. 11: 695—698.
Moll, R. H. and C. W. Stuber. 1974. Quantitative genetics. Adv. Agron. 26:
Mona, O. 1989. Population genetics in forest tree improvement. pp. 282—298. *In*: A. H. D. Brown, M. T. Clegg, A. L. Kahler and B. Weir (eds.). Plant Population Genetics, Breeding and Genetics Resources. Sinauer, Sunderland.
Monti, L. M. and L. Frusciante. 1984. Selection methods for yield improvement in faba beans. *In* P. D. Hebbilthwaite, T. C. K. Dawkins, M. C. Heath and G. Lockwood (eds). *Vicia faba*: Agronomy, Physiology and Breeding.
Mundt, C. C. 1990. Disease dynmaics in agroecosystems. pp. 263—300. *In*: C. R. Carroll, J. H. Vandeermer and P. M. Rosset (eds). Agroecology. McGraw Hill, New York.
Namkong, G., H. C. Kang and J. S. Brouard. 1989. Tree breeding: Principles and Strategies. Springer-Verlag, Berlin.
Nevo, E. 1983. Genetic resources of wild emmer wheat: structure, evolution and application in breeding. Proc. 6th Intern. Wheat. Genetics Symp. 421—431.
Power, A. G. and P. Kareiwa. 1990. Herbivorous insects in agroecosystems. pp. 301—328. *In*: C. R. Carroll, J. H. Vandermeer and P. Rosset (eds.). Agroecology. McGraw Hill, New York.

Qualset, C. O. 1975. Sampling germplasm in a center of diversity: an example of disease resistance in Ethiopian barley. *In* Frankel, O. H. and J. G. Hawkes (eds.) Crop Genetic Resources for Today and Tomorrow. Cambridge U. Press. Chapter 5: 81—96.

Ramage, R. T. 1987. A history of barley breeding methods. Plant Breeding Reviews 5: 95—138.

Rhodes, I. and W. Harris. 1979. The nature and basis of differences in sward composition and yield in ryegrass-white clover mixtures, pp. 55—60. *In*: A. H. Charles and R. J. Haggar (eds.) Changes in Sward Composition and Productivity. British Grassland Sco., Harley, England.

Schnell, F. W. 1981. Breeding Methods: Synthesis Proc. EUCARPIA Meeting, Quant. Genetics and Breeding Methods.

Schnell, F. W. 1989. Quantitative genetics in crop improvement. *In*: B. S. Weir, E. J. Eisen, M. M. Goodman and G. Nankoong, Eds. Quantitative Genetics, pp. 462—464. Sinauer Sunderland.

Simmonds, N. W. 1962. Variability in crop plants, its use and conservtion. Biol. Revs. 37: 422—465.

Simmonds, N. W. 1979. Principles of Crop Improvement. Longman, London.

Snaydon, R. W. 1984. Plant demography in an agricultural context. *In* R. Dirzo and J. Sarukhan (eds) — Perspectives on Plant Population Ecology, Sinauer, Sunderland. pp. 389—407.

Turkington, R. 1985. Variation and differentiation in populations of *Trifolium repens* in permanent pastures. *In*: J. White (ed) Studies on Plant Demography, Chapter 5, pp. 69—81.

Via, S. 1986. Quantitative genetic models and the evolution of pesticide resistance. *In* Pesticide Resistance: Strategies and Tactics for Management. Nat. Acad. Press, Washington, D.C.

Weber, W. E., C. O. Qualset, and G. Wricke. 1989. Selection strategies for the improvement of autogamous species. pp. 299—316. *In* A. H. D. Brown, M. T. Clegg, A. L. Kahler and B. S. Weir (eds.). Plant Population Genetics, Breeding and Genetic Resources. Sinauer, Suderland.

Witcombe, J. R. and P. J. Murphy. 1986. Covered and naked barleys from the Himalaya. 2. Why do they differ from each other so extensively? Theor. Appl. Genet. 71: 736—741.

Wolfe, M. S. and C. E. Caten (eds.) 1987. Populations of Plant Pathogens: Their Dynamics and Genetics. Blackwell, Oxford.

Wolfe, M. S. and J. Barrett. 1985.

Wricke, G. and W. E. Weber. 1978. Number of lines for synthetic varieties in rye. Ann. Amelior. Plantes 28: 667—674.

Wricke, G. and W. E. Weber. 1986. Quantitative genetics and selection in plant breeding. W. DeGruyter, Berlin.

Wright, A. J. 1973. The selection of parents for synthetic varieties of outbreeding diploid crops. Theor. Appl. Genet. 43: 78—82.

Zobel, B. J. and J. Talbert. 1984. Applied forest tree improvement. Wiley, New York.

7. The population dynamics and control of the parasitic nematode *Trichostrogylus tenuis* in red grouse in the North of England

A. P. DOBSON and P. J. HUDSON

Abstract

In this chapter we present evidence which suggests that the parasitic nematode *Trichostrongylus tenuis* is responsible for producing the patterns of cyclic abundance commonly observed in red grouse *Lagopus lagopus scoticus*. We first review the life history of the parasite, then outline some mathematical models that may be used to explore the population dynamics of this system. This model produces cycles similar to the observed cycles for reasonable parameter values. Different ways of controlling the parasites are discussed and other work on grouse populations are discussed in terms of our own work.

Introduction

In the last ten years, ecologists have shown considerable interest in the impact that parasites and diseases have on the population dynamics of their hosts (Anderson and May 1978, 1979, 1985; May and Anderson 1978, 1979). Where previously diseases might have been considered to appear sporadically in host populations and cause sudden bouts of density-independent mortality, it is now felt that parasites are continuously present in most wild and domestic host populations. Because their population growth rates are usually faster than those of their hosts, they possess considerable potential to act as powerful regulators of host population growth.

Anderson and May (1979) suggest a distinction between the macroparasites and the microparasites when considering the dynamics of different pathogens. The latter group, which includes pathogens such as viruses, bacteria and some protozoans, tends to produce sustained immunity in their hosts. Their dynamics may be modeled by dividing the host population into susceptible, infected, and immune classes. In contrast, the macroparasites, such as the parasitic helminths and arthropods, tend not to produce sustained immunity to reinfection. Models of their dynamics must consider the statistical distribution of parasites within the host population (Dobson 1989). Moreover, variation in responses of different individuals of the host populations to infection by the parasite has a large genetic component.

S. K. Jain and L. W. Botsford (eds), Applied Population Biology, 149—171.

Dobson (1989) reviewed several well-studied examples of parasitic helminths for a wide range of modelling efforts. In this paper we illustrate how a model that was initially developed by Anderson and May (1978), can be used to suggest management strategies for a parasitic nematode that effects the population density of game birds.

The problem

The red grouse (*Lagopus lagopus scoticus*) inhabits the heather moors of Britain; managed grouse moors cover 1.5 million hectares of Britain and provide an important and profitable land use in many upland areas. Under favorable conditions, red grouse can reach densities of up to 100 birds km^{-2}, though densities of around 40 paris km^{-2} are more common in the moors of northern England. Grouse shooting first developed as a fashionable sport for landed gentry in the early 19th century after the development of light flintlock guns. It was not until the development of breech-loading guns and ejectors in the 1860s that rapid shooting was encouraged and grouse were driven to sportsmen standing in butts. With the increase in demand for driven grouse, moorland management improved and more gamekeepers were employed to manage heather, the birds main food supply. Although grouse shooting was probably at its most intense in the early 1900s, the decline of lowland gamebirds such as partridge, due mainly to the effects of intensive modern agriculture, has resulted in increased demand for grouse shooting.

The main aim of grouse management is to produce a consistent surplus of birds that can be harvested each autumn. However, grouse populations do not remain constant, but show a tendency to fluctuate in an oscillatory manner. Years of large grouse bags are frequently followed by population crashes with few birds being shot. This causes serious cash-flow problems for commercial estates, hence the causes of the population cycles have concerned sportsmen and biologists for more than a century.

The population dynamics of the red grouse have been studied for many years (Jenkins *et al.* 1967; Watson *et al.* 1984; Hudson and Watson 1985). The idea that parasitic nematodes are important in determining grouse numbers was originally suggested by Edward Wilson in his contribution to the report "The Grouse in Health and Disease" (Lovat 1911). However, his ideas were abrogated by Lord Lovat, the chairman of the committee, who felt that heather quality was more important in determining grouse numbers.

Here we present empirical evidence which suggests that the nematode is of prime importance in determining the fecundity of red grouse. This contrasts to previous studies which have only considered the parasite as a cause of mortality. We then develop a mathematical model for the grouse-nematode system. Analysis of the properties of this model suggests that the parasites *are* likely to cause the cyclic patterns of abundance of red grouse observed on many grouse moors in the North of England. The model further suggests that many of the mechanisms previously suggested to cause the cycling patterns of abundance, may in fact be consequences of the presence of the parasite.

Observed life histories

The field work for the studies described here was carried out in the North of England. The long term sets of data on the numbers of grouse shot on different moors were collected by sending a standard enquiry form out to the present owners of the estates. A total of 150 sites were censused in the North of England and a further 250 are now being processed from similar sites in Scotland.

The numbers of grouse shot each year on a grouse moor typically show cyclic patterns of abundance (Fig. 1). In more than 80% of cases the tendency to oscillate with a regular period of 4 to 5 years is statistically significant (Potts *et al.* 1984).

Observed patterns

Adult *Trichostrongyle tenuis* live in the caeca of red grouse. Their eggs leave the host in the caecal faeces, where the larvae emerge and, after molting, migrate to the tips of heather buds. The time from egg release to the appearance of third stage larvae is about seven days, although eggs may remain unhatched on the pasture for several months. The grouse pick up the infective larvae when they feed on the heather tips. Young grouse begin to pick up the infection in the first few weeks of life; most are infected by the time they are two months old and worm burdens usually reach at least 500 worms per bird by the age of six months (Fig. 2). To determine the distribu-

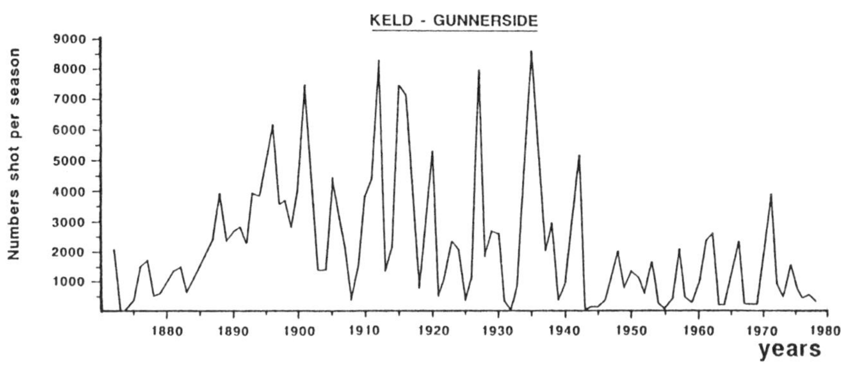

Fig. 1. Bag records for the Keld estate, Gunnerside in the North of England. The graph shows the number of grouse shot on the estate for the years 1870—1980.

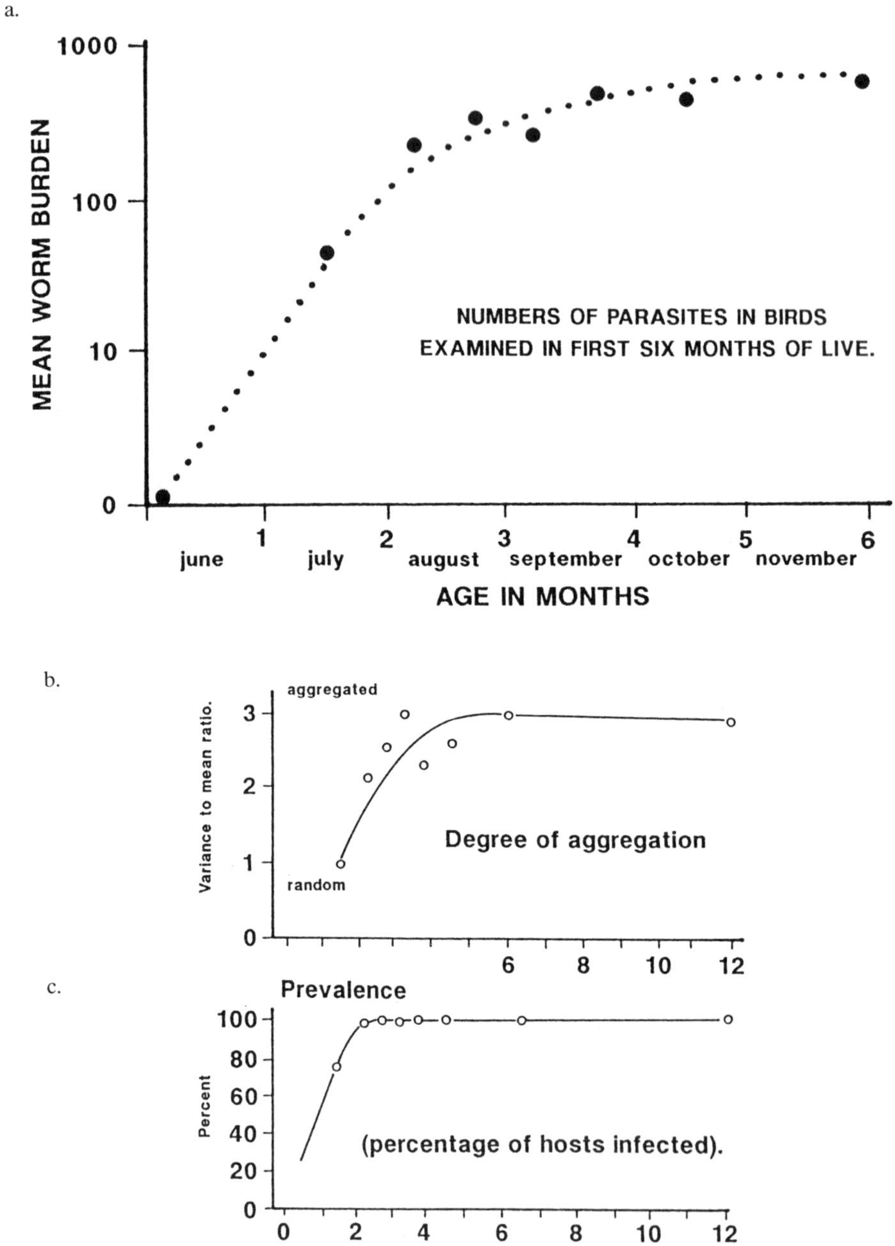

Fig. 2. (a) Age intensity curve for young grouse in their first six months of life. The chicks are usually born early in June. The data are pooled for four years on the same estate. (b) The degree of aggregation of the parasites in young grouse in the first year of life and (c) the proportion of grouse infected. The degree of aggregation is measured using the variance to mean ratios of parasite numbers per host.

tion of parasite numbers per host, the numbers of parasites in the caecae of grouse were counted as grouse were collected during the shooting season. Figure 3 compares the burdens in males and females. In both cases, the negative binomial distribution provides the best fit to the observed frequency distribution of parasites per host. (The binomial distribution is a probability distribution that is commonly used to fit parasite data. It may result from an aggregative process. The parameter k varies inversely with the degree of aggregation.) The distributions are essentially identical for each sex with mean parasite burden of 2200 and degree of aggregation, k, of 1.6. The degree of aggregation is relatively low, though this is partly a consequence of the high parasite burdens.

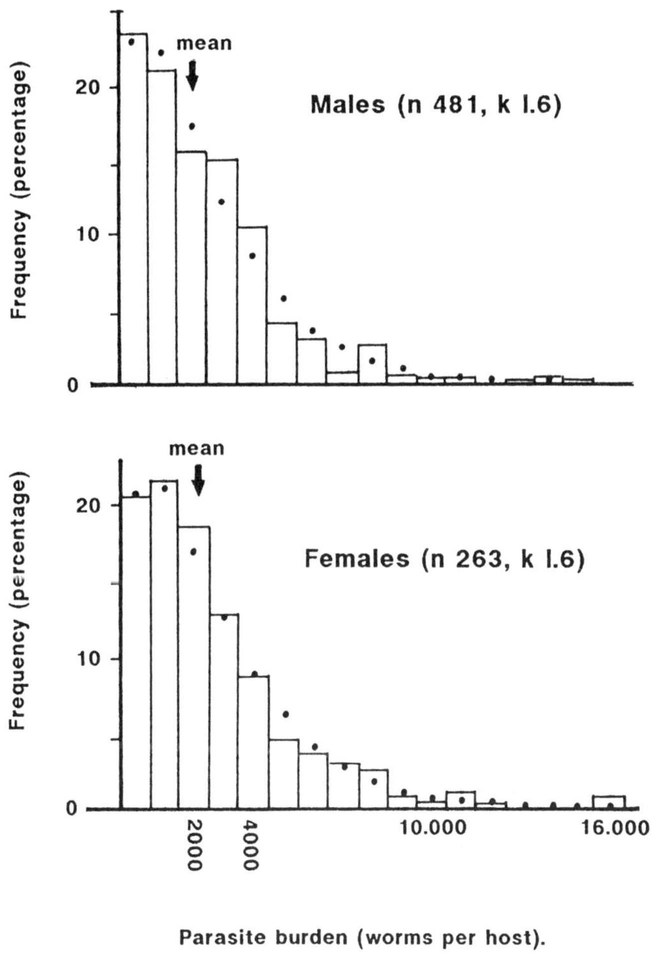

Fig. 3. Frequency distribution for the parasite burdens in 481 adult male grouse and 263 adult female grouse. The frequency classes increase in units of 1000 worms per host, the y-axis gives the proportion of worms in the population carrying this burden of worms.

The numbers of eggs produced per host is dependent on the parasite burden of the host (Fig. 4), as is the case with many helminths (Anderson and May 1982, 1985). The number of eggs released from each infected female seems to peak at around 100,000 eggs per gram of faeces once parasite burdens exceed 4,000 worms per bird.

Survival of the free-living egg and larval stages of *T. tenuis* is dependent on both the humidity and temperature of the habitat. Data from other related *Trichostrongyle* worms suggests that parasite life expectancy decreases with increasing temperature (Fig. 5). The study by Watson, Lee and Hudson (in press) suggests that the survival of *T. tenuis* may be somewhat less than that for other *Trichostrongyles.*

Preliminary studies suggest that adult worms may be at least as long lived as their hosts. In a study on captive grouse, Wilson (personal communication) found that adult worms continued to produce eggs for at least eighteen months. It is also possible that larvae possess the ability to arrest their development and enter a resting stage when they first enter the host (Fitzsimmons 1969; Smith, this volume). Under these conditions they remain temporarily inactive but attached to the lining of the host's caecae.

Parasite pathogenicity may operate in two different ways, affecting either host survival or host fecundity. The early work of Jenkins *et al.* (1963) suggested that males not holding territory had higher parasite burdens than

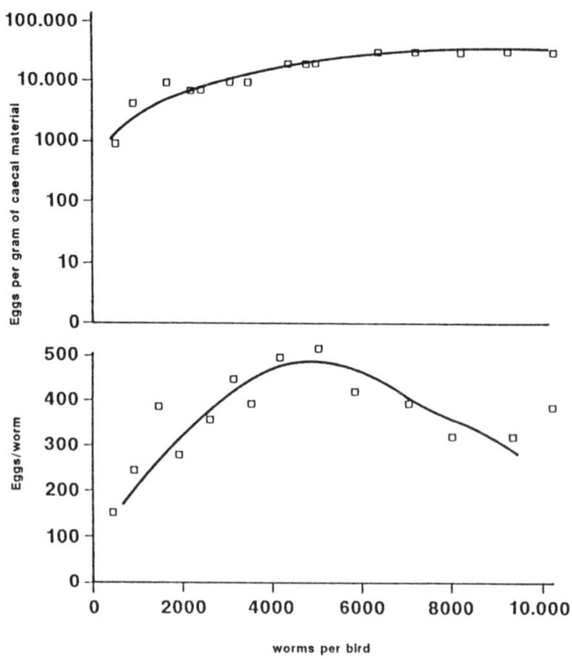

Fig. 4. The number of eggs produced per gram of faeces for a range of parasite burdens.

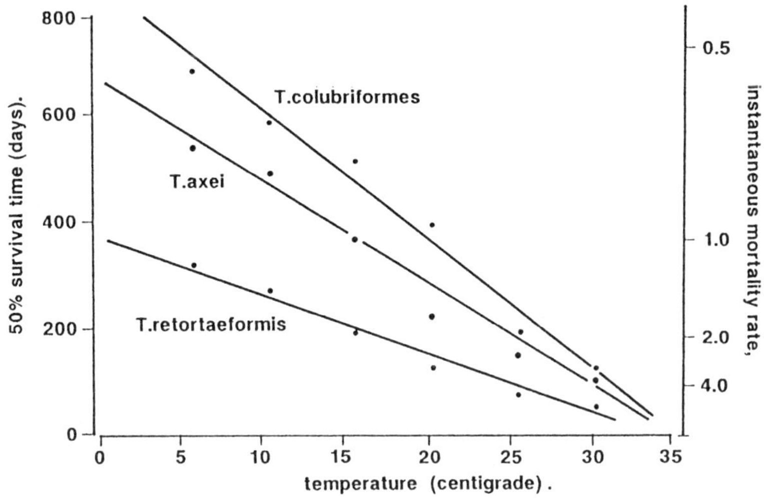

Fig. 5. Life expectancies of three other species of Trichostrongyle worms at a range of different temperatures. Life expectancies are illustrated on the left hand axis and instantaneous mortality rates on the right hand axis. The data used are from Boag and Thomas 1985.

territory holding birds. This suggested that the parasite may have some general effect on host vigor and ability to hold a territory. The more detailed physiological studies of Wilson and Wilson (1978) showed that parasitized birds developed a mild aneamia when infected with the worm. However, because peak parasite burdens tend not to coincide with periods of decline in cycling populations of Red Grouse, most earlier workers concluded that parasite-induced increases in host mortality are relatively insignificant in determining host survival (Lovat 1911; Jenkins *et al.* 1963).

The data of Jenkins *et al.* (1963) suggest that the fecundity of female grouse decreases as parasite burdens increase (Table 1). The more detailed experiments of Hudson (1986) illustrate that the parasite has a significant

Table 1. Breeding success of female red grouse in relation to parasite burden (after Jenkins *et al.* 1963).

		Parasite infection		Brood size	
Year	# Grouse	% Infected	Mean burden	Low	High
1956–7	52	47	245–2261	2.43	1.80
1958	79	85	1341–4102	1.05	0.97
1959	44	89	4168–7498	0.84	0.65

Data are presented for two study sites: one at the base of a hill (LOW site) and one on higher ground (HIGH site).

effect on all indices of female reproductive success (Fig. 6), while also suggesting that the survival of infected females may also be diminished (Table 2). The latter effect may be due to the parasite increasing the susceptibility of infected hosts to predation (Hudson and Dobson, in prep.).

Model framework

The population dynamic consequences of the interaction between the parasite and its host can be explored by developing a simple mathematical model of the life cycle. Here we will develop a model based upon those described in Anderson and May (1978) and May and Anderson (1978). The model considers the dynamics of three different populations: The grouse hosts, H, the adult parasites in the grouse caecae, P, and the free-living larval stages of the parasite, W. The various birth and death rates of these three populations and the flow rates between the different stages of the parasite's

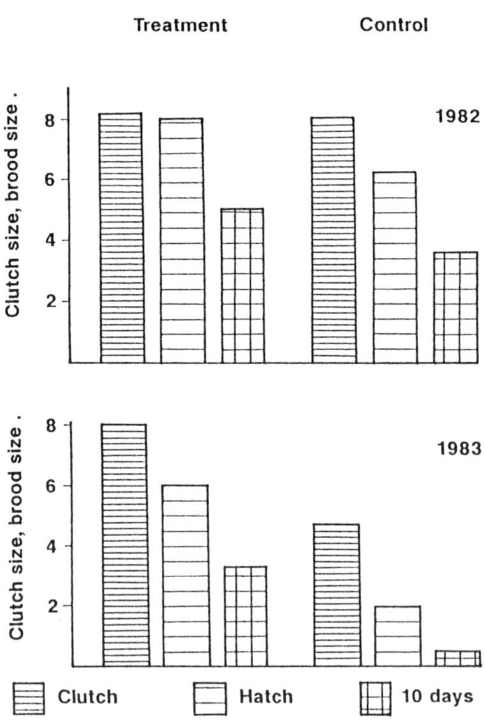

Fig. 6. Comparative fecundity of parasitized and treated female grouse. The treatment birds were given an aural dose of a standard anthelmintic, the control birds were treated with water (see Hudson 1986a for details). Fecundity is measured in three ways; the number of eggs laid, the number of eggs hatching and the number of chicks alive at age ten days. Data for two different years are illustrated.

Table 2. Parasite burdens and survival of treated and control hens (see Hudson 1986 for further details).

Year	Treated	Control	Significance
1982			
Number tagged	44	159	
Two months after treatment	4	9	
Nematode eggs/gram faeces	5.0 ± 2.6	87.2 ± 1.4	0.01
Number recaught (Winter 82)	12	30	
'Survival'	0.27	0.29	0.1
1983			
Number tagged	37	115	
Two months after treatment	7	8	
Nematode eggs/gram faeces	72.2 ± 1.4	223.4 ± 1.4	0.05
Number recaught (Winter 83)	4	9	
'Survival'	0.11	0.08	n.s.

life-cycle are illustrated in Fig. 7. Table 3 lists the parameters used to quantify each of these various rates.

Structure of the basic model

The dynamics of the host-parasite interaction may be described by the following set of three simultaneous differential equations:

$$dH/dt = (a - b)\,H - (\alpha + \delta)P \tag{1}$$

$$dP/dt = \beta WH - (\mu + b + \alpha)P - \alpha\,\frac{P^2}{H}\left(\frac{k+1}{k}\right) \tag{2}$$

$$dW/dt = \lambda p - \gamma W - \beta WH \tag{3}$$

This model assumes that the adult parasites are distributed in an aggregated manner and that this aggregation can be described by the negative binomial distribution. The parasites are assumed to be detrimental to both survival and fecundity of the host. This pathogenicity is assumed to act linearly. The model differs from that described in Hudson *et al.* (1985), in that we are here more fully considering the dynamics of the parasite's free-living stages.

A qualitative feel for how the various parameters affect the magnitude of each population can be gained by considering the equilibrium properties of the above equations. Setting $dH/dt = dW/dt = dP/dt = 0$, we obtain:

$$M^* = P^*/H^* = (a - b)/(\alpha + \delta) \tag{4}$$

$$H^* = \gamma\, d'/\beta(\lambda - d') \tag{5}$$

$$W^* = (a - b)\, d'/(\alpha + \delta)\beta \tag{6}$$

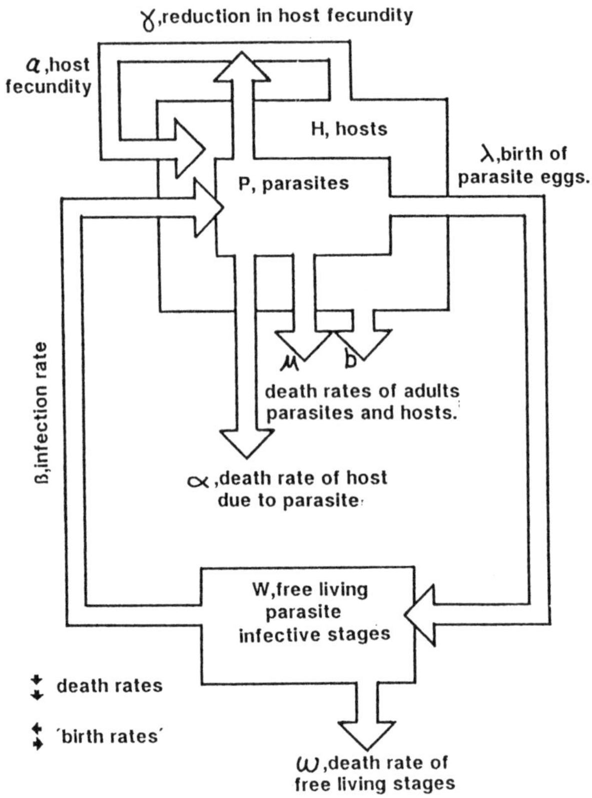

Fig. 7. Flow chart of the life cycle of *Trichostrongylus tenuis* in Red Grouse illustrating the parameters used in the model to mimic the flow rates through the different parts of the life cycle.

Here M* is the mean parasite burden of adult grouse at equilibrium and d′ $= \mu + a + b + [(a - b)\,(2\delta + \alpha)/(\alpha + \delta)]\,k'$, where $k' = (k + 1)/k$. This initial analysis suggests that mean parasite burdens vary inversely with the parasites' ability to affect host survival and fecundity. Host population density varies inversely with both parasite fecundity, λ, and the rate at which parasite larvae are ingested, β. The numbers of free-living larvae vary indirectly with the pathogenicity of the parasite $(\alpha + \delta)$.

As with other epidemiological studies, it is useful to obtain expressions for R_o, the basic reproductive rate of the parasite and for H, the threshold number of hosts required to sustain the infection (Anderson and May 1979). The former term is directly analogous to Fisher's R, the number of offspring produced by a female in her lifetime that survive to reproduce (Fisher 1930), while the latter expression was first used in the classic work of Kermack and McKendrick (1927). When the parasite is first introduced into the population, pathogenicity and other density-dependent constraints on its rate of

Table 3. Notations used to denote various population parameters.

Parameter	Description
a	Instantaneous birth rate of the grouse (/host/unit of time).
b	Instantaneous death rate of the grouse due to natural causes from all sources of mortality except the parasite (/host/unit of time).
α	Instantaneous death rate of host due to influence of parasites (/parasite/host/unit of time).
δ	Instantaneous reduction in grouse fecundity due to the presence of the parasite (/parasite/host/unit of time).
Δ	Density dependent reduction in grouse fecundity (/host/unit of time).
λ	Instantaneous birth rate of parasite eggs (/parasite/unit of time).
μ	Instantaneous death rate of adult parasite in the grouse due to natural causes or host induced (immunological) causes (/parasite/unit of time).
γ	Instantaneous death rate of the parasite's free-living egg and larval stages (/parasite/unit of time).
β	Instantaneous rate of ingestion of parasite free-living stages of the grouse (/parasite/host/unit of time).
k	Parameter of the negative binomial distribution which measures inversely the degree of aggregation of the parasite within the host population.

increase are effectively trivial. Collapsing equations (1), (2), and (3) under these conditions gives an expression for this initial growth rate:

$$R_0 = \frac{\beta \lambda H}{(\gamma + \beta H)(\mu + b = \alpha)} = T_1 M_1 M_2 \qquad (7)$$

Here T_1 = bkH, the transmission rate of eggs from mature adult female to establishment in another grouse, $M_1 = 1/(\gamma + \beta H)$, the life expectancy of a free living larvae, and $M_2 = 1/(\mu + b + \alpha)$, the life expectancy of an adult worm. Thus the basic reproductive rate of the parasite consists of the product of the numbers of new infections established by each worm and the life expectancies of the free-living and parasitic stages of the worm. Estimates of some of these parameters are available from either our own work or the published literature on *T. tenuis* and *L. l. scoticus* (Table 4). Estimates of the two parameters which determine the impact of the parasite on host fecundity and survival may be obtained from the experimental data described above and in Hudson (1986). As is usually the case with epidemiological studies, few data are available with which to estimate rates of transmission. Crude estimates of these parameters may be obtained by determining the threshold host density at which we would expect the parasite to just maintain itself in the host population.

Table 4. Population parameter estimates for *T. tenuis* and red grouse.

Parameter value	Symbol	
Grouse fecundity	a	0.8—2.5/grouse/year
Grouse mortality	b	0.55/grouse/year
Parasite fecundity	k	10^4—10^6 worm/year
Adult worm mortality	l	0.5/worm/year
Parasite pathogenicity	α	0.0001—0.00001 parasite/host/year
Parasite reduction in host fecundity	δ	0.0004/parasite/grouse/year
Aggregation of parasites in hosts	k	0.5—1.6
Mortality rate of free-living parasite larvae	γ	2—6/larvae/year
Rate of ingestion of free-living larvae by grouse	β	2×10^{-6}—4×10^{-5} larvae/grouse/year

The threshold number of grouse required for the parasite to establish may be obtained by setting R_0 to unity and rearranging Equation 7.

$$H_T = \frac{\gamma/\beta}{\left(\dfrac{\lambda}{m_2} - 1\right)} \tag{8}$$

The denominator of this expression will roughly be equal to λ (Table 4). Data from a number of studies suggest that *T. tenuis* is effectively absent from moors where grouse densities drop below 20 grouse/km^2. Thus a crude estimate of β, suggests a figure of the order of 10^{-6} (roughly, between 2×10^{-6} and 4×10^{-5}). Substitution of this estimate into equation 7 suggests a value for R_0 in the range of 5 to 10 for a typical grouse moor in the north of England.

Dynamical properties of the model

The dynamic properties of the model can be explored by local stability analysis. The details of this are given in Appendix 1. This analysis suggests that the dynamic behavior of the model is determined by four parameters: the degree of aggregation of the parasite in the definitive host, k, the life expectancy of the free-living larvae, $1/\gamma$, and the relative magnitudes of the parasite-induced reductions in host fecundity, δ, and survival α. When the life expectancy of the free-living stages is sufficiently long and parasite-induced reductions in host fecundity are greater than parasite-induced increases in host mortality, the system will tend to oscillate. Some numerical simulations are illustrated in Fig. 8. These oscillations may ultimately be sufficiently large to drive both the host and the parasite to extinction. However, several other mechanisms may be included in the model which reduce this tendency to produce unstable oscillations. These are the territorial behavior of the red grouse, which tend to put an upper bound on their numbers in any area, and the density dependent decline in parasite fecundity as worm burdens increase.

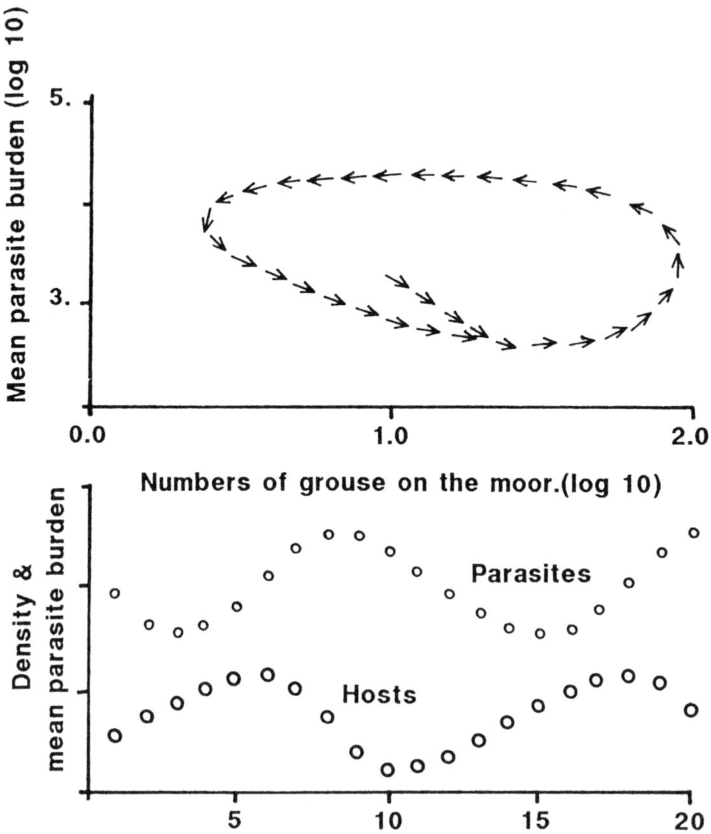

Fig. 8. Simulations of the basic model's behaviour. The upper diagram depicts the results as a phase plot with numbers of grouse on the x-axis and mean parasite burden on the y-axis. The lower graph depicts the same simulation as a time series. The parameter values used were a = 0.8, b – 0.5, k = 5000, u = 0.5, $\alpha = 0.0001$, $\delta = 0.0004$, $\gamma = 3.00$ and k = 1.6.

The territorial behavior of the grouse

The territorial behavior of the grouse may be included in the model by modifying equations 1 and 2 so as to include an additional term which accounts for increases in grouse mortality rates as their density rises. Such an effect would mimic the increases in numbers of birds unable to obtain territories at high population densities.

$$\frac{dH}{dt} = (a - b - \Delta H)\,H - (\alpha + \delta)P \tag{9}$$

$$\frac{dP}{dt} = \beta WH - (\mu + b + \alpha + \Delta H)P - \alpha\,\frac{p^2}{H}\left(\frac{k+1}{k}\right) \tag{10}$$

In the absence of the parasite, the grouse will stabilize at some carrying capacity K, where

$$K = (a - b)/\Delta = \gamma/\Delta \tag{11}$$

Inclusion of the territorial behavior of the grouse hosts into the model removes the tendency to increasing oscillations, and, instead leads to stable limit cycles (Fig. 9). The period of these cycles is primarily determined by the life expectancy of the free-living parasite stages and the intrinsic growth rate of the grouse. A simple life-table analysis for grouse using a conservative range of parameter values suggests that γ (= a − b) will be in the range of 0.1 to 1.0 (Table 5). The experiments of Watson, Lee and Hudson (in press)

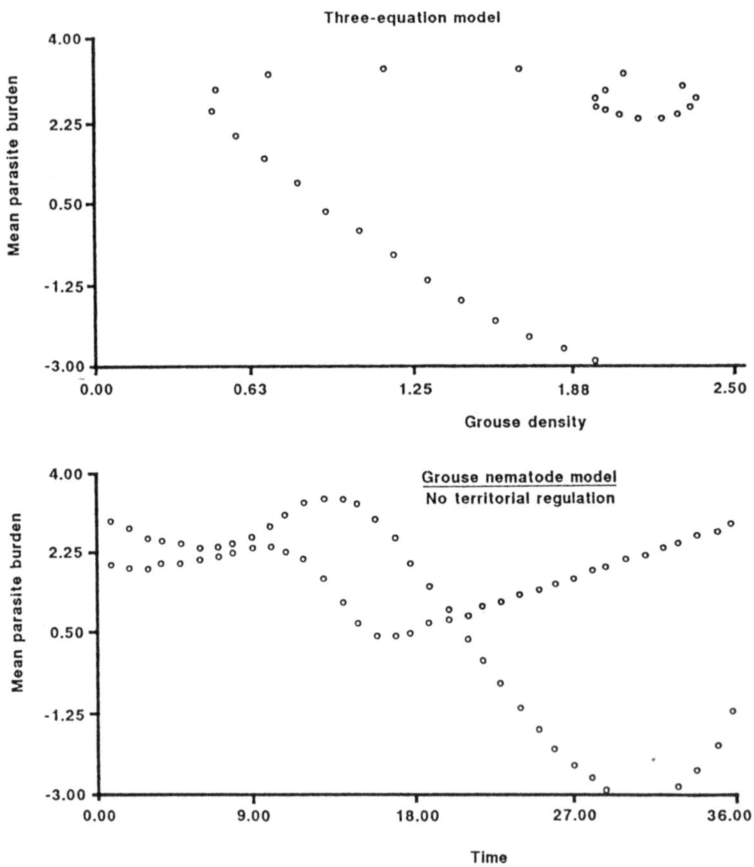

Fig. 9. Behaviour of the model after inclusion of a term for the territorial behavior of the grouse. The lower graph shows a time series of grouse numbers and mean parasite burdens. The top graph shows this data as a phase plane plot with numbers of grouse on the x-axis and mean parasite burden on the y-axis. The parameter values are as for Fig. 8 except that D now equals 0.005.

Table 5a. Survival of adult red grouse (from Jenkins *et al.* 1977).

Age group	Proportion surviving	Annual survival (%)
Between 1 and 2 years	278/819	33.9
Between 2 and 3 years	107/278	38.5
Between 3 and 4 years	26/98	26.5
Between 4 and 5 years	11/25	44.0
Between 5 and 6 years	5/9	55.6
Between 6 and 7 years	0/3	0.0
Overall mean	427/1232	34.7

Table 5b. Estimates of r for grouse for a range of published clutch sizes and immature survival rates and assuming adult survival of 34.7% per annum.

Clutch size	5	6	7	8	9	10
Immature survival						
0.2	0.000	0.000	0.000	0.000	0.000	0.011
0.3	0.000	0.000	0.031	0.085	0.134	0.177
0.4	0.000	0.000	0.148	0.204	0.254	0.299
0.5	0.102	0.177	0.242	0.299	0.350	0.395
0.6	0.177	0.254	0.320	0.378	0.429	0.475
0.7	0.242	0.320	0.386	0.445	0.497	0.543

suggest that the free-living larvae have a life expectancy of between 2 to 6 months. Data collected from other closely related *Trichostrongyles* suggest that under certain conditions survival rates may be considerably longer (Fig. 4, Boag and Thomas 1985). The expected behavior of the model for these ranges of parameter values is illustrated in Fig. 10. This graph suggests that grouse populations on moors, where grouse productivity is relatively large, will cycle with a period of around 4 to 5 years. In areas where the growth rate of the grouse populations is lower, the cycle period tends to increase. This may account for the longer cycle lengths we tend to see in the data from Scotland (Table 6). Decreases in the survival of the free-living stages of the parasite tend to increase the cycle length and may ultimately lead to the parasite being unable to sustain itself in the host population.

On one particular study site, we have been able to collect data from both numbers of grouse shot *and* mean parasite burdens. These data, illustrated in Fig. 11, exhibit essentially the same pattern of long term dynamics as are illustrated by the simulations of the model (Fig. 9). The most important point to emerge from both this graph and the simulation is that mean parasite

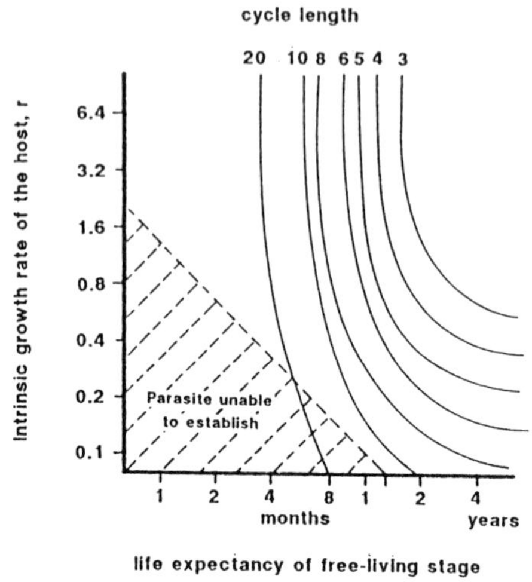

Fig. 10. Expected behaviour of the model for a range of values of the intrinsic growth rate of the grouse and the life expectancy of the free-living stage of the parasite. The rest of the parameters are as given in Table 4. In the shaded area at the bottom left portion of the graph, the parasite is unable to establish within the host population, this assumes a carrying capacity for the area modeled of 100 grouse. Increasing the value of this carrying capacity shifts this line down and to the left. Everywhere else on this plane the populations will exhibit stable limit cycles of numbers. The dotted contour lines give the expected cycle lengths for any particular combination of parameter values.

Table 6. Cycle lengths in different studies of grouse.

Authors	Geographical region	Cycle length (years)*
Barnes & Hudson (unpublished)	N.W. Highlands	6.3
	Central Highlands	6.6
	Southeast Scotland	6.6
	Southwest Scotland	4.9
Potts, Tapper & Hudson	Northern England	4.8

* Cycle length increases from south to north.

burdens do not crash until *after* the host population density has crashed. This temporal disparity could explain why previous workers have inadvertently tended to diminish the importance of the parasite as a mechanism leading to decreases in grouse density.

Discussion

This discussion falls into two parts: in the first half we compare different

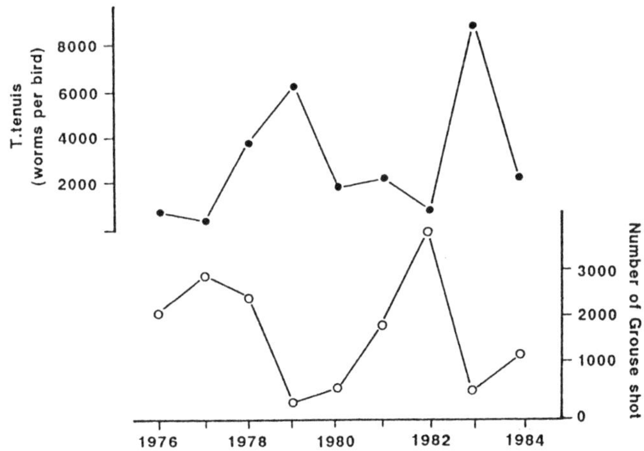

Fig. 11. Observed numbers of grouse shot and mean parasite burdens of these birds on a moor in the North of England. The data were collected by methods described in Hudson (1986c).

ways of controlling the parasite, while in the second, we discuss other mechanisms that have been suggested to account for the oscillatory patterns of abundance in grouse. The former exercise suggests ways of managing grouse populations to obtain more information that may be used to test the importance of the parasite as the mechanism leading to cyclic patterns of grouse abundance.

Methods of controlling grouse nematodes

Three methods may be adopted for controlling the parasitic nematode *T. tenuis* in red grouse. Each seeks to reduce the magnitude of one specific parameter of the parasite-host association. Methods of control which seek to reduce the life expectancy of the free-living egg and larval stages of the parasite are attractive to many land owners as they minimize direct interference with the grouse. However, the free-living larvae may be very numerous in years of high worm densities and they are likely to be spread quickly across the grouse moor and are probably washed down into the top soil. Reductions in survival of these stages could be achieved by chemical means or by draining the soil, but both of these actions are likely to be very expensive. Similarly, the efficacy of either technique is likely to be limited. In areas where rainfall exceeds 1—2 meters of rain per year, most chemicals applied to the soil surface will be quickly washed away. At the same time, very few drainage schemes will be able to handle this amount of rainfall and maintain the habitat in a sufficiently dry condition to affect parasite survival.

Neither of these two techniques is likely to produce significant decreases in the survival of parasite eggs and larvae.

Chemotherapy may be directed against the adult worms in the grouse. Two methods may be employed in this context. At the simplest level, it may be possible to treat supplementary food or grit with chemicals which the grouse will then seek out and ingest. A more complex method of treating the birds would employ selective chemotherapy of infected females utilizing a scaled up version of the experiments designed to demonstrate differences in the fecundity of infected and uninfected females (Hudson 1986). Here a proportion of the female grouse on each moor would be treated with a large dose of anthelmintic prior to the breeding season. This second method of chemotherapy is specifically targeted against the impact of the parasite on host fecundity. Numerical simulations suggest that treating 50% of the females reduces the tendency of the system to oscillate and leads to a mean grouse population density of around 35% of what it would be in the absence of the parasite. Treating 75% of the females completely stops the cycling and leads to grouse densities of around 40% of their value in the absence of parasites.

Selective chemotherapy targeted towards females is the more sensible of the two approaches since it should minimize the rate at which the parasite evolves resistance to the anthelmintic agent used. This contrasts to the case where chemotherapy is given at a low level to all grouse; in such cases, selection may operate fairly strongly and resistance evolves rapidly. In a recent review, May and Dobson (1986) show that while the rate at which resistance to chemotherapy evolves, it varies directly with the generation time of the pathogen or pest, but it varies only logarithmically with factors such as the initial frequency of the resistant allele and the selection intensity. As resistance to chemical control tends to appear within 10 to 50 generations for most organisms (May and Dobson 1986), treating all the birds at a low level is likely to exhaust the active utility of the chemical within 5 to 10 years. Treating only a proportion of the females ensures that selection only operates on a portion of the parasite population. This maintains susceptible genes within the worm population and should prolong the active life of the chemical agent used.

Alternative mechanisms for cycles in grouse abundance

Several different mechanisms have been suggested by a variety of workers to account for the oscillatory behavior of grouse populations. Three of the most significant of these will be discussed here.

Variations in food quality was favored as a potential cause of population cycles in grouse by Lord Lovat in the published proceeding of the committee on grouse disease (Lovat 1911). As variations in the amount and quality of available food have been observed at different points of cycles of grouse abundance, a series of experiments were undertaken to rigorously test this hypothesis on a number of moors in the east of Scotland (Miller *et al.* 1970; Moss *et al.* 1975). Although grouse may only feed on 2 to 10% of the available heather shoots, these experiments showed that food quality, and, in particular, the density of heather tips is important in determining breeding success of grouse. However, an experiment which attempted to stop the decline phase of a cycling grouse population by artificially raising the nutrient supply to the heather, failed to prevent the grouse population from declining (Watson and Moss 1980; Watson *et al.* 1984). This suggests that variations in the amount of food at different points of the population cycle may be a consequence of variations in feeding pressure, rather than a cause of the cycles. This effect would be particularly significant if parasite burden has a negative effect on the feeding rates of grouse, as is the case with many studies of helminths in mammals (Symmons 1969; Crompton 1984).

The "genetic hypothesis" has been proposed as a general theory to account for the oscillations of vertebrate species (Chitty 1967; Krebs 1978). This theory seems to have been developed as a compromise in response to earlier criticisms of the work of Wynne-Edwards (1962) on group selection and population egulation. Chitty's theory assumes that genetic variation in the level of aggression in a population is responsible for driving cycles of host abundance. Increases in the level of aggression as population density increases have been observed in grouse (Moss and Watson 1980; Moss *et al.* 1984), and these have been correlated with variations in the testerone levels of males (Watson 1970). Although variations in the frequency of a variety of alleles have been recorded in populations of grouse at different densities (Henderson 1977; Redfield 1973), genetic variation has not been followed in a population through a whole cycle. It may be that variations in the burdens of parasites harbored by the grouse may be responsible for producing different levels of aggression at different times of the cycle. Evidence from the experiments of Wilson and Wilson (1978) tend to support this, and we are undertaking further experiments to test this ourselves (Hudson and Dobson in prep.). Similarly, a mathematical analysis of some of the more general assumptions underlying the Chitty hypothesis has suggested that in the absence of some extrinsic driving factor, genetic variations in levels of aggression will not produce oscillations in population density (Stenseth 1981). If parasites are selecting for particular genetic strains of the host, it may be that the observed genetic variation in grouse populations is being driven by the parasites, rather than being the driving force of the cycle, a conclusion also reached by Moss *et al.* (1984).

At present a behavioral mechanism has achieved ascendancy as a cause of population cycles in grouse (Moss and Watson 1985). This operates in a similar manner to the Chitty hypothesis, except that differential dispersion is included into the model and it is assumed that birds disperse further in years when the population density is declining than in years of increases in population numbers (Watson *et al.* 1984; Moss and Watson 1980). However, many other studies of birds have shown that dispersal to a new breeding site occurs more frequently in years following a poor or unsuccessful breeding attempt (Lack 1966; Catchpole 1972; Harvey *et al.* 1979). As parasite burden plainly has an effect on breeding success, it may be that the differential dispersal observed at different times in the cycle may be a consequence of variations in parasite burden, rather than a cause of the cycles. Although Watson *et al.* (1984) develop some mathematical models to mimic these behavioral mechanisms of population regulation, these models only display patterns of sustained oscillations when primed with sets of oscillatory data. Analyses of the properties of the models suggests that they would give stable non-oscillatory dynamics when run for reasonable ranges of parameter values.

Conclusions

We feel that the evidence presented above and in other recent publications (Hudson 1986; Hudson and Dobson 1987) strongly supports the notion that the parasitic nematode *T. tenuis* is driving the cycles of abundance commonly observed in the North of England and Scotland. Many previously suggested mechanisms for cycles in abundance may, in fact, be consequences of variation in parasite and host abundance, rather than causes of population cycles. However, full resolution of this debate awaits further experimental testing. The full proof of the model suggested above for parasite driven cycles awaits the results of a study that attempts to produce a stable population from an oscillating one by considerably reducing parasite burdens. This experiment may also be designed to compare the different potential ways of controlling nematodes in grouse.

The study of *T. tenuis* in red grouse is also interesting in that its dynamics may be examined within the framework of a fairly basic mathematical model for a macroparasite and its host. The presence of one major parasite in a habitat containing predominantly one species of host is rather unique in field studies of parasite-host systems. In most other studies the presence of several species of parasite and two or more species of hosts, reduces the potential for straightforward experimental manipulation. In many cases information is only obtained about such systems following either natural perturbations or by bringing parts of the system into the laboratory.

Acknowledgements

We would like to thank Roy Anderson, Dick Potts and Bob May for many helpful conversations during the development of this work. Considerable stimulation was also provided by the Population Biology Group at Princeton University and the Applied Epidemiology Group at Imperial College,

London University. A. P. Dobson was supported by a NATO post-doctoral fellowship and a Rockefeller Foundation grant to R. M. May.

References

Anderson, R. M. and R. M. May. 1978. Regulation and stability of host — parasite population interactions: I. Regulatory processes. Journal of Animal Ecology 47: 219—249.

Anderson, R. M. and R. M. May. 1979. Population biology of infectious disease: Part I. Nature 280: 361—367.

Anderson, R. M. and R. M. May (eds.). 1982. Population Biology of Infectious Diseases. Dahlem Konferenzen No. 25, 314 pp. Springer Verlag, Berlin.

Anderson, R. M. and R. M. May. 1985. Helminth infections of humans: mathematical models, population dynamics, and control. Advances in Parasitology 24: 1—101.

Anderson, R. M. and R. M. May. 1986. The invasion, persistence and spread of infectious diseases within animal and plant communities. Phil. Trans. Roy. Soc. Lond. B. 314: 533—570.

Boag, B. and R. J. Thomas. 1985. The effect of temperature on the survival of infective larvae of nematodes. J. of Parasitology 71: 383—384.

Catchpole, C. K. 1972. A comparative study of territory in the reed warbler (*Acrocephalus scrirpaceus*) and sedge warbler (*A. schoenobaenus*). J. Zoology, London 166: 213—231.

Chitty, D. 1967. The natural selection of self-regulatory behaviour in animal populations. Proc. Ecological Society of Australia 2: 51—78.

Crompton, D. W. T. 1984. The influence of parasitic infection on food intake. Federation Proceedings 43: 239—245.

Dobson, A. P. 1989. The population biology of parasitic helminths in animal populations. Pp. 145—175 in T. G. Hallam, L. J. Gross and S. A. Levin (eds.). Applied Mathematical Ecology. Springer-Verlag, Berlin.

Fisher, R. A. 1930. The Genetical Theory of Natural Selection. Clarendon Press, Oxford.

Fitzsimmons, W. M. 1969. Pathogenesis of the Trichostrongyles. Helminthological Abstracts 38: 139—190. Harvey, P. H., P. J. Greenwood and C. M. Perrins. 1979. Breeding area fidelity of the great tit (*Parus major*). J. Animal Ecology 48: 305—313.

Henderson, B. A. 1977. The genetics and demography of a high and low density of Red Grouse *Lagopus l. scoticus.* J. Animal Ecology 46: 581—592.

Hudson, P. J. 1986. The effect of a parasitic nematode on the breeding performance of red grouse. J. Animal Ecology 55: 85—92.

Hudson, P. J. and A. P. Dobson. 1987a. Red Grouse. The Biology and Management of a Wild Gamebird. Brourne Press, Bournemouth.

Hudson, P. J. and A. P. Dobson. 1987b. The population dynamics of *Trichostrongylus tenuis* in Red grouse. Parasitology (in press).

Hudson, P. J. and A. Watson. 1985. Exploited animals: Red Grouse. Biologist 32: 13—18.

Hudson, P. J., A. P. Dobson, and D. Newborn. 1985. Cyclic and non-cyclic populations of red grouse: a role for parasitism. In: Ecology and Genetics of Host-Parasite Interactions. D. Rollinson and R. M. Anderson, eds., pp. 77—90. Academic Press, London.

Jenkins, D., A. Watson, and G. R. Miller. 1963. Population studies on red grouse, *Lagopus lagopus scoticus* (Lath.), in northeast Scotland. J. Animal Ecology 32: 317—376.

Jenkins, D., A. Watson and G. R. Miller. 1967. Population fluctuations in the Red grouse *Lagopus lagopus scoticus.* J. Animal Ecology 36: 97—122.

Kermack, W. O and A. G. McKendrick. 1927. A contribution to the mathematical theory of epidemics. Proceedings of the Royal Society of London, Series A, 138: 55—83.

Krebs, C. J. 1978. A review of the Chitty hypothesis of population regulation. Canadian J. of Zoology, 56: 2463—2480.

Lack, D. 1966. Population Studies of Birds. Clarendon Press, Oxford.
Lovat, L. 1911. The Grouse in Health and Disease: being the final report of the Committee of Inquiry on Grouse Disease. 2 vols. Smith, Elder and Co., London.
Mackenzie, J. M. D. 1952. Fluctuations in the numbers of British tetraonide. J. Animal Ecology 21: 128—153.
May, R. M. and R. M. Anderson. 1978. Regulation and stability of host-parasite population interactions. II. Destablising processes. J. Animal Ecology 47: 249—268.
May, R. M. and R. M. Anderson. 1979. Population biology of infectious diseases: Part II. Nature 280: 455—461.
May, R. M. and A. P. Dobson. 1986. Population dynamics and the evolution of pesticide resistance. In: Pesticide Resistance and Management, pp. 170—193, NAS-NRC publications, Washington.
Miller, G. R. and A. Watson and D. Jenkins. 1970. Responses of Red Grouse populations to experimental improvement of their food. Symposia of the British Ecological Society, 10: 323—334.
Moss, R. and A. Watson. 1980. Inherent changes in the aggressive behaviour of a fluctuating red grouse *Lagopus lagopus scoticus* population. Ardea 68: 113—119.
Moss, R. and A. Watson. 1985. Adaptive value of spacing behaviour in population cycles of red grouse and other animals. In: Behavioural Ecology, R. M. Sibly and R. H. Smith, eds., pp. 275—294. Blackwell Scientific Publications, Oxford.
Moss, R., A. Watson and R. Parr. 1975. Maternal nutrition and breeding success in red grouse (*Lagopus lagopus scoticus*). J. Animal Ecology 44: 171—190.
Moss, R., A. Watson and P. Rothery. 1984. Inherent changes in the body size, viability and behaviour of a fluctuating red grouse (*Lagopus l. scoticus*) population. J. Animal Ecology 53: 171—189.
Potts, G. R., S. C. Tapper and P. J. Hudson. 1984. Population fluctuations in red grouse: analysis of bag records and a simulation model. J. Animal Ecology 53: 21—36.
Redfield, J. A. 1973. Demography and genetics in colonizing populations of Blue Grouse (*Dendrtagapus obscurus*). Evolution 27: 576—592.
Stenseth, N. C. 1981. On Chitty's theory for fluctuating populations: the importance of genetic polymorphism in the generation of regular density cycles. J. Theoretical Biology 90: 9—36.
Symons, L. E. A. 1969. Pathology of gastrointestinal helminthiasis. International Review of Tropical Medicine, 3: 49—100.
Watson, A. 1970. Territorial and reproductive behaviour of red grouse. J. Reproduction and Fertility, Supplement, 11: 3—14.
Watson, A. and R. Moss. 1979. Population cycles in the Tetraonidae. Ornis Fennica 56: 87—109.
Watson, A., R. Moss, P. Rothery and R. Parr. 1984. Demographic causes and predictive models of population fluctuations in red grouse. J. Animal Ecology 53: 639—662.
Watson, A., R. Moss and R. Rothery. 1984. Effects of food enhancement on numbers and spacing behaviour of red grouse. J. Animal Ecology 53: 663—678.
Watson, A. and R. Moss. 1980. Advances in our understanding of the population dynamics of Red Grouse from a recent fluctuation in population numbers. Ardea 68: 103—111.
Wilson, G. R. 1983. The prevalence of caecal threadworms (*Trichostrongylus tenuis*) in red grouse (*Lagopus lagopus scoticus*). Oecologia 58: 265—268.
Wilson, G. R. and L. P. Wilson. 1978. Haematology, weight and condition of captive red grouse (*Lagopus lagopus scoticus*) infected with caecal threadworm (*Trichostrongylus tenuis*).
Wynne-Edwards, V. C. 1962. Animal Dispersion in Relation to Social Behaviour. Oliver and Boyd, Edinburgh.

Appendix

A linearized stability analysis of the models equilibrium is carried out as outlined in Anderson and May (1978) and May and Anderson (1978). Writing $H(t) = H^* + x(t)$, $W(t) = W^* + y(t)$ and $P(t) = P^* + z(t)$, and linearizing by neglecting terms of order x^2, xyy^2, etc., we get from eqns. (1), (2) and (3) in the main text:

$$dx/dt = c_1 x - c_3 z \quad \text{(A1)}$$

$$dy/dt = -(c_1/c_3)x - \lambda c_2 y + \lambda z \quad \text{(A2)}$$

$$dz/dt = (c_1/c_3)(d' + c_1 c_4)x + d' c_2 y - (d' + c_1 c_5)z \quad \text{(A3)}$$

Here for notational convenience, $c_1 = (a - b)$, $c_2 = \gamma/(\lambda - d')$, $c_3 = (\alpha + \delta)$, $c_4 = (k + 1)/k$, and $c_5 = c_4 \alpha/(\alpha + \delta)$. The temporal behavior of x(t), y(t) and z(t) then goes as exp (Kt), where the stability determining eigenvalues, K are given from eqns (A1), (A2), and (A3) by the cubic equation

$$K^3 + AK^2 + BK + C = 0 \quad \text{(A4)}$$

Here

$$A = \lambda c_2 + d' + c_1(c_5 - 1) \quad \text{(A5)}$$

$$B = \lambda c_1 c_2 (c_5 - 1) \quad \text{(A6)}$$

$$C = c_1 c_2 d' (\lambda - 1) \quad \text{(A7)}$$

The requirement for neighborhood stability is given by the Routh-Hurwitz criterion $A > 0$, $B > 0$, $C > 0$ and $AB > C$. $B > 0$ requires $c_3 > 1$, e.g., either k must be small or $a \gg d$.

For $AB > C$:

$$\lambda(c_5 - 1)(\lambda c_2 + d' + c_1(c_5 - 1)) > d'(\lambda - 1) \quad \text{(A8)}$$

or approximately $\gamma[\lambda/(\lambda - d')]^2(c_5 - 1) > d'$.

When either of these inequalities is transgressed, the system oscillates with period $2\pi/w$. Where $w = (s_1 - s_2)$ appraoches $\sqrt{3}/2$.

$$s_1 = [r + (q^3 + r^2)^{1/2}]^{1/3}, \text{ and } s_2 = [r - (q^3 + r^2)^{1/2}]^{1/3},$$

with

$$q = -1/3\,B - 1/9\,A^2, \; r = -1/6\,(AB - 3C) - 1/27\,A.$$

8. Population biology of helminth infections of veterinary importance and its relevance to control

GARY SMITH

Abstract

This chapter examines how mathematical models can be helpful in the design and evaluation of control programs directed against the debilitating and sometimes fatal diseases that arise when domestic ruminants are infected with helminth parasites. The models comprise hypotheses about the natural processes which control and regulate parasite abundance. The empirical description of regulatory processes is described in detail since these processes act in such a way as to confine parasite population density between certain bounds and so tend to work in opposition to disease control strategies.

It is demonstrated that models can rank strategies in order of efficacy — at least in parasitological terms — but currently give no good indication of how much better one strategy might be than another in terms of the things that most interest the farmer: the net return on his investment in parasite control, and the risk associated with that investment. It is argued that the proper use of the models described here is not to replace anthelmintic field trials but rather as tools to guide our thinking, a means of refining judgments about the strategies most likely to work, and a framework for the development of new methods of disease control. Models should reduce the number of field trials required to evaluate the possible permutations of some new strategy because we will already have some very good idea of which will work best.

Introduction

This chapter examines how mathematical models can be helpful in the design and evaluation of control programs directed against the debilitating and sometimes fatal diseases that arise when domestic ruminants are infected with helminth parasites (round worms, tapeworms and flukes).

The most successful strategies depend upon the use of drugs (anthelmintics) which can be used to prevent (chemoprophylaxis) as well as treat (chemotherapy) disease. A strategy is defined by a number of key features: the choice of anthelmintic, the timing and frequency of treatments, the method by which the drug is administered, and the age class of animals treated. The factors which have to be taken into consideration when deciding upon a strategy include the pharmacological half-life of the anthelmintic

S. K. Jain and L. W. Botsford (eds), Applied Population Biology, 173–195.

(which may range from a few hours to several days), the efficiency with which the drug kills parasites in a particular phase of the life-cycle, and the population biology of the parasite in question.

The population biology of the parasite is important for two reasons. First of all, the mean time spent in any of the free living stages is critically dependent upon the environmental temperature. It follows that the rate at which the ruminant host becomes infected usually varies with the season, and this has obvious implications for any strategy intended to *prevent* disease outbreaks. Second, there are a number of mechanisms by which the abundance of parasites is naturally regulated. These mechanisms operate in a density-dependent manner, and, being sensitive to changes in the mean parasite burden, interact with density-independent causes of parasite mortality (such as anthelmintics) so as to ameliorate their effect.

The principal difficulty, then, in choosing an appropriate strategy is finding a way to take into account each of the relevant factors without being overwhelmed by the complexity of the task. Mathematical models can help. The models dealt with here are summaries of a corpus of hypotheses about the population biology of the parasites considered. They provide a formal framework within which the consequences of those hypotheses can be investigated and a stable point of reference when novel strategies for disease control are being evaluated.

There is no pretense that these models are predictive in the sense that they will describe the exact outcome of a particular strategy on a particular farm. Nevertheless, it is the intention that they be sufficiently well constructed to generate and explain the epidemiological patterns typical of a given region or management practice and so, to that extent, can be used to evaluate and compare the consequences of a series of strategies.

The use of models is just one component of the process of choosing or designing a particular strategy, and the things that models cannot do should be attended to as carefully as those things they can accomplish. As we shall see, there is no simple relationship between parasite burden and the health or productivity of the host, and the models do not lead easily into a consideration of the *economic* aspects of control. Further, the models considered here are deterministic in construction. They take no account of the effect of chance fluctuations in parasite abundance, particularly at low parasite densities, and have only a limited capacity to indicate the confidence limits associated with any particular result. Finally, it should be remembered that the model reflects the extent of our knowledge concerning the population biology of the parasite. If our knowledge is flawed or incomplete, then so is our model.

The chapter is organized as follows. First, there is a brief history of modeling helminth diseases of veterinary importance. Next we consider a method of estimating optimum treatment times and show how models can be elaborated to include the impact of natural regulatory process. We then give an example of the way in which models can be used to screen prophylactic strategies, and finally discuss the role of models in the general decision making process.

The problem

The production of cattle and sheep is an important component of the economy in most countries of the world (Cole and Ronning 1974). In none of these countries are the stock entirely free from the risk of infection by digenean flukes or gastrointestinal nematode parasites, and there is a continuing effort to reduce the economic impact of these infections to more acceptable levels. The most successful strategies of control or prophylaxis depend upon the administration of anthelmintic drugs. However, the circumstances under which one disease control strategy might be more successful than another are greatly influenced by local conditions. This is due to the frequently highly seasonal nature of parasitic disease and the range and variety of livestock management systems encountered world-wide.

Disease forecasting systems

In the past, when anthelmintics were either unavailable or else relatively ineffective against the pathogenic phase of the parasite life-cycle, disease control depended on judicious grazing management policies which attempted to moderate the rate of infection (i.e. the number of infective stages ingested by each grazing animal). Such policies were guided by disease forecasting systems, the most successful of which used simple climatic indices to predict the period in each year when the livestock were at most risk of infection (Ollerenshaw 1966, 1974; Ollerenshaw *et al.* 1978; Thomas and Starr 1978; Gettinby and Gardiner 1980). Several of the more elaborate systems, although still ultimately dependent upon readily available meteorological records, were preliminary attempts to model the population biology of the parasite in question (Gettinby *et al.* 1979; Hope-Cawdery *et al.* 1978; Paton *et al.* 1984).

The usefulness of these systems diminished as safe and effective anthelmintic preparations became more widely available. However, it is conceivable that they might once more receive greater attention as the spread of anthelmintic resistance compels producers to rethink chemoprophylactic strategies which rely on repeated treatments carried out irrespective of the actual risk of disease in that year.

Models of parasite population biology

A wide variety of safe and effective anthelmintics is currently available, and as disease control policies have moved from chemotherapy (treatment of sick animals) to chemoprophylaxis (disease prevention), there has evolved an increasingly sophisticated understanding of the way in which the characteristics of the population biology of the parasite affect the outcome of any particular strategy. Nevertheless, choosing an optimum strategy of chemoprophylaxis remains a formidable task, and the literature is replete with accounts of field trials comparing various anthelmintic preparation with each other as well as various strategies with each other (see Donald 1985; Campbell 1986 for a general review). Not only do anthelmintic preparations differ in their mode of action, spectrum of efficiency, pharmacological half-life and means of delivery, but the epidemiology of helminth associated disease varies with climate, immunological status of the host, and management practices specific to a geographical region.

Mathematical models of helminth infections have proved to be especially useful in this respect and provide a formal framework for comparing the likely outcome of different control measures prior to their application in the field (Smith 1984 a,b,c; Smith and Grenfell 1985; Grenfell *et al.* 1987b; Smith *et al.* 1987b) as well as helping to explain why control strategies already in use did not have the expected effect (see Roberts *et al.* 1987 for a particularly good example involving tapeworms).

Two such models will be considered below. The first and simplest describes a population of flukes (*Fasciola hepatica*) in a flock of sheep kept on pasture during the fall and winter. The second describes the population biology of *Ostertagia ostertagi*, an important gastrointestinal roundworm parasite of cattle.

Fasciola hepatica

The model

The life cycle of the common liver-fluke is given in Fig. 1a. It is convenient to begin with this parasite for a number of reasons. First, sheep appear to develop no significant resistance to infection and so the mortality of the flukes is independent of the host's experience of the parasite. Second, if the sheep are moved onto a contaminated pasture during the late summer or early fall under the temperature conditions typical of northern Europe, we can assume that there will be no significant addition to the population of infective stages (metacercariae, sometimes called cysts) until the following year. Third, the number of infective, accessible metacercariae declines due to mortality or their displacement into the lower herbage mat at a constant average rate (Over and Dijkstra 1975).

Thus, the rate of change with respect to time (t) of the number of metacercariae (C(t)) on the pasture and the number of immature (I(t)) and

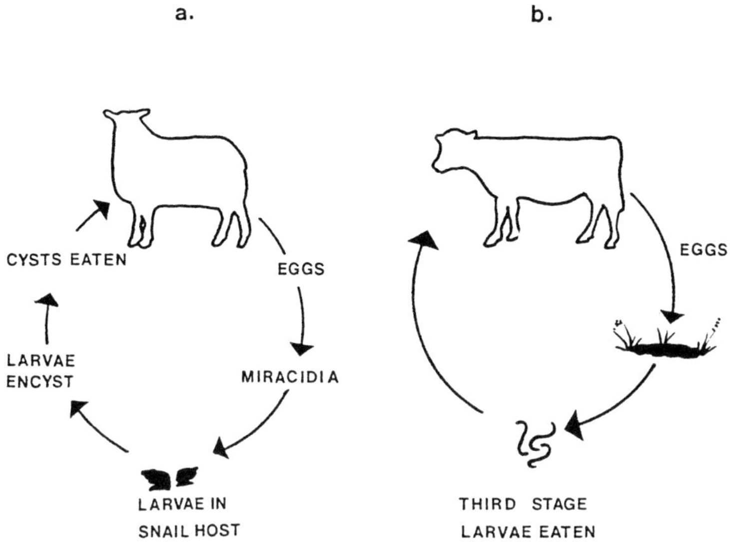

Fig. 1. (a) Life-cycle of *Fasciola hepatica*. (b) Life-cycle of *Ostertagia ostertagi*.

mature (M(t)) parasites within a flock of H(t) sheep can be represented as follows:

$$dC(t)/dt = -\mu_1 C(t) - \beta H(t)C(t)\theta(t-\tau_1) \tag{1}$$

$$\begin{aligned} dI(t)/dt = {} & \beta H(t)C(t)\theta(t-\tau_1) - \mu_2 I(t) - \\ & - \beta H(t-\tau_2)C(t-\tau_2)\exp[(-\mu_2\tau_2)\theta(t-\tau_1-\tau_2)] \end{aligned} \tag{2}$$

$$\begin{aligned} dM(t)/dt = {} & \beta H(t-\tau_2)C(t-\tau_2)\exp[(-\mu_2\tau_2)\theta(t-\tau_1-\tau_2)] - \\ & - \mu_3 M(t) \end{aligned} \tag{3}$$

All parameters are defined in Table 1. The model follows the fate of a cohort of metacercariae during the six month period of fall and winter. Sheep are turned out onto the pasture τ_1 weeks after the start of the simulation at $t = 0$. The model makes the conventional assumption that the rate of infection is proportional to the number of hosts and infective stages respectively (namely, $\beta H(t)C(t)$) and so equation (1) describes the rate of loss of infective stages from the pasture as they either die (at a net rate $\mu_1 C(t)$) or are ingested by the grazing sheep. Equation (2) describes the development of the immature stages within the hosts and assumes that those flukes which survive τ_2 weeks become mature egg-laying parasites (M(t)). These parasites die at a rate, μ_3, which is much smaller than that of the immature stages (μ_2). The functions, $\theta(z)$, are step functions which determine the time delays in the model (see Table 1).

An analytical solution to equations (1), (2), and (3) is presented in the

Table 1. Parameter definitions for model expressed by equations (1), (2) and (3) in the main text. All rates are instantaneous rates.

Population variables and parameters	Definition
t	Elapsed time in days
C(t)	Density of metacercariae (ha^{-1}) at time t
I(t)	Density of immature flukes (ha^{-1}) at time t
M(t)	Density of mature flukes (ha^{-1}) at time t
H(t)	Density of sheep (ha^{-1}) at time t
μ^1	Death rate of metacercariae ($cyst^{-1}$ day^{-1})
μ^2	Death rate of immature flukes ($fluke^{-1}$ day^{-1})
μ^3	Death rate of mature flukes ($fluke^{-1}$ day^{-1})
β	Average instantaneous rate of infection (day^{-1} $host^{-1}$)
τ_1	Number of days after start of simulation when sheep turned onto pasture
τ_2	Average maturation time of flukes (days)
$\theta(z)$	A step function such that $\theta = 1$ when $z \leqslant 0$, otherwise $\theta = 0$

Appendix, and Fig. 2 provides a qualitative comparison between the behavior of the model and field observations on the decline in metacercarial abundance and changes in the mean number of flukes per sheep [(I(t) + M(t)/H(t))].

Before the model can be used to evaluate and compare anthelmintic strategies, we must choose the criteria by which the strategies are to be judged. Suitable criteria would be the prevalence and intensity of infection at the end of a simulated trial, differences in the age structure of the parasite population in the treated flock compared with that in the untreated control group, or some index of the host's entire experience of infection such as the integral of the mean parasite burden throughout the whole period of the trial (i.e. the area under the intensity of infection curve in Figs. 2 and 3) (Smith 1982, 1984 a,b,c). In the example that follows, changes in the integral of the mean number of parasites per host (A_n) are used to determine optimal times for administering n treatments.

A simulated field trial

A well-maintained flock in those regions of northern temperate Europe where fascioliasis is endemic, is likely to be treated at least three times during the fall and winter. The timing of the third treatment is fixed by the lambing date; the timing of the two preceding treatments generally depends upon convenience. Given the multiplicity of possible treatment times, we could not field-test even that fraction of combinations which seemed intuitively reasonable without some further selection process. We therefore use the model to refine our ideas about which strategy is most likely to work.

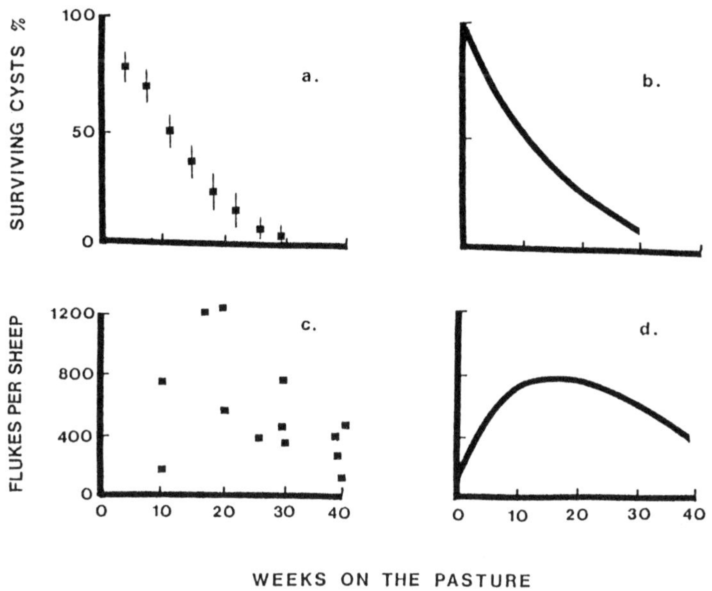

Fig. 2. Quantitative comparison between the predictions of the *Fasciola* model and observed changes in the density of metacercariae and abundance of flukes. (a) Observed survivorship curve for metacercariae (mean and standard error); (b) predicted metacercarial survivorship; (c) observed intensity of infection in a flock of lambs (obtained by serial slaughter); (d) predicted intensity of infection. (Data from Over and Dijkstra 1975; Ross 1967.)

We imagine a fluke-free flock of H sheep turned out onto pasture contaminated with C(t) cysts at time t ($t = \tau_1 = 0$). The flock is treated with a conventional anthelmintic (half-life < 1 day) on days t_1, t_2 and t_3. The trial finishes on day $t = t_3$. The efficacy of the strategy is measured by the effect it has on the integral (A_3) of the total parasite burden ($P(t) = I(t) + M(t)$). Since the number of hosts (H) is constant, P(t) is a simple direct index of the area under the intensity of infection curve from $t = 0$ to t_3, i.e.,

$$A_3(t_1, t_2) = \int_0^{t_3} P(t)\, dt. \tag{4}$$

The optimum treatment times ($\hat{t}_1$, $\hat{t}_2$) are those values of t_1 and t_2 which minimize this integral. The conditions

$$\left(\frac{\partial^2 A_3}{\partial t_1^2} \cdot \frac{\partial^2 A_3}{\partial t_2^2}\right) - \left(\frac{\partial^2 A_3}{\partial t_1 \cdot \partial t_2}\right)^2 > 0, \text{ and } \frac{\partial^2 A_3}{\partial t_1^2} > 0 \tag{5}$$

are satisfied when t_1 and t_2 are assigned their optimum values (Smith 1984b). A graphical example is given in Fig. 4.

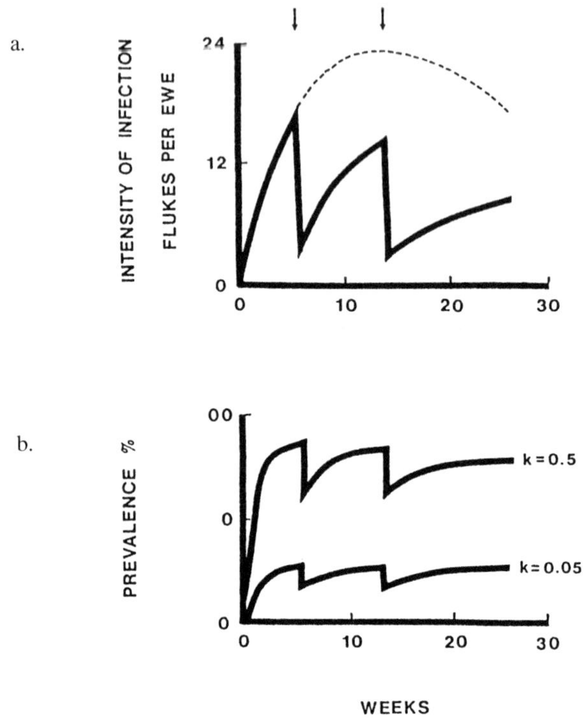

Fig. 3. Predicted effect (solid line) of two treatments (arrows) with a conventional anthelmintic 6 weeks and 14 weeks after a previously uninfected flock is turned out onto a pasture contaminated with *F. hepatica* metacercaria.

The choice of the integral, A_n, as the measure by which the optimum treatment times are selected satisfies the intuitively reasonable assumption that the detrimental effects of infection are a function of the host's *entire* experience of the parasite. Nevertheless, we are currently unable to describe the exact relationship between infection and production loss, and, until such time as we are able to do so, the choice of criteria by which the strategy is judged a success or failure will remain relatively arbitrary.

Natural control and regulation of *F. hepatica* populations

The model described so far deals with that free living stage least affected by microclimate variations and rests upon the important assumption that the death rate of the flukes is independent of the sheep's experience of infection. This is a reasonable assumption for a well managed flock in which fluke-related host deaths are avoided by judicious grazing management policies or effective chemotherapeutic strategies. However, it is no longer valid when

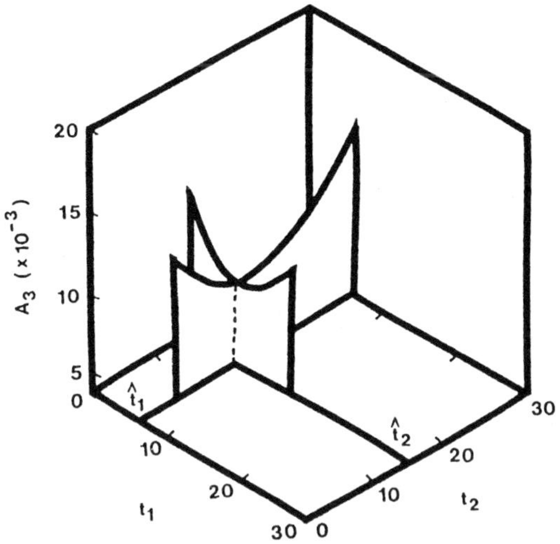

Fig. 4. A graphical illustration of the optimum treatment times (t_1 and t_2) and their effect upon the area (A_3 (t_1, t_2) under the intensity of infection curve. This area provides an index of the parasitemia to which the flock was subjected. The objective of parasite control is to reduce this area to the minimum possible, given the resources available. The bowl-shaped surface representing all possible combinations of t_1 and t_2 has been intersected in the figure at the optimum treatment times.

there is significant parasite induced host mortality since the probability of host death increases with the intensity of infection.

Accordingly, the application of the fasciola model in other than the rather narrow set of circumstances described above depends upon the identification and description of those natural processes which regulate and control the parasite population through the effect they have on parasite development and mortality. It is useful to distinguish between regulation and control here because some of the processes which affect parasite numbers do so in such a way as to confine population density between certain bounds (regulatory processes), while others may be regarded as mere perturbatory influences (controlling processes).

The importance of weather (a controlling factor) in determining the distribution and abundance of *F. hepatica* is well known (Wilson *et al.* 1982), but less attention has been given to processes which are truly regulatory in that they *promote the stability of the parasite population.* Natural regulatory processes render the parasite-host system more or less refractory to perturbations, at least in the long-term. Since the principal function of anthelmintics is to effect a profound and, if possible, lasting perturbation in one particular direction, it is worth considering the opposing influence of the natural regulatory processes in some detail. Two of these processes are summarized below together with a brief discussion of the methodological difficulties of representing them in the *Fasciola* model. We begin with the density-dependent regulation of fluke fecundity and then discuss parasite deaths attributable to parasite-induced host mortality.

Density-dependent fecundity

Experimental work has demonstrated that the *per capita* fecundity of *F. hepatica* in sheep is an inverse function of the number of flukes per host (Boray 1969). A good empirical model of this relationship is provided by an exponential function (Fig. 5a), i.e.

$$\lambda = \lambda_0 \cdot \exp(-\delta m) \qquad (6)$$

where λ is the average fecundity of the parasite subpopulation in a *single* host containing m mature flukes (λ_0 and δ are constants) (Smith 1984d). The net rate of egg production summed over the entire parasite population depends upon the way in which the parasites are distributed amongst their hosts. For a flock of H sheep, this rate is

$$\lambda_0 H \sum_{m=0}^{\infty} mp(m)\exp(-\delta m) \qquad (7)$$

where p(m) is the probability that a host will contain m mature flukes (Smith 1984d). Numerical evaluations of expression (7) for typical probability distributions indicate that we should expect the total daily egg output of *F. hepatica* populations to tend towards an asymptotic value as parasite abundance increases (Fig. 5b), but mere abundance is not the only thing that affects total egg output. It is usual to find that most of the sheep in a flock are infected, but that only a very few of them harbor substantial numbers of parasites. Parasite populations distributed like this are said to be aggregated. The overall severity of the density-dependent constraint on fecundity is greatest in the most aggregated populations because a greater proportion of the flukes is subjected to the full force of the constraint. Here we have assumed that the negative binomial distribution provided a good phenomenological mimic of the distribution of flukes in a flock of sheep. This distribution is especially useful because its exponent, k, provides an inverse index of the degree of aggregation. As the degree of aggregation increases, the maximum egg output assumes a lower value (Fig. 5b).

Parasite mortality due to parasite-induced host deaths

When an infected host dies, its parasite burden dies with it. If the probability of host death increases with the intensity of infection, then parasite deaths due to that cause are density-dependent and regulate the parasite population (Fig. 5c)

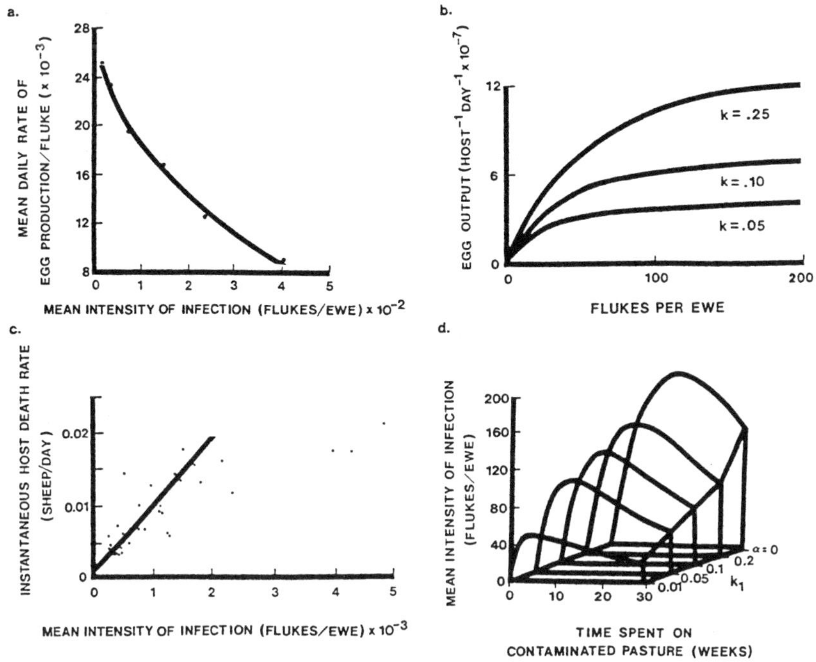

Fig. 5. (a) Density dependent fecundity in *F. hepatica*; (b) Predicted total daily egg production in a natural *F. hepatica* population assuming an aggregated distribution of parasites (k is the exponent term of the of the negative binomial distribution and is thus an inverse index of the degree of aggregation of mature flukes); (c) Host death rate in sheep infected once only with *F. hepatica*; (d) Predicted mean parasite burden in sheep infected with *F. hepatica* (when $\delta = 0$, the infection is assumed non-pathological, k_1 is the exponent term of the negative binomial distribution describing the distribution of mature and immature flukes) (Smith 1984d).

In the case of *F. hepatica*, the rate of host deaths due to parasite induced host mortality in a flock of H sheep can be represented by

$$\alpha H \sum_{i=0}^{\infty} i p(i) \tag{8}$$

where α is the average instantaneous rate of parasite induced host mortality (/parasite/host), and p(i) is the probability that a host contains a total of i parasites. The net rate of *parasite* deaths due to this cause follows directly:

$$\alpha H \sum_{i=0}^{\infty} i^2 p(i) \tag{9}$$

Smith (1984d) incorporated this expression into a model of an outbreak of fascioliasis among sheep in northern temperate Europe and obtained the results shown in Fig. 5d. The death of the infected hosts has a considerable impact on the mean intensity of infection particularly when the parasites are highly aggregated.

Methodological problems

The putative mechanisms by which density-dependent processes regulate parasite populations are as follows.

1. Intra-specific competition for finite resources.
2. Pathological or non-specific immune responses to the infection creating an environment inimical to parasite survival or reproduction.
3. Pathological or immuno-pathological responses leading to host deaths.
4. Acquired immune responses.

The models embodied by expressions (7) and (9) are merely empirical descriptions of observed relationships derived from single infection experiments. Nevertheless, once incorporated in a model of parasite population biology they become implicit hypotheses about the mechanism of regulation. For example, an important limitation on both expressions is that all of the processes operate instantaneously: there is an immediate effect of *current* parasite density on fluke fecundity or the probability of host death. Neither model contains an anamnestic component (i.e. there is no reference to past experience of infection) and it is assumed that the density-dependent process is mediated, without any time delay, by intra-specific competition (in the case of fecundity) or some pathological host response (parasitic deaths due to parasite induced host mortality).

When the actual mechanism of regulation is unknown and an empirical mimic is the only resort, it seems reasonable that it should be based on long-term trickle infection experiments carried out over a range of infection intensities in order to minimize the risks of extrapolation. Trickle infection experiments involve the repeated infection of host animals with known numbers of infective stages over a long period. The procedure mimics the infection process in the field but lends itself to a very detailed analysis of the dynamics of parasite demography. This was the approach used in the next example, a mathematical model of the population biology of *O. ostertagi.*

Ostertagia ostertagi

The parasite

Ostertagia ostertagi is one of the most prevalent and harmful gastro-intestinal nematode parasites of cattle in the temperate world. It has a simple direct life cycle (Fig. 1b). The cattle become infected when they ingest the third stage larvae. After a brief period of development, the young adult stages emerge from the gastric glands of the abomasum and there is a profound disruption of mucosal architecture and function (Armour and Ogbourne 1982). If the intensity of infection is sufficiently high, the clinical signs of ostertagiasis appear: loss of appetite, diarrhea, reduced live weight gain or weight loss, and, in some cases, death.

The development and survival of the free-living stages of the nematode are largely determined by density-independent factors such as soil-surface moisture and temperature (Smith *et al.* 1986; Grenfell *et al.* 1986), but the immunological status of the host dominates the population dynamics of the parasitic stages (Grenfell and Smith 1983; Smith and Grenfell 1985; Grenfell *et al.* 1987a; Smith *et al.* 1987a).

The acquired immune response of the host

Smith and Grenfell (1985) reviewed a number of trickle and single infection experiments with *O. ostertagi* and concluded that the host's immunological response to this parasite expressed itself in three ways:

1. A reduction in the *per capita* fecundity of the female worms (Fig. 6a).
2. A reduction in the rate of establishment of the ingested third stage larvae in terms of the proportion of larvae that manage to penetrate into the abomasal mucosa (Fig. 6b).
3. An increase in the death rate of mature worms (Fig. 6c).

The empirical representation of each of these processes is described below.

Parasite fecundity

A good empirical model of the changes in fecundity (λ) in trickle infection experiments was provided by:

$$\lambda = \lambda_0 \exp(-\delta \mathrm{mt}) \tag{10}$$

where λ_0 and δ are constants (Smith *et al.* 1987a). This model differs from equation (6) which describes the regulation of the fecundity of *F. hepatica* in that fecundity here is represented as declining with time as well as parasite abundance. The rate of change in fecundity depends upon the product of the number of mature roundworms (m) and the duration of infection (t). This is a is a crude approximation to the integral of the worm burden through time (Smith *et al.* 1987a) and was necessary to take into the account the observation that the fecundity of *O. ostertagi* varies with the hosts entire experience of infection rather than merely the current parasite burden.

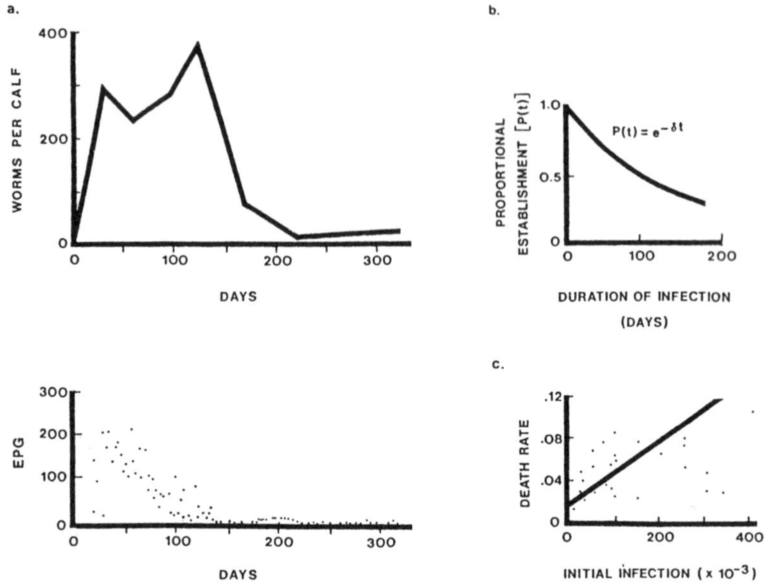

Fig. 6. (a) Comparison of the worm burden (upper graph) and fecal egg count (EPG, eggs per gram) (lower graph) in calves infected with 1500 third stage *O. ostertagi* larvae per day (data from Michel 1963); (b) Predicted decline in the rate of establishment of third stage *O. ostertagi* larvae; (c) Death rate of fifth stage worms in calves infected once only with *O. ostertagi* (Data from Smith and Grenfell 1985).

Changes in the rate of establishment

In order to begin to develop to maturity, the worm must first penetrate the lining of the host stomach (abomasum). Not all of the worms manage to do this and, as the duration of the infection increases, fewer and fewer of them manage to become established; eventually, the host becomes almost entirely refractory to new infections. The mechanism by which the rate of establishment is controlled is unknown, but a declining exponential function of the duration of the infection satisfactorily explains the observed changes in fourth stage larval abundance in trickle infection experiments (Grenfell *et al.* 1987a) (Fig. 6b). This pleasingly simple model, which has exact analogues in other host-parasite systems (Smith 1988), is in fact rather perplexing. A valuable evidence indicates that the rate of establishment is in part governed by immunological processes. If that is true, why is establishment dependent only on the duration of infection? We might reasonably expect it to vary with the intensity of infection also — but it does not (Grenfell *et al.* 1987a; Smith 1988).

Death rate of the mature parasites

Early analyses of the survival of *O. ostertagi* in calves indicated that the death rate of the mature stages is an increasing linear function of the total cumulative number of third stage larvae ingested (Anderson and Michel 1977; Grenfell *et al.* 1986), although there is now some suggestion that the death rate tends to an asymptote as the hosts experience of infection increases (Fig. 6c). This makes sense, since we would not expect the death rate to become infinitely large!

A simulated field trial

Grenfell and Smith (1983) and Grenfell *et al.* (1987b) incorporated each of the regulatory processes described above into a complete model of the life cycle of *O. ostertagi* (Fig. 7). The model was used to screen five anthelmintic strategies against ostertagiasis by simulating a field trial in which 60 previ-

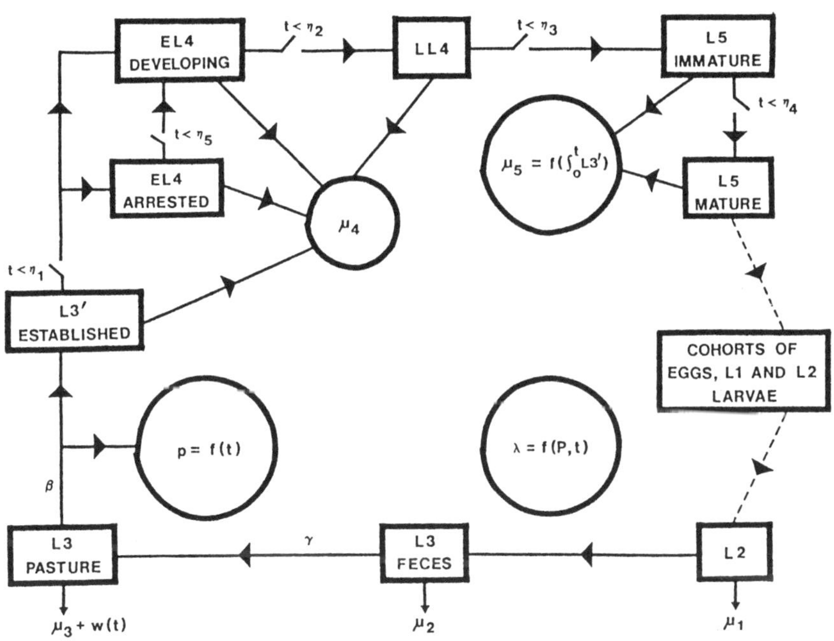

Fig. 7. Basic architecture of the *Ostertagia* model. Differential equations are used to model changes in abundance of those life-cycle stages represented by the nine smaller rectangles. Difference equations are used for those stages represented by the larger rectangle. Explicit time delays in the parasitic phase of the life cycle are represented as a series of switches which remain open as long as the corresponding statement, $t < 3_i$ ($i = 1$ to 5), remains true. Other parameters: p is the proportion of larvae that become established, P is the total number of parasites per host, and λ is the *per capita* fecundity (Grenfell *et al.* 1986c).

ously uninfected calves were divided into six groups of ten calves each. The groups were maintained on adjacent paddocks under climatic conditions typical of those in southeastern England. The simulation covered a period of 18 months commencing one month before turnout (April 1st) and described pasture larval contamination, fecal egg out-put and worm burdens throughout the entire grazing season and subsequent housing period (Smith *et al.* 1987b). The calves in paddock A were left untreated. The calves in the other paddocks received (hypothetical) anthelmintics as outlined in Fig. 8. Drug 1 was effective against all of the parasitic stages, a single treatment killing 90% of the larval forms and more than 99% of the fifth stage worms. Drug 2 was effective against only the adult worms but was delivered via an intra-ruminal sustained release device administered to each calf on the day before turnout. The device had first order release characteristics (so that the actual amount of drug released declined in proportion the amount remaining in the capsule) and effective drug levels were maintained for at least 90 days after administration. Unlike drugs 1 and 2, which had assumed half-lives of a few hours only, drug 3 had a half life of 4 days. Drug 3 brought about a 99% reduction in all parasitic stages measured one week after a single treatment.

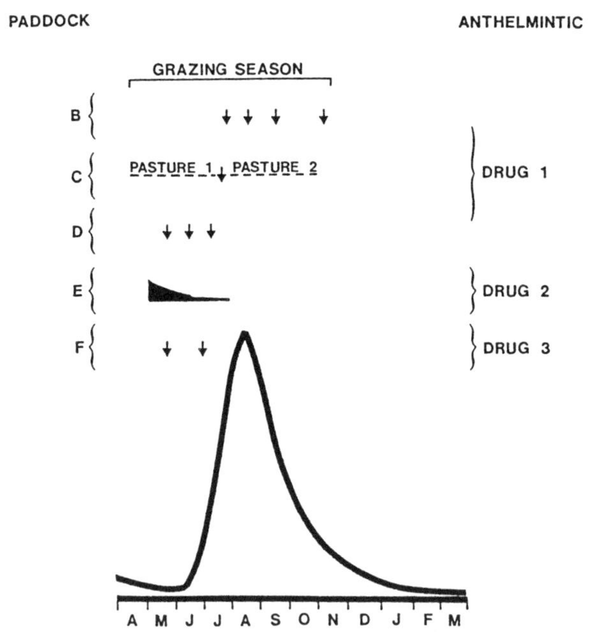

Fig. 8. Chemotherapeutic strategy B and chemoprophylactic strategies C to F against *O. ostertagi.* The expected pattern of pasture larval contamination (L3 larvae per kilogram dry herbage) on the untreated paddock A (solid line) is shown with treatment times (arrows) on each of the other paddocks. Protocol E uses an intra ruminal sustained release device, and protocol C involves two paddocks ("dose and move").

The strategies

The protocol used in paddock A was a simple therapeutic strategy in response to the onset of clinical ostertagiasis. The prophylactic protocols used in paddocks D, E and F were designed to curtail the midsummer rise in pasture larval contamination and so retard or present the sudden increase in the intensity of infection in the calves which is its usual sequel. In each case, the strategy was to keep the number of mature parasites as low as possible during that period in the first half of the grazing season when the level of host-resistance to reinfection is relatively meager. Later in the season, the microclimatic conditions at the pasture surface were less favorable (hotter, drier) to the development of the free living stages, and the parasitic stages survived for a shorter period in increasingly resistant hosts. By this time the calves have had sufficient experience of infection for the rate of establishment to be very low, and the death rate of mature worms to be approaching its asymptotic value.

The "dose and move strategy" in paddock C does not prevent the midsummer rise but rather seeks to avoid damaging levels of infection by a judicious mixture of anthelmintic treatment and grazing management.

The results of the simulation

The efficacy of each strategy was measured in terms of the effect on pasture larval contamination (Fig. 9) and the integral of the mature worm burden over the period of the simulation (Table 2).

The simple chemotherapeutic protocol on paddock B had a negligible effect on larval contamination of pasture and the calves were rapidly reinfected after each treatment. All of the prophylactic protocols (C to F) had a profound effect on the magnitude of the midsummer rise in larval abundance, a fact reflected in the integral of the mean mature worm burden for each paddock. Even the least effective of the prophylactic protocols (C) resulted in a 75—80% reduction in the mature worm burden when compared with the untreated control group on paddock A.

Assessing the likelihood of disease and the role of the model in the decision making process

Experimental studies of calves and sheep infected with *O. ostertagi* and *F. hepatica* respectively indicate that the severity of the signs and lesions associated with the infection is broadly related to the number of infective stages administered (Anderson *et al.* 1966; Hawkins and Morris 1978). There is considerable circumstantial evidence that this is true of a variety of other helminth infections (Barger 1982). However, there seems to be no

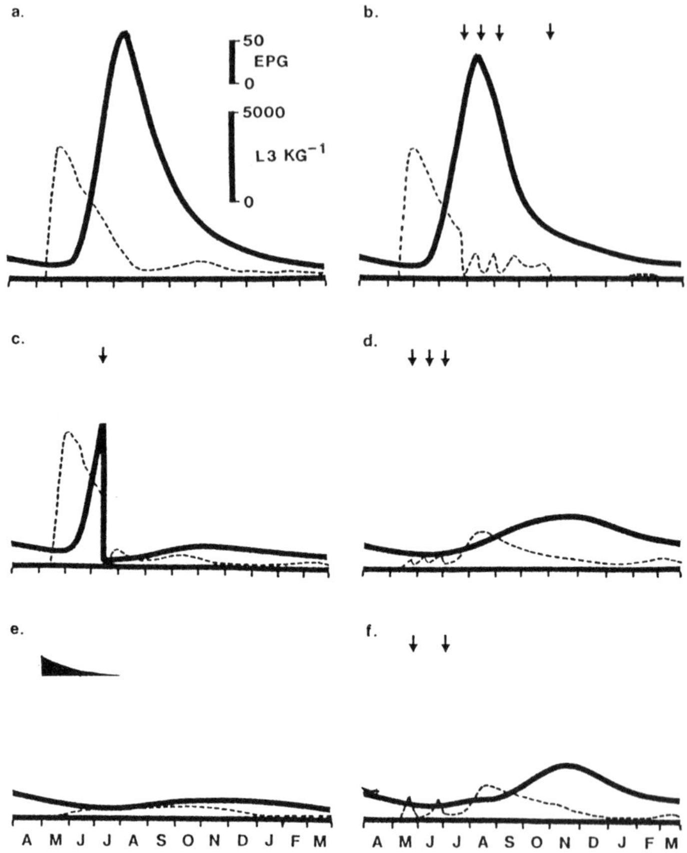

Fig. 9. Predicted patterns of parasite abundance on Paddocks A to F ((a) to (f)). Dashed line: eggs per gram feces (EPG); solid line: pasture larval contamination (L3 KG^{-1}). All figures drawn to the same vertical scale.

simple relationship between the current intensity of infection and host morbidity or mortality and it is intuitively reasonable to assume that this is because the impact of a parasite on the disease status and productivity of a herd or flock is related to the hosts' entire experience of the infection. This is the reason that the integral of the parasite burden was chosen as an index of anthelmintic efficacy in the examples given above. Even so, the models do not yet take into account other confounding features such as the nutritional status of the hosts (Berry and Dargie 1976), and although we may make the general assumption that changes in parasitological parameters will be directly reflected in the performance of the herd or flock, we are unable to say how great a change is necessary before the difference would be detectable in terms of production parameters like milk yield, weight gain or wool crop.

Table 2. Outcome of a simulated field trial comparing various strategies against bovine ostertagiasis.

Anthelmint	None	Drug 1			Drug 3	Drug 2
PROTOCOL	Control group	3 therapeutic treatments & 1 housing dose	"treat and move"	3 early season prophylactic treatments	2 early season prophylactic treatments	Intraruminal sustained released of drug during early season
PADDOCK	A	B	C	D	F	E
Integral of mature worm burden (worm days × 10^{-5})	38.23	15.79	8.23	7.42	5.66	2.33
Peak mean worm burden	166893	126945	37783	29715	24458	28850

This is a general problem in veterinary parasitology. There are many studies which demonstrate a production advantage when treated animals are compared with untreated controls (e.g. Gaafar 1983) but very few which directly demonstrate a relationship between the intensity of infection and the degree of production loss. Furthermore, such studies as there are indicate that the relationship is markedly non-linear (e.g. Hawkins and Morris 1978).

It appears then, that models can rank strategies in order of efficacy — at least in parasitological terms — but currently give no good indication of how much better one strategy might be than another in terms of the things that most interest the farmer. Even if one feels justified in inferring that reductions in the mean intensity of infection will automatically translate into production advantages, one should not be tempted to recommend that strategy which yields the greatest probable production benefit unless it has already been shown that this strategy will also provide the greatest economic return. Control strategies are often expensive and one is justified in spending increasing amounts on control only for as long as the net return continues to increase. Experience shows that the point of maximum profit (where the marginal cost equals the marginal return) is typically rather below the level of control required to achieve maximum possible production and will vary according to market prices (Morris 1969; Anderson *et al.* 1976). Furthermore, recommendations to producers based on a rank order derived from simulations such as those described here embody the implicit assumption that the variance associated with each return is identical for all interventions (Smith and Galligan 1988). In practice, this is not true — as every farmer knows. Each strategy has a certain risk (variance in outcome) associated with it and experience shows that farmers choose strategies based on their perceptions of the risks involved as well as the average return they might expect on their investment in control.

So, what is the value of modeling the population biology of helminth parasites? The first point to make is that the criticisms directed at models in the preceding paragraphs could just as easily, and with as much justification, be directed against field trials of anthelmintic strategies (Smith and Galligan 1988). It follows that we cannot dismiss the models on the grounds that we have some other mechanism that does the job better. It might even be tempting to argue that models are a cheaper method of obtaining the same result — but this would be to misunderstand the nature of models. The models described here are not intended to replace field trials, as this would be to eschew the experimental testing of the hypothesis that the model represents. The models are intended to provide the "first look" at a problem. They are tools to guide our thinking, a means of refining our judgments about the strategies most likely to work in any particular circumstance, and a framework for the development of new methods of disease control. Properly used, models should reduce the number of field trials required to evaluate the possible permutations of some new strategy because we will already have some very good idea of which will work best.

Summary and speculation

Mathematical models of the population biology of helminth parasites of veterinary importance provide a useful framework for the evaluation of chemotherapeutic and chemoprophylactic strategies, at least in terms of the effect these strategies have on parasite numbers. Nevertheless, it should be remembered that disease and production loss are multicausal phenomena. The mere presence of infection does not guarantee disease, nor do changes in parasite burden necessarily leading to correspondingly large changes in the production efficiency of the flock or herd. Furthermore, veterinary practitioners in the field will often be called upon to deal with multispecies assemblages of parasites whereas the models discussed above describe only single infections. In short, since the ultimate criteria for the success of a control strategy are economic, the models are instructive but not definitive.

As for the future, the most serious impediment to the use of anthelmintics is the spread of drug-resistant phenotypes amongst helminth parasite populations. The emergence of predominantly resistant populations is inevitable and will become an increasingly important consideration in the design and management of anthelmintic strategies. Mathematical models provide a particularly suitable tool for the investigation of strategies designed to keep production losses to an acceptable minimum while simultaneously impeding the spread of resistant forms.

Literature cited

Anderson, N., J. Abrmour, R. M. Eadie, W. F. H. Jarrett, F. N. Jennings, J. S. D. Ritchie, and G. M. Urquhart. 1966. Experimental *Ostertagia ostertagi* infections in calves: results of single infection with five graded dose levels of larvae. American Journal of Veterinary Research 27: 1259—1265

Anderson, N., R. S. Morris, and I. K. McTaggart. 1976. An economic analysis of two schemes for the anthelmintic control of helminthiasis in weaned lambs. Australian Veterinary Journal 52: 174—180.

Anderson, R. M. and J. F. Michel. 1977. Density dependent survival in populations of *Ostertagia ostertagi.* International Journal for Parasitology 7: 321—329.

Armour, J. and C. P. Ogbourne. 1982. Bovine Ostertagiasis: a review and annotated bibliography. Miscellaneous publication No. 4., Commonwealth Institute of Parasitology.

Barger, I. A. (1982) Helminth parasites and animal production. In, Biology and Control of Endoparasites. (ed. by L. E. A. Symons, A. D. Donald, J. K. Dineen) Academic Press, Sydney. pp 133—156.

Berry, C. I. and J. D. Dargie. 1976. The role of host nutrition in the pathogenesis of ovine fascioliasis. Veterinary Parasitology 2: 317—332.

Boray, J. C. 1969. Studies on experimental infections with *Fasciola hepatica* with particular reference to fascioliasis in sheep. Advances in Parasitology 7: 95—210.

Campbell, W. C. 1986. The chemotherapy of parasitic infections. Journal of Parasitology 72: 45—61.

Cole, H. H. and M. Ronning. 1974. Animal Agriculture. W. H. Freeman and Co., San Francisco.

Donald, A. D. 1985. New methods of drug application for control of helminths. Veterinary Parasitology 18: 121—137.
Gaafar, S. M. (ed.). 1983. Morantel Sustained Release Bolus. Special Issue of Veterinary Parasitology 12: 215—362.
Gettinby, G., K. Bairdon, J. Armour, and C. Benitez-Usher. 1979. A prediction model for bovine ostertagiasis (*Ostertagia ostertagi*) Veterinary Record 105: 57—59.
Gettinby, G. and W. P. Gardiner. 1980. Disease incidence forecasts by means of climatic data. Biometeorology 7: 87—103.
Grenfell, B. T. and G. Smith. 1983. Population biology and control of ostertagiasis in first year grazing calves. Proceedings of the Society for Veterinary Epidemiology and Preventive Medicine (Southampton 12—13 April, 1983) pp 70—77.
Grenfell, B. T., G. Smith, and R. M. Anderson. 1986. Maximum likelihood estimates of survival and migration rates of the infective stages of *Cooperia oncophora* and *Ostertagia ostertagi.* Parasitology 92, 643—652.
Grenfell, B. T., G. Smith, and R. M. Anderson. 1987a. The regulation of *Ostertagia ostertagi* in calves: The effects of past and current experience of infection on proportional establishment and parasite survival. Parasitology 95: 363—372.
Grenfell, B. T., G. Smith, and R. M. Anderson. 1987b. A mathematical model of the population biology of *Ostertagia ostertagi* in calves and yearlings. Parasitology 95: 389—406.
Hawkins, C. D. and R. S. Morris. 1978. Depression of productivity in sheep infected with *Fasciola hepatica.* Veterinary Parasitology 4: 341—351.
Hope-Cawdery, M. J., G. Gettinby, and J. N. R. Grainger. 1978. Mathematical models for predicting the prevalence of liver fluke disease and its control from biological and meteorological data. In. Weather and Parasitic Animal Disease. W.H.O. Technical Note No. 159. (ed. T. E. Gibson), 21—39.
Michel, J. F. 1963. the phenomenon of host resistance and the course of infection of *Ostertagia ostertagi* in calves. Parasitology 53: 63—84.
Morris, R. 1969. Assessing the economic value of veterinary services to the primary industries. Australian Veterinary Journal 45: 295—300.
Ollerenshaw, C. B. 1966. The approach to forecasting the incidence of fascioliasis over England and Wales 1958—1962. Agricultural Meteorology, 3: 35—63.
Ollerenshaw, C. B. 1974. Forecasting liver fluke disease. In. The effects of Meteorological Factors upon Parasites. Symposium of the British Society for Parasitology Vol. 12. (ed. A. R. Taylor. and R. Muller) Blackwell Scientific Publications.
Ollerenshaw, C. B., E. G. Graham, and L. P. Smith. 1978. Forecasting the incidence of parasitic gastroenteritis in lambs in England and Wales. Veterinary Record 103: 461—465.
Over, H. J. and J. Dijkstra. 1975. Infection rhythm in fascioliasis, In Facts and Reflections II (Lelystad Workshop on Fascioliasis) (ed. H. J. Over and J. Armour).
Paton, G., R. J. Thomas, and P. J. Waller. 1984. A prediction model for parasitic gastroenteritis in lambs. International Journal for Parasitology 14: 439—445.
Roberts, M. G., J. R. Lawson, and M. A. Gemmel, M. A. (1987) Population dynamics in echinococcosis and cysticercosis: mathematical model of the life-cycles of Taenia hydatigena and T. ovis. Parasitology 94, 181—197.
Ross, J. G. 1967. An epidemiological study of fascioliasis in sheep. Veterinary Record 80: 214—217.
Smith, G. 1982. An analysis of variations in the age structure of *Fasciola hepatica* populations in sheep. Parasitology 84: 49—61.
Smith, G. 1984a. Analysis of anthelmintic trial protocols using sheep experimentally or naturally infected with *Fasciola hepatica.* Veterinary Parasitology 16: 83—94.
Smith, G. 1984b. Chemotherapy of ovine fascioliasis: use of an analytical model to assess the impact of a series of discrete doses of anthelmintic on the prevalence and intensity of infection. Veterinary Parasitology 16: 95—106.
Smith, G. 1984c. The impact of repeated doses of an anthelmintic on the intensity of infection

and age structure of *Fasciola hepatica* populations in sheep. Veterinary Parasitology 16: 107—115.

Smith, G. 1984d. Density dependent mechanisms in the regulation of *Fasciola hepatica* populations in sheep. Parasitology 88: 449—461.

Smith, G. 1988. The population biology of the parasitic stages of *Haemonchus contortus.* Parasitology 96: 185—195.

Smith, G. and B. T. Grenfell. 1985. The population biology of *Ostertagia ostertagi.* Parasitology Today 1: 76—81.

Smith, G. and D. T. Galligan. 1988. Mathematical models of the population biology of *Ostertagia ostertagi* and *Teladorsagia circumcincta*, and the economic evaluation of disease control strategies. Veterinary Parasitology 27: 73—83.

Smith, G., B. T. Grenfell, and R. M. Anderson. 1986. Development and survival of the non-infective free living stages of *Ostertagia ostertagi.* Parasitology 92: 471—482.

Smith, G., B. T. Grenfell, and R. M. Anderson. 1987a. The regulation of *Ostertagia ostertagi* populations in calves: density dependent control of fecundity. Parasitology 95: 373—388.

Smith, G., B. T. Grenfell, R. M. Anderson, and J. B. Beddington. 1987b. Population biology of *Ostertagia ostertagi* and anthelmintic strategies against ostertagiasis in calves. Parasitology 95: 407—420.

Thomas, R. J. and J. R. Starr. 1978. Forecasting the peak of gastrointestinal nematode infection in lambs. Veterinary Record 103: 465—468.

Wilson, R. A., G. Smith, and M. R. Thomas. 1982. Fascioliasis. In, Population Dynamics of Infectious Diseases: Theory and Applications (ed. by R. M. Anderson). Chapman and Hall, London, pp. 262—304.

Appendix

The solution to equations (1) to (3) in the main text is as follows:

Assumptions: $\tau_1 = 0$, $H(t) = H$, a constant.

$$C(t) = C(0)\exp(-\mu_1 t) \tag{A1}$$

$$I(t) = \frac{\beta HC(0)\exp(-\mu_2 t)\,[1 - \exp(\mu_1 + \beta H - \mu_2)t]}{(\mu_1 + \beta H - \mu_2)} \quad \text{for } t < \tau_2 \tag{A2}$$

$$I(t) = \frac{\beta HC(0)\exp(-(\mu_1 + \beta H)t)\,[\exp((\mu_1 + \beta H - \mu_2)\tau_2) - 1]}{(\mu_1 + \beta H - \mu_2)} \quad \text{for } t > \tau_2 \tag{A3}$$

$$M(t) = \frac{\beta HC(0)\exp((\mu_1 + \beta H - \mu_2)\tau_2 - (\mu_1 + \beta H)t)\,[\exp((\mu_1 + \beta H - \mu_3)(t - \tau_2) - 1]}{(\mu_1 + \beta H - \mu_2)} \quad \text{for } t > \tau_2 \tag{A4}$$

9. Ecological theory and biological control

WILLIAM W. MURDOCH

Abstract

Successful classical biological control in long-lived ecosystems occurs when an imported natural enemy keeps the density of the alien pest insect below the density at which it causes economic damage. It has been generally accepted for some 75 years that such enemies establish a low stable equilibrium pest density and maintain it by imposing density-dependent mortality on the pest. In recent years it has been suggested that success typically involves aggregation by the enemy, perhaps to those patches that contain more pests. This theory is embodied in Nicholson-Bailey models of the parasitoid-host interaction.

This chapter reviews this body of theory and compares it with real systems. It appears, at least in some cases, that the basic premise of the theory — a stable interaction on the small scale — does not hold. In the few examples studied, it also appears that the mechanisms proposed in the models, including aggregation, do not account for control. In the one apparently stable system studied — red scale on citrus controlled by the parasitoid *Aphytis* — a refuge may be the key stabilizing factor; otherwise this interaction also has features implying instability. In another case the maintenance of a pest weed (ragwort) in the face of enemies driving it extinct may depend upon an invulnerable pest stage (the seed bank).

Few real systems have been analyzed and it should not be assumed that the above results hold in general. Stability may be common in some circumstances, and it seems likely that aggregation to local pest density will be important in some instances.

There is a strong trade-off between stability and degree of pest suppression in the Nicholson-Bailey models discussed. The trade-off is associated with the lack of within-generation dynamics in these models: stability arises from density dependence in the parasitoid (decreasing efficiency with increasing parasitoid density) rather than in the pest. This process operates and the trade-off is severe even when the parasitoid aggregates to patches that (initially) contain more pests. By contrast, such aggregation in a model that allows the parasitoids to redistribute themselves in response to the continually-changing pest distribution results in better control, but can also be destabilizing.

The consequences of these results for biological control in theory and practice are discussed. More work is needed to expose the mechanisms operating in real examples of successful control. For systems in which control involves local instability, further modeling of ensemble dynamics should help pinpoint possible key features that may allow both regional persistence of the system and consistently low pest densities.

S. K. Jain and L. W. Botsford (eds), Applied Population Biology, 197—221.

Introduction

Successful biological control occurs when a "natural enemy" is able to drive the density of a pest species below the level at which it causes economic damage, and keeps it consistently below that density. In insect pests, enemies are often parasitoids (insects that develop in or on a single member of the pest species) or predatory insects, but they may also be fungi, nematodes, bacteria, or other types of organisms; in weed pests, the natural enemy is usually a fungus or an herbivorous insect. Here I will be concerned only with insects as natural enemies, and will concentrate mainly on parasitoids, for which most of the theory has been developed.

We will also concentrate on long-lived agricultural ecosystems such as orchards, forests and pastureland, even though biological control agents occur or are widely used in other agricultural situations. This restriction is necessary because the theory of population ecology hinges on the concept of mathematical stability, which is a long-term phenomenon. The models thus deal implicitly with populations that persist through many generations.

Both practicing entomologists and ecologists have written extensively over the past 75 years or so on the theoretical basis for successful control and, surprisingly, they have almost always agreed on a principle that ostensibly underlies successful control: enemies control pests by reducing their density to a low stable equilibrium level. Excursions of the pest from this density are curtailed by the enemy, whose attacks tend to return the pest population to its equilibrium. To explain successful control, therefore, we need to explain the stability of the interaction.

In the past decade or so a second principle has been added to the theory: successful enemies are those that aggregate in patches that have the most pests (Beddington *et al.* 1978; Hassell 1978). That predators and parasitoids might behave in this way is an eminently sensible idea that is supported by observations of individual behavior in a number of predatory insects.

In this chapter I will examine these ideas, together with some other possible explanations for successful control, by looking at evidence from real systems. First, however, I present a brief history and explication of the theory, in which I will survey only major models that have been tested against data from field studies of biological control. At the end of the presentation of each model, I will describe the type of field data predicted by the model, and against which it can be tested. It is a pleasure to record that many of the published results to be reviewed resulted from work done in collaboration with Drs. John Reeve, and Peter and Jean Chesson.

Theory of biological control

Density dependent parasitism in time

There has been a close connection between classical biological control and ecological theory from their inception. Two entomologists first introduced the idea of density dependence (Howard and Fiske 1911), and another gave the idea its name (Smith 1935). Howard and Fiske observed that, as a pest population varies through time, it will increase without limit unless the fraction killed by the natural enemy increases with pest density. Smith gave the name density dependence to this positive relationship.

Howard and Fiske's verbal model is easy to test in the field. We expect the fraction of the pest population parasitized in any given generation to be higher if the pest is more dense in that generation. (In some cases this relationship might contain a time delay.)

Parasitoids evenly distributed in space

The next major advance was the creation of the basic model of parasitoid-pest interactions (Nicholson and Bailey 1935). A general form of their equations is

$$H(t+1) = FH(t)f[P(t)]$$

$$P(t+1) = H(t)\{1 - f[P(t)]\}, \qquad (1)$$

where H(t) and P(t) are the number of hosts and parasitoids in generation t, F is the factor of increase of the pest per generation in the absence of the parasitoid, and f(P) (a number that varies between 0 and 1) is the fraction of the host population escaping parasitism. Each parasitized host is assumed to give rise to a single new parasitoid (hence the fraction giving rise to a parasitoid is 1 − f). The parasitoid and pest are assumed to have non-overlapping and discrete generations, which is perhaps best approximated by annual insects in temperate climates.

The specific model proposed by Nicholson and Bailey is the simplest that one might consider, being a discrete analog of the well-known Lotka-Volterra model. It is

$$H(t+1) = H(t)Fe^{-aP(t)}$$

$$P(t+1) = H(t)\,(1 - e^{-aP(t)}), \qquad (2)$$

where *a* is the attack rate of the parasitoid, or equivalently the vulnerability of the pest, or a combination of both. The fraction of pests that escape parasitism, $f(P) = e^{-aP(t)}$, can be derived in a number of ways. It is very common to assume that attacks are distributed among hosts at random, according to a Poisson distribution, $e^{-aP(t)}$ being the zero term of that distribution, i.e. the fraction without attacks. The parasitoid is assumed to require a negligible time to attack, "handle" and parasitize each pest, and is assumed not to run out of eggs.

There is no direct density dependence in the model, in either the pest or the parasitoid population. The parameter a is the instantaneous death rate per head of the pest, per parasitoid; it also determines the instantaneous rate of production of new parasitoids, per parasitoid; and it is a constant that does not vary with the density of either the pest or the parasitoid. Notice also that the fraction of pests surviving parasitism, f, does not depend on current pest density, only on that of the parasitoid.

Virtually all later models in the Nicholson-Bailey family have this same general structure. Only the term, f(P), describing the fraction of pests escaping parasitism varies, allowing the introduction of density dependence in the parasitoid population, and henceforth we will write only this term. Other versions do exist in which density dependence of the pest is introduced by making F a function (e.g. logistic) of pest density. Here we omit these models as our interest is in how an otherwise unlimited pest is controlled or stabilized by parasitism. This seems appropriate for most pests since few are likely to be self-limited (e.g. via the food supply) over the range of densities acceptable as successful biological control. The pest density might also directly affect the fraction surviving parasitism, i.e., f = f(H, P); in most models, however, there is no dependence on pest density.

The dynamics of the models are examined by local stability analysis, i.e. by perturbing the densities a small distance from equilibrium and asking whether the populations tend to return to equilibrium. The Nicholson-Bailey model is unstable, the populations oscillating with increasing amplitude until both go extinct. Because of this, the model has always been rejected as an explanation for successful control, since success has been equated with stability. Ironically, Nicholson was a major proponent of the idea that populations in nature are stable (Varley *et al.* 1973).

The Nicholson-Bailey model can be tested against the pattern of parasitism seen in field samples, such as fruit, twigs or branches. In each sample we determine the fraction of pests that is parasitized. The assumptions of the model predict that the frequency distribution of the fraction parasitized will be binomial (i.e. it will have a humped shape similar to, although not exactly the same as, the taller histogram in Fig. 1b), because if all pests are visited by the same number of parasitoids but attacks succeed or fail at random then each sample is a set of independent Bernoulli trials (Murdoch *et al.* 1984). Thus the Nicholson-Bailey model is consistent with the assumption that parasitoids are distributed evenly in space.

Parasitoid attacks concentrated in some pest individuals

In 1962 Nicholson and two colleagues produced versions of the model that predicted stable populations (Bailey *et al.* 1962). As noted by May (1978), the biological implications were not emphasized in this milestone paper, which therefore has been given less attention than it deserves.

The basic idea in the 1962 models is that pest individuals vary in their susceptibility to parasitism. The authors suggested variation in the quality of

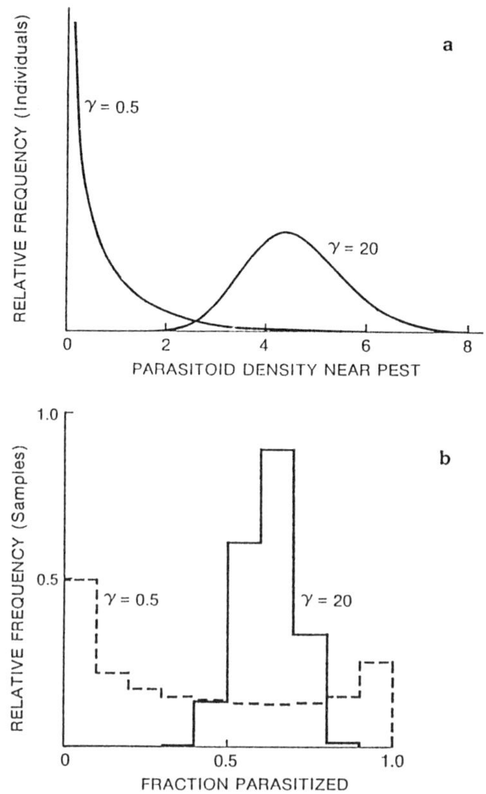

Fig. 1. Distribution of risk among hosts, and of parasitism among samples, when parasitoids aggregate independently of local host density. (a) is the theoretical distribution of parasitoid density in the vicinity of individual hosts, when the distribution is gamma (see text). When τ = 20, most of the host population lives near an intermediate density of parasitoids; when τ = 0.5, a few hosts are close to high parasitoid densities while most hosts are virtually unvisited by parasitoids. (b) is the expected frequency of samples with a given fraction of parasitized hosts.

hiding places as a mechanism, but one could think of many others, such as a distribution of phenotypes (rather than just one), or of local parasitoid density (May 1978), or any process that produces variation in relative risk among hosts (Chesson and Murdoch 1986). They modeled variation in vulnerability, without reference to spatial considerations, by letting vulnerability (i.e. a) vary among pest individuals. In one version a has a gamma distribution, whose shape is controlled by a single parameter, τ.

The same model arises if we assume instead that the density of parasitoids in the vicinity of each pest is a random variable, which we designate by X, the mean local density being P(t) (Fig. 1a). Here the parasitoids are distributed independently of the pests. When the parameter is large, $\tau > 1$, the distribution is bell-shaped (e.g., $\tau = 20$ in Fig. 1a), so that most pest individuals are near an intermediate number of parasitoids, as might be expected by chance. When $\tau < 1$, however, the distribution becomes extremely skewed (e.g., $\tau = 0.5$ in Fig. 1a) so that most pest individuals have a very low probability of being attacked, while a few pest individuals receive many attacks. The fraction of the pest population escaping parasitism is now obtained by summing over individual pests:

$$f[P(t)] = S\, e^{-aX} g(X) dX, \qquad (3)$$

where g(X) is a gamma distribution, with τ its "shape" parameter and P(t) its mean.

This model is stable provided $\tau < 1$. Stability arises, however, because of direct density-dependence in the parasitoid, not the pest population. Taylor (1988) stresses the generality of this feature in current Nicholson-Bailey models. We can think of it arising because the previously constant attack rate, *a*, is replaced by a new attack rate (involving *a* and g(X)) that is a decreasing function of P(t). Thus, as parasitoid density increases over time, each individual parasitoid becomes less efficient (because it is re-attacking already parasitized pests more frequently than it does in the basic Nicholson-Bailey model). The pest population thus gains an increasing degree of protection from the average parasitoid as parasitoid density increases. (The rate of production of parasitoids, per parasitoid, also declines.) This has been called "pseudointerference" because it operates as though parasitoids waste increasing amounts of time actually interfering with each other as their density increases (Free *et al.* 1977).

May (1978) pointed out that the model can be rewritten by recognizing that, with g(X) a gamma distribution, f(P) is the zero term of the negative binomial distribution, which can be written

$$f(P) = [1 + aP(t)/\tau]^{-\tau}, \qquad (4)$$

where τ now measures the degree of aggregation when attacks are distributed among pests according to the negative binomial.

The model is difficult to test with field samples under some circumstances. It predicts the risk of parasitism run by individual hosts, but that risk is hard to measure because in practice each host is either parasitized or not. The problem is solved if each attack leaves a record (e.g. a single egg), but this is not typical of parasitism. Where hosts that are close together tend to run a similar risk, however, we can estimate variation in individual risk by estimating the fraction parasitized in a local sample of hosts. This condition will be met if, for example, an uneven distribution of parasitoids, or of physical refuges or other spatial features, is the source of heterogeneity. Reeve and Murdoch (1985) showed that when c is very large, the fraction of parasitized pests in samples should approximate a binomial distribution, as in the Nicholson-Bailey case. When $\tau > 1$ but not very large, the distribution will usually be humped, although not exactly binomial (e.g., $\tau = 20$ in Fig. 1b). In the stable case, when $\tau < 1$ (and the attack rate is not very low) we expect a peculiar, U-shaped, frequency distribution that is the "opposite" of the binomial — the fraction parasitized in many samples is low, in some samples it is high, and it is intermediate in relatively few samples (e.g., $\tau = 0.5$ in Fig. 1b).

Density dependent parasitism in space

A more biologically appealing model was developed by Hassell and May (1973) (these authors also explore interesting alternative formulations in Hassell and May 1974). Some pest individuals are again more vulnerable than others, but in this case vulnerability depends on the spatial distribution of both the pest and the parasitoid: pests in high-density patches are more vulnerable.

The pest is assumed to occur in patches. It is further implicitly assumed that, at the start of the generation, the parasitoid evaluates patches and spends its time preferentially in those patches that initially contain more pests. In particular, the fraction of the parasitoid population searching in a patch is assumed to increase with the initial fraction of the pest population that is found there. (This requires the parasitoid to evaluate not only the density of hosts in a patch, but also to compare this with the average density). The result is that the fraction parasitized in a patch increases with initial pest density in the patch.

In this case,

$$f(P) = \sum^{n} [\alpha_i e^{-\beta_i aP(t)}], \qquad (5)$$

where α_i and β_i are, respectively, the fractions of pests and parasitoids in patch i, and $\Sigma\alpha_i = \Sigma\beta_i = 1$. β_i is an increasing function of α_i. Equation (5) indicates that the fraction of the pest population escaping parasitism is now obtained by summing over all n patches in the habitat.

The model may or may not produce a stable interaction, depending on its detailed assumptions, and unfortunately the local stability properties are known only for a specific form of α_i and β_i. Stability is more likely, however, if the pest is strongly clumped in space and the fraction of parasitoids found in a patch is not only an increasing, but is a rapidly accelerating function of the initial fraction of the pests that occur in the patch (see Hassell and May 1973, Figs. 12 and 13). Stability, as in the previous model, arises because of density dependence in the parasitoid population: "pseudointerference" increases with parasitoid density (Free *et al.* 1977).

It has sometimes been assumed that the more general but perhaps less realistic model described by equation (4) can be taken to describe the results of the aggregative behavior described in equation (5). Chesson and Murdoch (1986), however, have shown that this is not usually valid; model 4 assumes that parasitoid attacks are distributed independently of pest distribution, while model 5 assumes that attacks are concentrated in patches with more prey.

Hassell (1984) and, using a different approach, Chesson and Murdoch (1986), show that parasitism decreasing with host density in the patch (i.e. inversely density dependent parasitism in space) can have precisely the same effect on stability as direct spatially density dependent parasitism. This serves to emphasize two general points: (1) in these models it is the distribution of parasitism, not of the parasitoids, that is the key to stability (Chesson and Murdoch 1986); (2) there is no simple link between spatial density dependence and temporal density dependence, or stability.

The model of equation (5) makes a clear assumption that can be tested in the field. Samples from the habitat should show strong spatial density dependence (or inverse density dependence) in the parasitism rate, provided the samples are not all taken within one or a few patches or, alternatively, do not lump together many patches in a sample. The model does not explicitly address what a patch is (indeed the model need not have a spatial interpretation). Presumably, however, a patch must be small enough that the average parasitoid could visit at least several in its lifetime, but large enough to signal to the searching parasitoid an initial reward rate that differs from other patches.

Notice that, while the models of equations (4) and (5) rest on ideas about the behavior of parasitoids, they can be tested using data that are much easier to obtain, namely the distribution of parasitism. By contrast, Taylor (in press) has developed an interesting model of competition among parasitoid larvae in which the distribution of the number of attacks, not merely of the fraction parasitized, influences stability and needs to be determined.

Density dependent parasitoid sex ratio

Hassell *et al.* (1983) show that several models gain stability if the fraction of females in the parasitoid population declines as either the parasitoid density or the ratio of parasitoids to pests increases. We can test this model against field data simply by looking at these relationships in samples taken at different times.

A useful aspect of this model is that it makes explicit the temporal density dependence in the parasitoid population that is fundamental to the two models dealing with aggregation: because only female parasitoids parasitize, the instantaneous attack rate of the average parasitoid decreases as parasitoid density increases.

Other mechanisms

Other stabilizing mechanisms have been incorporated into the Nicholson-

Bailey framework, but they have not been considered important to biological control. Refuges have long been recognized in both field and modeling studies in ecology as a potentially important stabilizing factor. In Nicholson-Bailey models a fixed number of spatial refuges for the host is strongly stabilizing Hassell (1978), and Beddington *et al.* (1978) suggested it as a possibility worth exploring in successful cases of biological control. Time wasted by parasitoids when they are sufficiently abundant can act to stabilize models by causing actual (rather than "pseudo") interference to increase with increasing parasitoid density (Hassell and Varley 1969). It is not generally thought, however, that this mechanism is important in the field (Free *et al.* 1977).

Finally, "transit time" is known to be stabilizing in a range of both parasitoid-host and predator-prey models (Hassell and May 1974; Murdoch and Oaten 1975; Oaten 1977). It arises when time is required for the parasitoid to move from one patch to another, and can be stabilizing when the parasitoid stays longer in patches that have more hosts, thus wasting less total time in transit when average host density is higher.

Aggregation and pest density

Parasitoid attacks concentrated in some pest individuals

While aggregation can be a powerful stabilizing mechanism in Nicholson-Bailey models, it has the unfortunate effect of greatly increasing the pest equilibrium density. That this should be the case is most intuitively obvious in model (3), where the parasitoid concentrates its attacks on some pest individuals regardless of their local density. For stability the model requires $\tau < 1$, and for τ markedly less than 1 the parasitoid tends to ignore most pest individuals; not surprisingly, the pest equilibrium becomes extremely high (May 1978, Fig. 1; Hassell 1978; Murdoch *et al.* 1984). The pest equilibrium can be depressed to realistic levels by adding linear density dependence (i.e. a logistic term) to the pest population (Hassell 1978), but such self-damping by the pest is probably not common near economically acceptable pest densities.

Thus, while aggregation independent of local pest density may contribute to stability in real systems, we would not expect it to be the major process operating in the interaction between an enemy species and the pest under its control.

A critical feature of the Nicholson-Bailey framework underlying the inefficiency in this and other aggregation models is the absence of within-generation dynamics; that is, parasitoids distribute themselves according to the *initial* host distribution (or host type, or environmental feature, etc.) at the start of each generation, and do not redistribute themselves to attack new hosts as those being visited accumulate superfluous attacks. (See Comins and

Hassell 1979 for an alternative approach). This process is more easily seen to operate in aggregation to local host density, which we explore next.

Density dependent parasitism in space

Although it may seem paradoxical, in the Nicholson-Bailey framework parasitoid aggregation to areas where pests are relatively abundant also appears to cause pest equilibrium density to increase. The basic mechanism causing this relationship is that, as aggregation becomes stronger and stronger, more and more of the parasitoids are confined for the entire generation to a small fraction of the (initially most dense) patches, and only a tiny fraction is available to attack pests in the remaining patches. Even though this can result in the virtual extermination of the pests in the dense patches, those in the more sparsely populated patches are almost exempt from attack, and a very large parasitoid population is required to prevent them from increasing. This in turn requires a large (but non-increasing) pest population to sustain the parasitoids.

The operation of this mechanism can be seen most easily from an analysis of a particular case (Hassell and May 1973). Although the details of this example are complicated, the basic message is clear: there is a severe trade-off between stability and suppression of the pest, such that as parasitoid aggregation is increased to achieve stability, pest density increases enormously (Fig. 2).

To obtain an analytically tractable form of model (5), Hassell and May (1973) assumed a particular function relating the fraction of parasitoids in a patch (β_i) to the fraction of pests there (α_i). They assumed, specifically, that a fraction, α, of the pests occur in one high-density patch and that each of the remaining (n − 1) patches contains the fraction, $\alpha_i = (1 - \alpha)/(n - 1)$, of the remaining pest population. The high density patch is in turn assumed to contain a fraction, α, of the parasitoids, the remaining low-density patches each containing a fraction $\beta_i = \varepsilon\beta$ of the remaining parasitoid population, so that $\beta = [1 + \varepsilon(n - 1)]^{-1}$. Aggregation was obtained by ensuring that the fraction of parasitoids in the high-density patch is greater than the fraction in each of the low density patches, by setting $\varepsilon = (\alpha_i/\alpha)^l$, where $\mu > 0$ is the index of aggregation; $\mu = 0$ implies that all patches contain the same fraction of parasitoids (in which case we recover the basic Nicholson-Bailey model), while larger values of μ imply that a larger fraction of the parasitoids is concentrated in the high-density patch.

Hassell and May (1973) found that stability is greater the larger μ is. Stability is also greater the more aggregated the *pest* is. Aggregation in the pest can be increased simply by increasing the number of patches (n) in the environment, thereby increasing the relative difference in pest (and parasitoid) density between the high- and low-density patches. Then, for each particular set of values of α, μ and (n − 1), the model is stable for a particular range of values of the pest rate of increase, F.

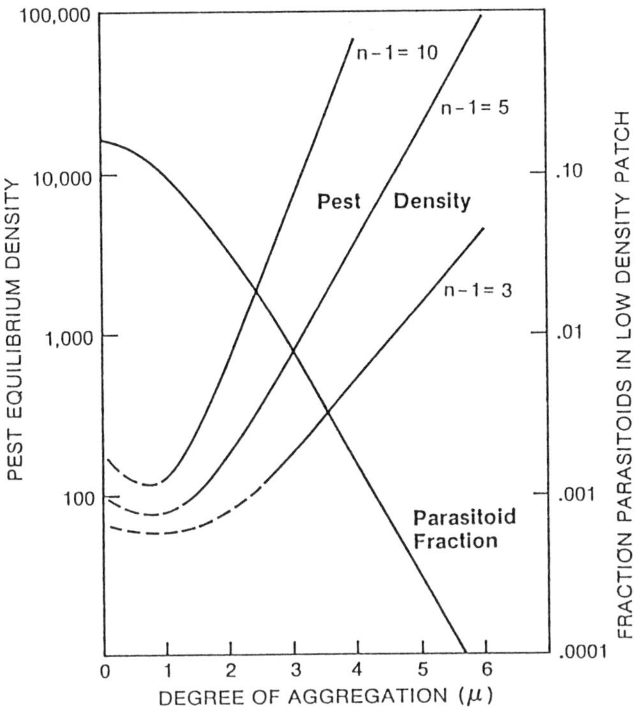

Fig. 2. Trade-off between stability and pest control in aggregating parasitoids. Equilibrium pest density increases with degree of aggregation by the parasitoid, and with level of pest aggregation (n − 1 = 3, 5 and 10), while fraction of parasitoids in the average low-density patch decreases with aggregation (for n − 1 = 5). Note the logarithmic scale in both cases. Dashed curve indicates unstable, solid curve indicates stable pest equilibrium densities. $\mu = 0$ is the basic Nicholson-Bailey case. See text for fuller explanation.

Hassell and May (1973) found that stability is greatest for an intermediate case in which half the pest population is in the high density patch ($\alpha = 0.5$). Stability in this case can be achieved, for F lying between 1 and about 4, for various combinations of values of μ and n − 1. In particular, as pests become more concentrated in space (n − 1 increases), stability can be achieved with less aggregation by the parasitoid (smaller μ), for a particular value of F. For example, for (n − 1) lying between 3 and 10, the model is stable provided μ has a minimum value that lies between 1 and 7, depending on the particular values of F and (n − 1) (Hassell and May 1973, Fig. 12).

Figure 2 illustrates the trade-off between stability and pest suppression for the case of F = 3, which is the value of the pest rate of increase that makes it easiest to achieve stability. Each curve shows that the pest equilibrium actually declines slightly as μ increases initially from 0. (It reaches a minimum at $\mu = 1$, when the ratio of pests to parasitoids is the same for all patches.) These declines, however, typically occur for conditions in which the system is unstable (dashed curves). As μ increases, the pest equilibrium eventually becomes stable, but also soon increases exponentially with μ, thus illustrating the trade-off between pest suppression and parasitoid aggregation. The figure also shows that increased pest density is also the cost of stability obtained as pest aggregation (n − 1) increases.

The reason for the trade-off is analogous to that in the case of aggregation independent of local host density. As μ increases, the fraction of the parasitoids in the high density patch increases and rapidly approaches 1; correspondingly, the fraction in each sparse patch approaches 0 (the decreasing curve in Fig. 2). Since there is no rearrangement of parasitoids within each generation regardless of parasitoid density, the fraction of hosts escaping parasitism in the dense patch asymptotically approaches 0, while that in the sparse patches approaches 1. An enormous total parasitoid population is therefore required so that enough parasitoids exist in the low-density patches to keep the pest population there from expanding without limit. An enormous total pest population is therefore needed to maintain the parasitoids. Again, as n − 1 increases, the fraction of the parasitoids in each low-density patch is decreased and this allows an even greater fraction of pests to escape parasitism, for a given degree of parasitoid aggregation.

The underlying processes can be seen clearly in the following analysis. The equilibrium densities, H and P, are obtained, by iteration, from

$$F^{-1} = \alpha e^{-a\beta P} + (n-1)\alpha_i e^{-a\beta_i P}, \qquad (6)$$

and H = PF/(F − 1).

As μ increases (for $\alpha > 1$), aP necessarily becomes large and the first term in (6) goes to 0 (because ε goes to 0 and hence β goes to 1). The parasitoid equilibrium is therefore determined solely by the second term in (6). At the same time, $\beta_i = \varepsilon\beta$ goes to ε. The parasitoid equilibrium therefore closely approximates $P = c\varepsilon^{-1}$, where c is the positive constant $[(\mu n F^{-1})/ - \alpha]$; i.e., $P = c(\alpha/\alpha_i)^{\mu}$, and therefore grows exponentially with μ at rate $\alpha/\alpha S_i$.

Aggregation by parasitoids in response to absolute pest density in a patch, rather than to relative density there, may lead to a weaker trade-off between pest suppression and parasitoid aggregation (Reeve 1988).

Tests of the theory

Four mechanisms discussed above have been considered important in promoting stability in successful biological control systems: parasitism density dependent through time, parasitism highly concentrated in a small fraction of pest individuals, parasitism density dependent in space, and (less importantly) parasitoid sex-ratio density dependent through time. We are now armed with the predictions needed to test whether, in particular cases of biological control, these mechanisms operate as suggested in the models. First, however, we need to determine whether the basic premise of the theory holds: do parasitoids achieve success through creating a stable interaction with the pest?

Stability

This question was examined by Murdoch *et al.* (1985) who examined seven "classic" cases of successful biological control plus some results from control of mosquitoes. We looked for evidence of a stable (or even an unstable, e.g. cyclic) equilibrium. If neither was present we asked if there was evidence of non-equilibrium dynamics: random walk or trends in the populations, including trends towards zero, or evidence of the pest or parasitoid going extinct locally. The answer is that there was almost no evidence for stability, there was good evidence for non-equilibrium dynamics, and in some cases good evidence for local extinction or near-extinction (Table 1).

One case was an exception: there was moderately convincing evidence that red scale controlled by *Aphytis* in citrus groves in California is a stable system. Notice that the olive scale, one of whose parasitoids is also an *Aphytis,* seems by contrast to be unstable (Table 1). We will return to this point.

Table 1. Dynamic behavior of successful biological control systems. See text for explanation. Based on Murdock *et al.* (1985).

Pest (agent)	Place	Stable equilibrium	Unstable equilibrium	Non-equilibrium	Extinction
Mosquitos (*Gambusia*)	California			+	+
Mosquitoes (*Notonects*)	California			+	+
Cottony cushion scale (*Vedalia*)	California			+	+
Larch sawfly (*Olescicampa*)	Canada			+	+?
Olive scale (*Aphytis)*	California			+	?
Walnut aphid (*Trioxys*)	California		?	?	
Winter moth (*Agrypon, cyzaenis*)	Canada	?	?	?	?
Red scale (*Aphytis*)	California	+			
Red scale (*Aphytis*)	Australia	No data			

These conclusions must be tempered with the caution that stability and instability are difficult to establish in the field, and that there are very few good population data available. There is also the issue of what we mean by "local" extinction: obviously there is a spatial scale so small (e.g. the individual) that extinction is both inevitable in a short ecological time and uninteresting, and a scale that is so large (the ensemble of all the populations of a species) that extinction is extremely unlikely and stability (or at least relative constancy) is highly probable. The solution is that we need to look for stability on the same spatial scale as that implied by the mechanisms in the models which, in this case, involve the behavior of individual parasitoids (Murdoch *et al.* 1985).

Mechanisms

The red scale-*Aphytis* system seems stable and hence is an appropriate place to test the four hypotheses contained in the theory. They were tested by sampling the scale and the parasitoid in a lemon grove in southern California over a period of two years, or about eight generations (Reeve and Murdoch 1985, 1986). We confirmed that the system in this grove was stable. Nevertheless, all four of the hypotheses were rejected.

1. Parasitism varied through time, but was independent of pest density; furthermore, it was a source of instability since there was random variation in parasitism through time, unrelated to pest density (Reeve and Murdoch 1986).
2. Parasitism rates in red scale not only were non-skewed as required by the Bailey *et al.* model (equation 3), their distribution among samples was invariably hump-shaped, indicating that $\tau > 1$. Indeed, the distributions were either binomial or close to it, indicating a fit to the simple and unstable Nicholson-Bailey model. The same circumstances held in the successful control of olive scale by *Aphytis pseudomaculicornis* and the parasitoid *Coccophagoides utilis* (Table 1) (Murdoch *et al.* 1984).
3. There was no tendency for parasitism rates to be higher where the pest was denser, regardless of the spatial scale on which the hypothesis was tested. Occasionally there was inversely density dependent parasitism at the smallest spatial scale, but this was always weak. Smith and Maelzer (1987) studied the same system on Citrus in Australia and showed that parasitism was not spatially density dependent there either (although there was weak spatial density dependence in the distribution of parasitoids). An analysis of data from the olive scale showed spatial density dependence in only one case out of eight (Murdoch *et al.* 1984).
4. The sex ratio of *Aphytis* on red scale was also not density dependent.

These results present something of a paradox since they are consistent with the unstable Nicholson-Bailey interaction, yet the population dynamics provide evidence for stability. Some other factor must be causing stability in the red scale system. We considered two.

An invulnerable age class

Immature stages of the red scale are vulnerable to parasitism but the adult reproductive females are not. This might stabilize the interaction since the adult stage might operate rather like a refuge, continuously leaking young (crawlers) into the interaction, and thus sustaining both the scale and the parasitoid.

This possibility was explored by Murdoch *et al.* (1987) in a model with the following structure. Adult scale produce immature scale (crawlers) which may mature to become adults in T_1 days (probably about 30 in summer in our grove), or they may be parasitized by randomly-searching parasitoids, or they may die from some other cause. Once adult, they die off at a constant rate, but do not live beyond T_A days (again, probably about 30). Parasitized scale constitute the immature parasitoid population, which again dies off at a constant rate, survivors maturing to adults in T_2 (around 10) days. Adult parasitoids die off at a constant rate.

Unlike Nicholson-Bailey models, this one assumes that generations overlap, and it recognizes adult and juvenile stages. Hastings (1983, 1984) also explored models with this general structure, emphasizing how they can give rise to complex dynamic behavior. The model without an invulnerable age class, however, has the same basic assumption, also in the Lotka-Volterra model, concerning randomly-searching parasitoids. Like the Lotka-Volterra, the model without an invulnerable class is neutrally stable, the oscillations around equilibrium neither increasing nor dying away with time.

The system is described by four simultaneous, time-delayed differential equations. Local stability analysis shows that the invulnerable class tends to add stability, in the sense that no stability is possible if this stage is absent whereas stability is possible when the stage is present. However, stability requires that the invulnerable stage be long relative to the parasitoid's time lag ($T_A > T_2$). A "guesstimate" of the parameter values obtaining in our grove suggests that there the adult stage does not last long enough to confer stability (cycles are predicted). The model does suggest, however, that the system is more stable than it would be without the invulnerable class. As with the Nicholson-Bailey models, there is a trade-off between control and stability: the longer the invulnerable class lasts the higher is the equilibrium pest density.

A physical refuge

Towards the end of the lemon grove study we discovered a physical refuge. Scale were usually sampled by the standard procedure of selecting at random from an exterior hemisphere made up of twigs bearing leaves and fruit. We noticed, however, that the unsampled trunk and interior scaffolding branches appeared rough, and on close inspection they turned out to be covered with live and dead scale. Sampling in this interior region, and in the outside of the tree where standard samples are taken, showed that the scale was several orders of magnitude more dense in the inside, and that parasitism there was negligible.

These observations suggest the hypothesis that stability is caused by the existence of a refuge population leaking crawlers from the interior to the outside of the tree, which otherwise would support an unstable interaction between *Aphytis* and scale. Reeve and Murdoch (1986) provide some evidence to support the hypothesis, and several models in ecology demonstrate that physical refuges can be strongly stabilizing (e.g. Murdoch and Oaten 1975; Hassell 1978). [McNair (1986), however, shows that this is by no means a universal property of refuges, which may even be destabilizing in some circumstances.]

Argentine ant runs were common on the interior of the tree, which suggested that the ants might be interfering with *Aphytis,* and hence creating the refuge. This hypothesis is being tested in a grapefruit grove in the same area (unpub. data of W. Murdoch, R. Luck and J. Reeve). Ants were removed from some blocks of trees and left in others. Preliminary results show that the density of scale is higher in trees with ants than in those without ants, providing support for our hypothesis that ants protect the scale from one or more natural enemies. However, the situation does not appear to be straightforward: the effect of ants on *Aphytis* seems small, and some other factor(s) must play a role in creating and maintaining the refuge.

In a second experiment we removed the refuge population in some trees, by scraping the interior surfaces. In subsequent months the *exterior* population was lower in trees with a reduced refuge population, thus supporting the idea that the refuge population feeds the exterior population.

These results, of course, do not establish that the refuge causes the system to be stable, although they are consistent with that hypothesis.

Control of ragwort

The control of the ragwort weed, *Senecio jacobaea,* in Oregon is being studied in an interesting experimental program by Dr. Peter McEvoy, who kindly provided the following detailed account (see also McEvoy 1987). This case provides comparisons with both red scale and other insect examples discussed above.

Ragwort is a weed of pastures introduced from Europe to the western U.S. in the early 20th century. The cinnabar moth, *Tyria jacobaea,* whose caterpillars strip the above-ground shoots in summer, was introduced from France in 1959. It is efficient at attacking the large reproductive individuals, although these typically develop regrowth shoots, but frequently misses smaller vegetative plants. The moth was unable to depress the weed to acceptably low levels, and subsequently the flea beetle, *Longitarsis jacobaea,* was introduced from Italy in 1969. Its larvae tunnel into the petioles, stems and roots in winter, and its actions are thus complementary to those of the cinnabar moth. The beetle attacks all plants, regardless of their size. The two species have achieved successful control in many pastures in Oregon.

The moth was released at McEvoy's study site, a very densely infested pasture, in 1978, and the beetle in 1979—80. By 1983 the moth and beetle had increased enormously and the ragwort population had been driven to a very low density. Perhaps the most striking fact has been that in the pasture in each of the four years since 1983 no plant has survived to become reproductive. The population has persisted, nevertheless, because of the existence of an invulnerable and long-lived stage — the seed bank with dormant seeds that can live below ground for many years and germinate when the soil is disturbed. These have produced plants each year but, as noted, none has survived to reproduce. McEvoy created a dense stand of the plant in 1985, by disturbing the soil, and showed that the plants were quickly exterminated by the flea beetle. Persistence of the system, at least in this particular pasture, thus clearly depends on the invulnerable stage.

An interesting question is whether this and other cases of control of ragwort are stable local interactions between the weed and its herbivores, or whether the ragwort is being maintained by dormant seeds (which is presumably not a sustainable situation in the long run), or by immigration of seeds from other areas. Dempster (1982) believes there is local extinction and reinvasion of ragwort and cinnabar moth populations in England.

McEvoy searched for evidence of spatial density dependence in the herbivores' attacks on reproductive plants. There was no tendency for attacks to be concentrated where the plants were most dense, and all were destroyed. There thus appears to be no let up in attacks as the reproductive plants become rare. Indeed McEvoy believes that a key feature of this successful system is that the herbivores spread their attacks broadly in space so that plants cannot escape.

Success on ragwort in Oregon, as in the olive scale (Table 1), involves two complementary enemies. It is not clear, however, whether both herbivores are needed, or whether the beetle could control ragwort on its own.

Records have been kept of ragwort density in other areas where control has been carried out, and they show that the plant in many cases has continued to occur at very low densities, for up to nine years in one situation.

Implications for biological control

Stability and control

It seems that the major premise of classical biological control theory does not necessarily hold such that stability does not appear to be essential to control. The results do not show, however, that stability is never the key to control; it may well be in some situations, and only more detailed studies of many examples of biological control will resolve this issue.

The stabilizing mechanisms discussed here actually interfere with the suppression of pest density, at least in theories. In addition, in red scale, the mechanism that seems to cause stability (the physical refuge) also interferes with control. Red scale is only just under control in many groves, and probably the pest would be suppressed to much lower densities if there were no refuge (as our experiment showed) and if there were no invulnerable class. Olive scale provides a picture of what red scale might be like in that case, since the olive scale's two parasitoids are complementary, leaving no stage unparasitized, and it apparently has no physical refuge (although no one has searched for a refuge in that system). Olive scale densities are typically several hundred-fold lower than those of red scale (cf. Murdoch *et al.* 1984, and Reeve and Murdoch 1985).

The other major mechanism that has been proposed to explain stability is parasitoid aggregation. Here again, as noted above, the cost of attaining stability in Nicholson-Bailey models is much higher pest densities, even when aggregation is to local (fractional) pest density. This trade-off is associated with the absence of within-generation dynamics in the Nicholson-Bailey framework; aggregation in the Lotka-Volterra framework (Murdoch and Stewart-Oaten in prep.) has different dynamic consequences (see below).

Mechanisms of control

The mechanisms that have been proposed previously to explain stability in parasitoid-host systems (temporal density-dependence in parasitism, or aggregation of one sort or another) do not operate in the red scale system nor, to my knowledge, have they been shown to be key in any other example of successful control. Again, this should not be taken to mean that aggregation is never important in biological control. A review by Walde and Murdoch (1988) found aggregation of parasitism in areas of higher host density in about 25% of the cases examined, although none of them presented an example of successful biological control. (In some of these cases spatial density-dependence may have arisen from a tendency for parasitoids to remain in the area of their birth, rather than from aggregative behavior. This mechanism suggests the existence of ensemble dynamics, discussed below.)

Almost certainly, aggregation is important to control of some pests that are themselves extremely aggregated. Aphid pests come to mind, and there is abundant evidence that their major predators — ladybird beetles — show strong aggregation. Such behavior has been established for many years (e.g. Banks 1957), and Kareiva (1984) has shown it to occur in field experiments. Insects like aphids are so aggregated that it is not likely that a specific predator like the ladybird could persist unless it were able to seek out the dense patches. Again, while such aggregation must play a role in reducing aphid density, it remains to be demonstrated that it is a critical factor leading to stability. Indeed, to the extent that the dynamics are closer to Lotka-Volterra than Nicholson-Bailey, the aggregation is likely to be destabilizing (see below).

In our empirical studies we tested the mechanisms in the form they take in particular models. Perhaps in reality, however, these mechanisms merely contribute to but are not solely responsible for stability, and hence do not need to take on the strong forms assumed in the models. In particular, stability sometimes can be obtained with $\tau > 1$ if it is combined with an additional mechanism that is also not stabilizing on its own. For example, Bailey *et al.* (1962) showed that a model with $\tau > 1$ could be stabilized by the presence of a fractional host refuge, although the effect held only for $\tau < 2$. Taylor (1988) provides another example. (The olive and red scale results, however, suggest that $\tau >> 2$ in the field.)

The mechanisms that appear to lead to stability in red scale (the refuge and the invulnerable class) are likely to play a stabilizing role in other biological control situations. Physical refuges may be common, given the heterogeneity of most agroecosystems involving long-lived perennials. For example, Embree (1966) suggested that apple trees in towns of Nova Scotia provide a refuge for the winter moth, a pest in oak woods, from parasitism by *Cyzenis.* Also common in insect pests are adult stages that are invulnerable to parasitism, although the key question is how long such stages live relative to other time lags in the system. The importance of an invulnerable stage to at least temporary persistence of a weed has been demonstrated in the ragwort (McEvoy 1987), as discussed above.

Aggregation and stability

I noted earlier in this chapter that aggregation stabilizes Nicholson-Bailey models by increasingly reducing the efficiency of parasitoids as they increase in density, so causing density dependence in the parasitoid population. This in turn derives, at least in part, from the peculiar assumption in these models that parasitoids respond to the initial density in a patch (for example), and do not reassort themselves among patches (or individual hosts) as conditions change through the generation.

Murdoch and Stewart-Oaten (1989) investigated aggregation to patches with higher prey density, within the framework of Lotka-Volterra predator-prey model. Here there are no time lags, and in particular the predators are assumed to respond instantly and continuously to changes in prey distribution, so that the fraction of predators in a patch is always an increasing function of the fraction of prey there (an analogue of equation (5)).

In this model, aggregation is efficient for the predators, which kill more prey per unit time than in the absence of aggregation. As a consequence, the prey equilibrium density is reduced. However, while this fits better with our intuition than the Nicholson-Bailey result, it can cause an otherwise neutrally stable Lotka-Volterra model to become unstable. Aggregation can be stabilizing in this context, especially if the spatial variance of the host distribution is a rapidly increasing function of its total density.

Lotka-Volterra models, like Nicholson-Bailey models, are caricatures of the "real world", and it can be rightly argued that predators cannot respond instantly or with omniscience to a changing prey distribution. In fact, the truth probably lies somewhere between this extreme and the equally unrealistic Nicholson-Bailey parasitoid, whose wings, so to speak, drop off when she encounters her first host. Further modeling could usefully pursue the direction begun by Comins and Hassell (1979) who studied the Nicholson-Bailey parasitoids that did respond to the changing distribution of unparasitized hosts, but the results depend on specific numerical assumptions and a more general treatment should prove rewarding.

Local instability, ensemble dynamics, and biological control

Finally, suppose we accept that stability may in some cases militate against successful control, and that our aim in those cases should be to use natural enemies that tend to cause local extermination of the pest. What are the practical implications of such a goal? (I concentrate below mainly on properties of the enemy, but clearly in any real system we need to characterize not just the enemy, but the pest-enemy interaction.)

The following might be appropriate tactics:

- Avoid invulnerable classes (implies complementary enemy species or reduce seed banks by proper weed management)
- Avoid major spatial and temporal refuges
- Seek high search and attack rates

Most obviously, we want to remove those features that enhance stability by directly protecting the pest. Obvious examples include spatial or temporal refuges, and long-lived invulnerable classes, especially the adult stage. Removing refuges is likely to be merely a gardening problem of little subtlety. Avoidance of long-lived invulnerable classes implies the use of complementary species of natural enemies, i.e. those that attack different stages of the pest, since it is rare for a single species of parasitoid to attack all stages, although true predators sometimes do.

We may also want to avoid protecting the pest via indirect stabilizing mechanisms. As noted above, within the Nicholson-Bailey framework, both types of aggregation, and a density dependent sex ratio, stabilize the system by making the *parasitoid* population density dependent, thus protecting pests when the parasitoid is abundant.

The potential role of spatial aggregation is especially interesting. On the one hand, if dynamics follow the Nicholson-Bailey model, the trade off between stability and control suggests that aggregation is to be avoided. On the other hand, if the dynamics are closer to Lotka-Volterra, aggregation to patches of high pest density is desirable since, although sometimes destabilizing, it leads to lower pest density (Murdoch and Stewart-Oaten 1989).

The obvious argument *against* a control strategy based on local pest instability and a tendency to local eradication, is that pest outbreaks are more likely in unstable situations. In particular, there is the danger that the enemy will become globally extinct and the pest will escape control. This possibility cannot be ignored, but the problem is perhaps less serious than it seems.

First, local stability in a model at a low equilibrium density says nothing about the size of population excursions that are possible; it simply says that the system has a tendency to return to equilibrium when perturbed some (perhaps small) distance from it. It does not guarantee return, and certainly not rapid return (Murdoch *et al.* 1985). It also may fail to predict the response to frequent and large perturbations (Goh 1980).

Second, instability and a tendency to extinction at the local spatial level do not imply that a population composed of an ensemble of sub-populations will approach extinction. On the contrary, there is now a substantial body of theory showing that a population occupying even small numbers of loosely-coupled patches in a heterogeneous environment will typically persist indefinitely (e.g. Murdoch and Oaten 1975; Hastings 1977; Crowley 1981; Nachman 1987a). The population in such an ensemble of patches is likely to appear stable, even when no stabilizing mechanisms operate at the level of the patch and movement between patches is not density dependent. Indeed, Chesson (1978) shows that the dynamics of the ensemble may be described by a standard, stable, deterministic model, even though this description is only phenomenological and does not contain the local mechanisms that drive the dynamics. Thus the spatial scale we choose to model, and whether the resultant population appears stable or not, is to some extent a matter of choice. That choice, in turn, may depend upon whether we seek an explanation or merely a description.

Such ensemble dynamics appear to explain the global persistence of interacting populations of spider mites and predatory phytoseiid mites, which inevitably become locally extinct (Sabelis and Laane 1986). It may also explain the persistence at low densities of systems such as the cottony cushion scale and its natural enemies (Thorarinsson, pers. comm.) and the olive scale and its parasitoids (Murdoch *et al.* 1985). In fact, such dynamics may be important in the key example discussed here — the red scale-*Aphytis* system. The scale and parasitoid interact quite differently on leaves, twigs, branches and fruits, so the whole system is actually a set of sub-systems, each with different parameter values (especially parasitism rate and scale fecundity), all linked by the movement of first instar scales and parasitoids.

While mere persistence of the enemy may be a non-problem, its ability to keep the pest below the economic threshold is certainly not guaranteed by stable ensemble dynamics. Little theoretical work has been directed at this question. Important features, in addition to those mentioned above, are likely to include the degree of spatial heterogeneity and vagility in the pest and the natural enemy (e.g. Nachman 1987b), and the degree of polyphagy of the latter.

Spatial heterogeneity should help create semi-isolated subpopulations with different vital life history attributes, and hence should prevent the entire system from developing regional cycles. An intermediate amount of coupling between patches is important since too much is likely to lead to "single-patch" dynamics and hence extinction of the ensemble (Murdoch and Oaten 1975).

Polyphagy can be a useful feature in enemies that drive their prey extinct (Murdoch *et al.* 1985), although the pest needs to be a sufficiently preferred food such that it is attacked vigorously when present. Other factors needing investigation include the effects of seasonal forcing, overlapping versus discrete generations, and the degree of synchrony between pest and enemy.

Theory of ensemble predator-prey dynamics

The general vagueness of these remarks points to the need for theoretical developments in this area. While there has been some stochastic modeling of presence and absence for multi-species systems in ensembles of patches, unfortunately there has been little or no work done on explicit dynamics of the density of interacting predators and prey in a stochastic framework. Yet this may be needed for biological control that does not fit the classical mold of stability. Chesson (1978, 1984a,b) has made a crucial contribution here by introducing the concept of stochastic boundedness, which implies that densities of bounded populations remain most of the time within limits and do not wander off towards zero or infinity; there seems a clear analogy between upper stochastic bounds to pest density, and the lower ones defining an economic threshold (Murdoch 1979; see also Strong 1984). Chesson has developed conditions for stochastic boundedness for models with only one species (Chesson 1984b) and with competing species (Chesson 1984a), but conditions for parasitoid-host systems are not yet established. For the near future we may need to rely on simulation of pests and enemies in ensembles of patches, as exemplified by the work of Nachman (1987a,b). Thus, in this important area of applied population biology, both theory and experimental tests have continually yielded new insights and critical research questions.

Acknowledgements

I am grateful to Drs. Robert Luck, Peter McEvoy, Andy Taylor, Sandy

Walde, and John Reeve for critical comments on this manuscript. John Reeve kindly generated the histograms in Fig. 1. The research was supported by NSF grant BSR 831 5235.

Literature cited

Bailey, V. A., A. J. Nicholson, and E. J. Williams. 1962. Interactions between hosts and parasites when some host individuals are more difficult to find than others. J. Theor. Biol. 3: 1—18.

Banks, C. J. 1957. The behaviour of individual coccinellid larvae on plants. British J. Anim. Behav. 5: 12—24.

Beddington, J. R., C. A. Free, and J. H. Lawton. 1978. Characteristics of successful enemies in models of biological control of insect pests. Nature 273: 513—519.

Chesson, P. L. 1978. Predator-prey theory and variability. Ann. Rev. Ecol. Syst. 9: 323—347.

Chesson, P. L. 1984a. The storage effect in stochastic population models. In Lecture Notes in Biomathematics (Ed. by S. A. Levin and T. G. Hallam), 54: 76—89. Springer-Verlag, New York.

Chesson, P. L. 1984b. Persistence of a markovian population in a patchy environment. Z. Wahrscheinlichkeitstheorie und verwandte Gebiete 66: 97—107.

Chesson, P. L. and W. W. Murdoch. 1986. Aggregation of risk: relationships among host-parasitoid models. Amer. Nat. 127: 696—715.

Clausen, C. P. (ed.) 1978. Introduced parasites and predators of arthropods and weeds: a world review. U.S.D.A. Agricultural Handbook No. 480.

Comins, H. N. and M. P. Hassell. 1979. The dynamics of optimally foraging predators and parasitodis. J. Anim. Ecol. 48: 335—351.

Crowley, P. H. 1981. Dispersal and the stability of predator-prey interactions. Amer. Nat. 118: 673—701.

Dempster, J. P. 1982. The ecology of the cinnabar moth, *Tyria Jacobaeae L.* (Lepidoptera: Arctiidae). Advances in Ecological Research 12: 1—36.

Embree, D. G. 1966. The role of introduced parasites in the control of the winter moth in Nova Scotia. Can. Ent. 98: 1159—1168.

Free, C. A., J. R. Beddington, and J. H. Lawton. 1977. On the inadequacy of simple models of mutual interference for parasitism and predation. J. Anim. Ecol. 46: 543—554.

Goh, B. S. 1980. Management and Analysis of Biological Populations. Elsevier, New York.

Hassell, M. P. 1978. The Dynamics of Anthropod Predator-Prey Systems. Princeton University Press, Princeton.

Hassell, M. P. 1984. Parasitism in patchy environments: inverse density dependence can be stabilizing. IMA J. Math. Appl. Med. Biol. 1: 123—133.

Hassell, M. P. and R. M. May. 1973. Stability in insect host-parasite models. J. Anim. Ecol. 43: 567—594.

Hassell, M. P. and R. M. May. 1974. Aggregation in predators and insect parasites and its effect on stability. J. Anim. Ecol. 42: 693—736.

Hassell, M. P. and G. C. Varley. 1969. New inductive population model for insect parasites and its bearing on biological control. Nature 223: 1133—1136.

Hassell, M. P., J. K. Waage, and R. M. May. 1983. Variable parasitoid sex ratios and their effect on host-parasitoid dynamics. J. Anim. Ecol. 52: 889—904.

Hastings, A. 1977. Spatial heterogeneity and the stability of predator-prey systems. Theor. Pop. Biol. 12: 37—48.

Hastings, A. 1983. Age-dependent predation is not a simple process. I. Continuous time models. Theor. Popul. Biol. 23: 347—362.

Hastings, A. 1984. Delays in recruitment at different trophic levels: effects on stability. J. Math. Biol. 21: 35—44.
Howard, L. O. and W. F. Fiske. 1911. The importation into the United States of the parasites of the gipsy-moth and the brown-tail moth. Bull. of Bureau of Ent., U.S. Dept. of Agriculture 91: 1—312.
Kareiva, P. 1984. Predator-prey dynamics in spatially structured populations: manipulating dispersal in a coccinellid-aphid interaction. In Lecture Notes in Biomathematics (ed. by S. A. Levin and T. G. Hallam), 54: 368—389. Springer-Verlag, New York.
May, R. M. 1978. Host-parasitoid systems in patchy environments: a phenomenological model. J. Anim. Ecol. 47: 833—844.
McEvoy, P. B. 1987. Depression in ragwort *Senecio jacobaea* abundance following introduction of *Tyria jacobaea* and *Longitarsus jacobaea* on the central coast of Oregon. Proc. VI Int. Symp. Biol. Contr. Weeds, Vancouver.
McNair, J. N. 1986. The effects of refuges on predator-prey interactions: a reconsideration. Theor. Pop. Bio. 29: 38—63.
Murdoch, W. W. 1979. Predation and the dynamics of prey populations. Fortschritte der Zoologie 25: 295—310.
Murdoch, W. W. and A. Oaten. 1975. Predation and population stability. Advances in Ecological Research 9: 1—131.
Murdoch, W. W. and A. Stewart-Oaten. 1989. Aggregation by parasitoids and predators: effects on equilibrium and stability. Amer. Nat. 134: 288—310.
Murdoch, W. W., J. D. Reeve, C. B. Huffaker, and C. E. Kennett. 1984. Biological control of scale insects and ecological theory. Amer. Nat. 123: 371—392.
Murdoch, W. W., J. Chesson, and P. L. Chesson. 1985. Biological control in theory and practice. Amer. Nat. 125: 344—366.
Murdoch, W. W., R. M. Nisbet, W. S. C. Gurney, and J. D. Reeve. 1987. An invulnerable age class and stability in delay-differential parasitoid-host models. Amer. Nat. 129: 263—282.
Nachman, G. 1987a. Systems analysis of acarine predator-prey interactions. I. A stochastic simulation model of spatial processes. J. Anim. Ecol. 56: 247—265.
Nachman, G. 1987b. Systems analysis of acarine predator-prey interactions. II. The role of spatial processes in system stability. J. Anim. Ecol. 56: 267—281.
Nicholson, A. J. and V. A. Bailey. 1935. The balance of animal populations. Proc. Zool. Soc. Lond.: 551—598.
Oaten, A. 1977. Transit time and density dependent predation in a patchily distributed prey. Amer. Nat. 111: 1061—1075.
Reeve, J. D. 1988. Environmental variability, migration, and persistence in host-parasitoid systems. Amer. Nat. 132: 810—836.
Reeve, J. D. and W. W. Murdoch. 1985. Aggregation by parasitoids in the successful control of the California red scale: A test of theory. J. Anim. Ecol. 54: 797—816.
Reeve, J. D. and W. W. Murdoch. 1986. Biological control by the parasitoid Aphytis melinus, and population stability of the California red scale. J. Anim. Ecol.
Sabelis, M. W. and W. E. M. Laane. 1986. Regional dynamics of spider-mite populations that become extinct locally because of food source depletion and predation by phytoseiid mites (Acarina: Tatranychidae, Phytoseiidae). In J. A. J. Metz and O. Diekmann (Eds). Dynamics of Physiologically Structured Populations. Springer-Verlag, Berlin.
Smith, A. D. M. and D. A. Maelzer. 1987. Aggregation of parasitoids and density independence of parasitism in filed populations of the wasp Aphytis melinus and its host, the red scale *Aonidiella aurantii.* Eco. Entomol., in press.
Smith, H. S. 1935. The role of biotic factors in the determination of population densities. J. Econ. Ent. 28: 873—898.
Strong, D. R. 1984. Density-vague ecology and liberal population regulation in insects. Pp. 313—329. *In* P. W. Price, C. N. Slobodchikoff and W. S. Gaud (eds.). A New Ecology: Novel Approaches to Interactive Systems. Wiley, New York.

Taylor, A. D. 1988. Parasitoid competition and the dynamics of host-parasitoid models. Amer. Nat. 132: 417—436.

Varley, G. C., G. R. Gradwell and M. P. Hassell. 1973. Population dynamics and pest control. *In* D. Price Jones and M. E. Solomon (eds.), Biology in Pest and Disease Control.

Walde, S. J. and W. W. Murdoch. 1988. Spatial density — dependence in parasitoids. Ann. Rev. Ent. 33: 441—446.

10. Habitat fragmentation, species diversity, extinction, and design of nature reserves

GEORGE R. ROBINSON and JAMES F. QUINN

Abstract

We review theories commonly offered to predict the consequences of habitat fragmentation for community- and population-level diversity, and their potential use in conservation strategies. Exploring community-level effects of habitat subdivision, we demonstrate why island biogeography theory is not applicable to the problem of determining the best reserve size for maximum diversity. Evidence from field studies and experiments tends to argue for higher levels of diversity with some moderate degree of habitat subdivision, provided that the amount of total habitat area, whether intact or subdivided, is the same. We examine the relationship between population subdivision and extinction. In theory, large populations are resistant to extinction due to demographic or genetic stochasticity, whereas population subdivision permits escape from chance environmental threats. It follows that a conservation advantage might be gained from preserving multiple populations, depending on the relative importance of different extinction mechanisms in nature. Empirical evidence for this is mixed, varying among different organisms in different habitats. We conclude that strategies emphasizing reserve size to the exclusion of other concerns are not well supported.

Introduction

Many species of plants and animals need protection from human activities. While the list of candidates is growing, not all species can be accorded full protection, simply because not all habitats can be sheltered. Consequently, appropriate methods for ranking conservation alternatives are necessary (Thiebodau 1983; Roberts 1988). Preservation programs are typically formulated under political and economic constraints (e.g., Salwasser 1987), and a prudent ecological perspective dictates that the planning processes for many parks and other natural areas include evaluations to estimate the potential of a given area for preserving the resident biota. The standard conclusion to date, including that adopted by the International Union for the Conservation of Nature (IUCN 1980), has been that large preserves are necessary, and in the extreme case that only large preserves will be adequate (Sullivan and Shaffer 1975; Wilcox 1980; Soulé 1980; Wilcox and Murphy 1985; Newmark 1987).

S. K. Jain and L. W. Botsford (eds), Applied Population Biology, 223—248.

Given the limited resources available, this approach should rest on firm ecological foundations, because it sets the stage for a program of land use that cannot easily be revised once implemented. Any decisions will necessarily have to be made with imperfect knowledge, because rapid habitat destruction demands action before satisfactory field studies can be conducted for every population of concern or interest. In this chapter, we will study the foundations for the belief that conservation is best served by a limited number of large reserves rather than a greater number of smaller reserves, by examining theoretical, empirical and experimental investigations relevant to that conclusion.

A large nature reserve is generally regarded as better able to preserve more and larger populations, and thus greater biotic diversity, than an equal area subdivided into a collection of small reserves (Sullivan and Shaffer 1975; Diamond 1976; Terborgh 1976; Whitcomb *et al.* 1976; Terborgh *et al.* 1978; Lovejoy *et al.* 1984; Wilcox and Murphy 1985). This is in part because contiguous areas are better able to preserve intact communities of interdependent species (Gilbert 1980) and to maintain viable populations of those species that typically occur at low population densities (Wilcox and Murphy 1985; Shaffer and Samson 1985). To the extent that parks and preserves are fragments of formerly larger habitats, it is further argued that larger fragments are better choices, since they better represent the historic natural landscape (Pickett and Thompson 1978).

On the other hand, natural populations, communities and habitats generally form a subdivided landscape. Preserving natural dynamics of the species involved requires an adequate representation of subhabitats they normally occupy. Protecting entire species assemblages requires at a minumum that some representation of each species and habitat type be maintained. Consequently the number and spatial relationships between protected areas, as well as their average sizes, are bound to be critical components of successful conservation strategies.

The issue of how habitat structure affects biological communities is a more general and fundamental question in ecology. A discussion of the importance of spatial heterogeneity in population dynamics can be found in any elementary textbook, but a predictive theory of spatial structure has been elusive. Experiments that directly investigate the effects of spatial habitat subdivision were attempted only recently (Simberloff 1976; Simberloff and Abele 1982; Karieva 1987; Quinn and Robinson 1987; Robinson and Quinn 1988). If regular patterns emerge from these lines of inquiry, they may promote better understanding of natural communities. Furthermore, if a predictive theory of spatial structure of populations emerges, it could be very useful in planning conservation efforts, particularly for the majority of species that are poorly studied, and whose specific habitat and resource requirements are unknown. Because parks and reserves serve a variety of objectives, including recreation, scenic values, and preservation of air and water quality, species conservation will never completely dictate land acquisition policy. Nevertheless, it is useful to be able to assess the relative effectiveness of competing conservation strategies in protecting threatened populations.

Our own investigations, employing several kinds of evidence, have yielded little support for the position that only large parks and preserves can be effective reservoirs of biological diversity, or for the underlying ecological proposition that habitat subdivision should depress diversity and promote extinction. A number of authors have arrived at similar conclusions, based on both theoretical considerations (e.g., Boecklen 1986; Boecklen and Gotelli 1984; Gilpin and Diamond 1980; Simberloff and Abele 1976, 1982) and empirical evidence (e.g., Higgs and Usher 1980; Jarvinen 1982; Simberloff and Gotelli 1984; Zimmerman and Bierregaard 1986). At issue is not only how large parks must be, but how many are needed. Future planning will inevitably include decisions with trade-off between enlarging existing conservation areas and acquiring new ones.

Habitat subdivision

Natural habitat fragmentation is a continuous process that operates on various spatial and temporal scales. A dramatic example is the formation of land bridge islands, when continental land masses are invaded by rising seas, with only areas of higher elevation escaping submersion. At their initiation, these new islands each hold a subset of mainland populations, at least some of which are likely to go extinct (Preston 1960; MacArthur and Wilson 1967; Willis 1974; Karr 1982). Irregular recruitment from other islands or the mainland, coupled with differential survival of remnant populations, results in altered community structures. Long-term evolutionary processes characteristic of small, isolated populations (Darwin 1859; Mayr 1963; Wright 1970) may contribute to further changes, with the result that land bridge islands are often quite unlike the mainland with which they were once contiguous. Under this kind of subdivision, biological diversity can be both lost and gained. If those populations which fail to survive have no counterparts in other locations, then one form of diversity — species richness — is reduced. If, on the other hand, changes in community structure permit new evolutionary pathways, then genetic and adaptational variation may increase.

An analogous process occurs in terrestrial habitats when combinations of climate changes and tectonic events result in the insularization of patches of formerly connected habitat, as in Great Basin mountain ranges (Brown 1971; Brown and Kodrick-Brown 1977). Less dramatic examples are abundant, and most populations are spatially subdivided to some extent (Levin 1976).

There is no agreement in the ecological literature on how the degree of fragmentation of natural habitats should affect the dynamics of single species, the ability of interacting species to coexist, or the diversity or stability of entire assemblages. By and large, students of strong negative interactions between species — i.e., competition, predation, or disease — predict that distributing the interactions among multiple patches enhances coexistence, diversity, and the dynamical stability of communities (see DeAngelis and Waterhouse 1987 for a recent review). Perhaps the best evidence for stabilizing and diversifying effects of habitat subdivision comes from laboratory studies of predator-prey interactions (Huffaker 1958; Huffaker *et al.* 1963). By introducing spatial heterogeneity, which restricted predator movement relative to its prey, Huffaker demonstrated that subdivided prey populations could be maintained for much longer than observed in freely mixing populations. Armstrong (1976) found similar results in laboratory cultures of competing fungi.

Students of complex positive interactions among organisms, including social behavior within species and coevoloved mutualisms between them, tend to argue that fragmenting populations will disrupt the normal functioning of communities, and lead to degeneration of structure and loss of species (reviewed by Gilbert 1980). It is this second perspective on habitat fragmentation, its negative effects on diversity, that is stressed in recent conservation biology literature (Frankel and Soulé 1979; Soulé and Wilcox 1980; Schonewald-Cox *et al.* 1983; Soulé 1987.) From this perspective, Lovejoy and colleagues (Lovejoy and Oren 1981: Lovejoy *et al.* 1984; Lovejoy *et al.* 1986) have suggested that there is a "minimum critical size" of ecosystems, below which the trophic or structural integrity is no longer maintained.

Island biogeography theory and reserve design

Oceanic islands have provided an important metaphor and model for the development of perspectives on reserve system fragmentation. One of the more predictable patterns in ecology is the regular relationship between the size of an island and the number of species it harbors. The relationship is typically described by the equation

$$S = cA^z,$$

where S is the number of species, A is the area, and c is a scaling constant. The exponent, z, indicates the extent to which increasing numbers of species are found on increasingly large islands (Fig. 2a). Observed species richness typically doubles with a ten-fold increase in area, i.e., $z = 0.25$ (MacArthur and Wilson 1967; Schoener 1974). Many species groups on terrestrial habitat "islands", such as mountaintops (Brown and Kodric-Brown 1977) and forest fragments (Harris 1984; McLellan *et al.* 1986) have similar species-area curves (see Connor and McCoy 1979 for an extensive review). Parks and nature reserves may be expected to show similar patterns, especially as they become increasingly isolated and insularized by human encroachment around their margins.

Regular species-area relationships were first explained by MacArthur and

Wilson (1967) as the result of a dynamic balance between immigration and local extinction (Fig. 1). In the simplest terms, they suggested that as more species are found on an island, fewer potential immigrant species are left in source areas, and overall immigration rates (species per unit time) will decline. At the same time, there are more species subject to extinction, thus extinction rates (species per unit time) will increase. A stable balance will be reached when the extinction rate balances the immigration rate. Logically, extinction risks for any given species are likely to be greater on smaller islands, because of smaller population sizes and more limited resource bases. As a result, the MacArthur-Wilson theory predicts that smaller islands should have lower equilibrium species counts, greater average extinction rates and higher rates of species turnover.

From this body of theory followed the straightforward proposition that larger islands support more species, and so by analogy larger parks and preserves are likely to be more effective reservoirs of diversity than smaller ones (Diamond 1975a; May 1975; Sullivan and Shaffer 1975; Wilson and Willis 1975). The species-area relationship has been extended to suggest that, within a given region, smaller islands draw from a subset of species found on larger islands. From this it is inferred that smaller parks and preserves necessarily contain species subsets of larger parks and reserves (Lovejoy *et al.* 1984; Wilcox and Murphy 1985; Patterson and Atmar 1986; Patterson 1988). If this description were generally true, then no combination of small reserves would contain as many species as one large reserve. The optimum conservation strategy would be clear: diversity would be maximized and extinction rates minimized by safeguarding the largest possible contiguous area.

Island biogeography theory, however, does not require that the species on small islands be a subset of those on the larger islands, and as an empirical matter, they generally are not. If the biotas of the small islands are sufficiently

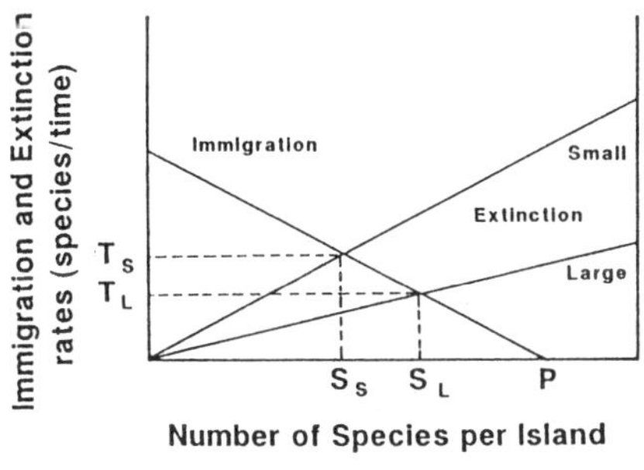

Fig. 1. Diagram of the proposed relationship between island size, turnover rates and species richness. S_S and S_L are predicted equilibrium species counts for small and large islands, respectively. P is the maximum number of species possible, i.e., the available mainland species pool. T_S and T_L are predicted turnover rates for small and large islands, respectively. (Source: MacArthur and Wilson, 1967).

dissimilar, collections of small islands may be expected to have more species than the same area on a single large island (Simberloff and Abele 1976, 1982; Higgs and Usher 1980; Boecklen and Simberloff 1986).

Simple, plausible biological scenarios may be used to predict that highly fragmented landscapes should support lower diversity, higher diversity, or similar diversity, compared to their unfragmented counterparts. For purposes of illustration, consider a hypothetical fragmented landscape in which one-third of the total habitable area is composed of fragments of each of three sizes: 15 fragments of one unit area; three fragments of five units area; and one fragment of 15 units area. Suppose each fragment has a total species richness dictated by the typical (z = 0.25) species-area curve shown in Fig. 2a. These fragments could represent possible parks or reserves, but they may

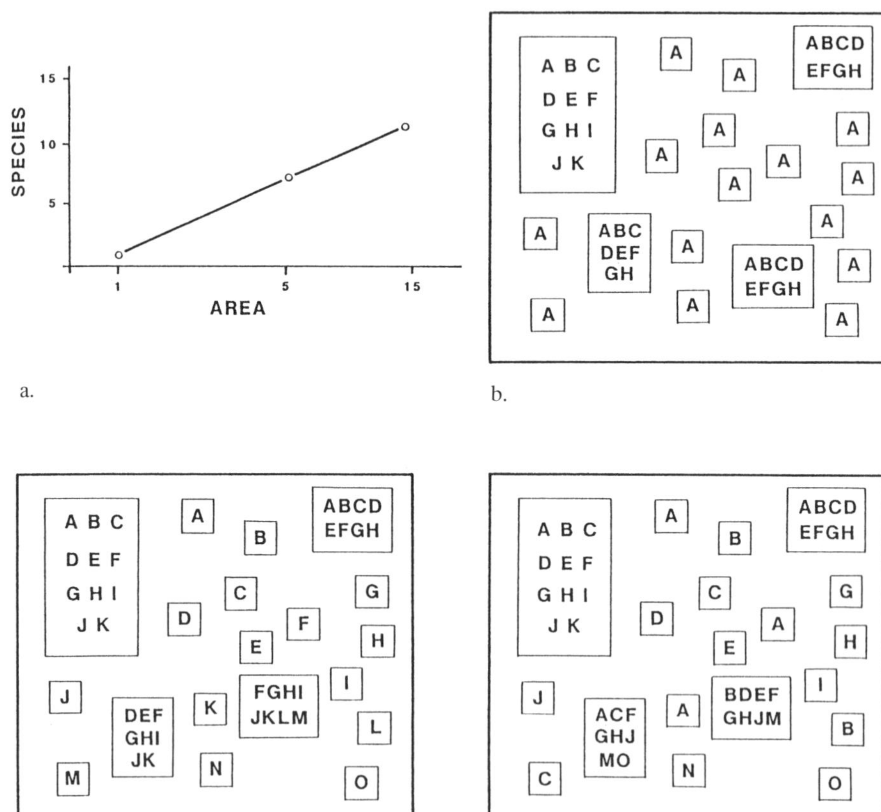

Fig. 2. (a) A hypothetical species-area curve with slope z = 0.25. (b–d) Three possible species distributions among hypothetical habitat fragments, each of which corresponds to the curve in (a). Each letter represents a unique species.

also represent islands or other island-like fragments in a naturally subdivided landscape. Assigning a letter of the alphabet to each potential resident species, several different scenarios can be depicted (Figs. 2b-2d).

The first scenario (Fig. 2b) corresponds to the kind of biology envisioned by Diamond (1975b) in his description of bird communities in oceanic archipelagoes. The smallest fragments support only a single weedy ("super-tramp") species, with minimal territory requirements and perhaps considerable dispersal abilities. The larger fragments harbor additional species with larger spatial requirements or needs for more extensive or reliable resources, such that the biota of the largest fragment is a superset of those of any collection of smaller fragments. Obviously if only a portion of this landscape were to be preserved, incorporating the largest fragment would be required to maintain the maximum number of species. This situation may be expected, for example, with fruit-eating primates in Amazon forest fragments (Lovejoy *et al.* 1986) or interior-dwelling bird species in temperate forest remnants (Whitcomb *et al.* 1981; Nilsson 1986). Such sensitive taxa are likely to disappear from habitats below a minimum area, and any number of small refuges will be insufficient to preserve them.

The second scenario (Fig. 2c) portrays a very different situation, in which a collection of the smallest fragments sample a larger number of species than any combination of larger fragments. This case might represents a situation in which there is competitive exclusion, such that the first species to occupy a fragment was able to prevent subsequent invasions, as is likely when colonization is random and infrequent, and early colonists adapt to local conditions. Lack (1976) argues for this kind of interaction between Darwin's finches on the Galapagos Islands. A number of examples in which several small islands contain more species than single larger islands have now been described (Simberloff and Abele 1982; Wilcove *et al.* 1986; Quinn and Harrison 1988).

In a third scenario (Fig. 2d), it is possible that the degree of subdivision does not affect overall diversity. Any randomly chosen subset of the landscape will contain approximately the same number of species, regardless of spatial structure. In this case, habitat area is the only important factor, and as long as the smallest fragments add up to the total area of a single large one, there will be no differences in species diversity, regardless of the degree of subdivision. This corresponds to a "null" model (Simberloff and Abele 1976, 1982), in which species distributions are primarily the results of chance events. Table 1 summarizes the species counts for each of our three scenarios.

Thus, island biogeography theory provides no unambiguous predictions on how many species can be incorporated into nature reserve systems following different tradeoffs between reserve size and reserve numbers. It is indeed likely that within the same habitat taxa, different will respond differently to alternate reserve acquisition strategies (Jarvinen 1982; Lahti and Ranta 1985). For example, large animals, or those of high trophic status, are

Table 1. Results of three levels of habitat subdivision under three different theoretical models.

	Model 1 (Fig. 2b)			Model 2 (Fig. 2c)			Model 3 (Fig. 2d)		
Number of fragments	1	3	5	1	3	5	1	3	5
Habitat area per fragment	15	5	1	15	5	1	15	5	1
Species per fragment	11	8	1	11	8	1	11	8	1
Total species	11	8	1	11	13	15	1	11	11
Unique species	3	0	0	0	0	2	1	1	1

likely to have larger minimum fragment sizes on which they can be maintained than do smaller or more sedentary species (Lovejoy and Oren 1981; Shaffer 1981; Belovsky 1987). Similarly, territorial species may require sufficient contiguous or connected habitat to permit natural breeding and dispersal (Lande 1987).

Field evidence for the effects of subdivision on diversity

A number of authors have now noted that the theory of island biogegraphy does not predict the best size, number, or spacing of nature reserves, particularly within the restraints posed by limited total resources for conservation (Boecklen and Gotelli 1984; Game and Peterkin 1984; Soulé and Simberloff 1986; Zimmerman and Bierregard 1986). At a more fundamental level, it is not established how habitat subdivision affects the diversity of natural communities in general. Since reasonable assumptions can predict almost any pattern, it becomes an empirical matter how habitat structure actually affects ecological assemblages of interest. While there is no reason that all taxa or communities should exhibit identical patterns, evidence from a large number of study systems at various spatial scales indicates that at least moderate amounts of spatial subdivision are associated with increased species diversity.

Much of this evidence comes from pairwise comparisons between the species richness on multiple small islands and that on one or a few larger islands. For example, Higgs and Usher (1980) made such comparisons among sixteen small nature reserves in Yorkshire chalk pavements, ranging in size from two to five hectares. They determined that all possible combinations of two smaller reserves held more vascular plant species than one larger reserve of equivalent total area. A similar conclusion was reached by Game and Peterkin (1984), who found that single large woodlands in Lincolnshire supported fewer plant species than most combinations of two smaller woodlands of comparable area. In a somewhat different approach, Gilpin and Diamond (1980) determined that among thirteen New Hebrides islands, two small islands held more species than one larger island, provided that species overlap between the two small islands was no greater than 75 percent. Further examples are found in Simberloff and Abele (1982).

The degree to which these examples are characteristic of patterns in nature is difficult to know. In most archipelagos, an enormous number of possible pairwise tests could be made between the richness of various collections of small and large islands. Many of those pairings would not be independent, and the fact that individual pairs can be found to support any given outcome is not surprising. Such observations disprove the naive prediction that subdivision necessarily depresses species diversity, but they do not make a strong case for the generality of alternative patterns in nature.

Similar pairwise comparisons can be made for existing park systems. For example, Newmark (1986) compiled mammal species lists for 24 medium to large parks in western North America. Roughly half the area studied (31,000 km^2) is in the two largest park complexes (The Banff-Jasper-Yoho-Kootenay complex in Alberta and the Yellowstone-Grand Teton complex in Wyoming). Twenty-two smaller parks total to 28,760 km^2. The two larger parks harbor 74 species of mammals, whereas the 22 smaller parks have 163 species in slightly less total area. Only one species (the Kennicott-Uinta ground squirrel) found in the large parks is absent from the collection of smaller parks. In contrast, the collection of smaller parks contains many habitats missing from the larger parks, and intersects more species ranges. As a result, at least 90 mammal species are found among the smaller parks but not in the two largest (Quinn 1990).

To overcome the problems of multiple pairwise comparisons, McClellan *et al.* (1986) designed an analysis that examines the contributions of fragmented habitats to diversity, with all fragments in a landscape considered simultaneously. They found that an area composed of small forest fragments contributed more to the overall diversity of birds in New Jersey forests than did comparable areas composed of larger fragments. Quinn and Harrison (1988) describe a simpler archipelago-wide analysis. It proceeds by ranking all islands or habitat fragments in order of area, largest to smallest, then calculating the cumulative number of species at each step. For each value of cumulative area, the cumulative species number then represents the species count in the subsample of the archipelago or landscape with the largest average island (or fragment) size. The same calculation is repeated, but with the islands (fragments) ranked from smallest to largest. Each species-area value then represents the richness of the most subdivided, possible subsample — i.e., maximizing the number of islands or fragments.

Quinn and Harrison (1988) apply this technique to data from 30 surveys of the biotas of islands and terrestrial habitat "islands". In no case does species richness saturate faster if the large islands are taken first (i.e., average fragment size maximized), and in most, collections of the smallest islands have consistently more species than in comparable areas of the larger islands.

Similar patterns emerge with data from national parks. We applied this technique to mammal data from 16 National Parks in California (Cook *et al.*

1990). Results of the analysis are summarized in Fig. 3, and show that smaller parks consistently sample as many or more species than the larger parks. Comparable results have been found for plants, birds, and mammals in parks in North America, Europe, and Africa (Quinn and Harrison 1988).

Cumulative species-area comparisons, while they offer potentially useful insights for conservation strategies, remain problematical for two main reasons. First, comparisons between levels of subdivision are necessarily indirect, because small subunits in fragmented natural systems, or for that matter existing natural reserve systems, are never replicated to include the same total area as the larger subunits to which they are compared. Second, habitat heterogeneity may make any such comparisons difficult to interpret. For example, in the park data summarized in Fig. 3, the combination of many smaller parks covers more habitat types than found in the few largest ones. On the other hand, large oceanic islands contain many more habitats (e.g., alpine zones, lakes and rivers) than found on collections of small islands. The similarity in pattern between the parks and islands may therefore mask very different causal mechanisms.

Experimental studies

Comparatively few experiments have investigated the effects of the degree of habitat subdivision on species diversity. Several have followed the lead of Huffaker (1958), and investigated connected and unconnected patches, approximating an unfragmented versus a fragmented landscape. (Armstrong 1976; Hanski and Ranta 1983; Karieva 1987). These studies have generally confirmed the predictions from "patch" models (Levin 1974; Slatkin 1974;

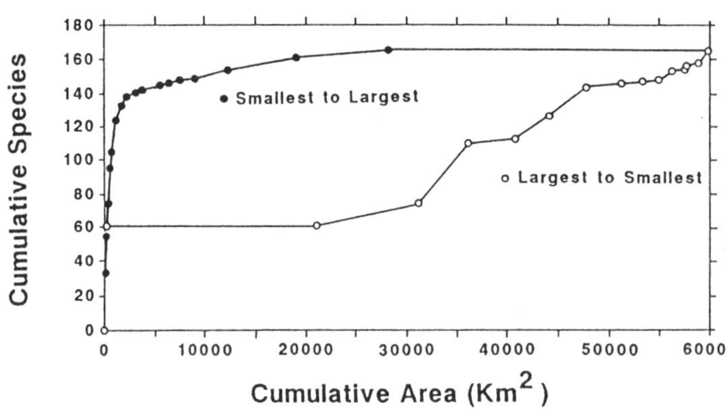

Fig. 3. Cumulative species-area curves for mammals from 24 western North American parks. Closed symbols (upper curve) begin with the smallest park and proceed through the largest. Open symbols begin with the largest park and proceed through the smallest. (Data from Newmark, 1986).

Caswell 1978; Hastings 1980) that species competitors or predators and prey are better able to coexist in patchy habitats than in homogeneous habitats. Armstrong (1976) found that fungal competitors were able to coexist in multiple serial cultures in the laboratory, when single cultures were invariably overtaken by single species. Hanski and Ranta (1983) examined a naturally-fragmented system which closely resembled experimental conditions. Three *Daphnia* species, all confined to small (less than 20 m in diameter) rock pools, were censused on fourteen Baltic Sea islands. Only on islands with a large number of pools did all three species occur, whereas islands with a small number of pools had only the same single species. This was explained as the result of differences in colonizing and competitive abilities, which became less important when more habitats were available. In contrast, Karieva (1987) found that a goldenrod aphid and its principal predator were less able to coexist in a fragmented meadow than they were in a continuous stand.

Fewer experimental studies have been done on community-level effects of fragmentation. Simberloff (1976) studied species colonization rates on eight artificially defaunated mangrove islands. His results indicated that smaller islands at equilibrium need not be limited to a species subset of larger islands, and this offered a reasonable basis for concluding that habitat subdivision could lead to increases in species diversity (Simberloff and Abele 1982).

An ambitious experiment on a much broader scale is underway in the Amazon Basin (Lovejoy *et al.* 1984, 1986). There, a large team of biologists is investigating the effects of tropical deforestation on remnant, uncut rain forest, by studying in detail the biota of forest fragments varying in size from one to 10,000 hectares. At the early stages of this work, there is evidence that: a) on smaller fragments, some species depart because they cannot survive throughout the year on the less predictable food supply (e.g., fewer fruiting trees); b) species typically associated with forest light gaps arrive to colonize the new edge habitat; and c) other species may be lost or reduced in numbers due to the incursion of edge (light-gap adapted) species along the perimeter. Thus, in this case diversity in a fragmented landscape appears to be both augmented by the addition of edge species and reduced by the loss of species with large home-range requirements. The experiment investigates mechanisms of extinction and "relaxation" of the biotas of single fragments following isolation, but it does not directly address the question of how the degree of subdivision of a given area affects total diversity. This is because the effects of "relaxation" on the entire biota depend upon the overlap in identity among the species disappearing from multiple small fragments. To measure this directly, the small fragments must be replicated sufficiently to equal the combined areas of the larger fragment-size treatments (as in the hypothetical examples in Fig. 2). For logistic reasons, the Amazon forest experiment cannot do this (It would require 10,000 of the small fragments to equal the area of the largest planned plot!) Similarly, past experimental island biogeography studies looking at island size effects (Schoener 1974; Sutherland 1974; Osman 1977) have not replicated small "islands" sufficiently to add up to the area of large islands. Ideally, an experiment to address this question should compare diversity between treatments of the same total area, composed of different numbers of patches.

We have conducted field experiments that employ such a balanced design on small spatial scales. In the first of these experiments, we subdivided a 1.25 ha California annual grassland into a series of fenced plots, 2 plots each of 32 m^2, 8 plots of 8 m^2, and 32 plots of 2 m^2, for a total of 64 m^2 in each treatment. The plots were randomly interspersed, isolated from each other by intensive sheep grazing, and for several years both flowering plants and arthropods were censused (for details, see Quinn and Robinson 1987; Robinson and Quinn 1988). Over five growth seasons, the most fragmented treatment (32×2 m^2), consistently had 9 to 11 (44 to 50 percent) more flowering plant species than the least fragmented treatment (2×32 m^2), to with the intermediate fragmentation treatment intermediate in species richness. Similar results were found for arthropods. Among 6 pitfall-trap censuses spread over two years, the richness of the most fragmented treatment was 19 to 39 percent greater than the least fragmented treatment, with counts from the intermediate treatment again falling between the two extremes. Among the plants, the patterns appear to result at least in part from interspecific interactions. In all treatments, one or a few species were able to dominate the canopy area of a given plot. However the identity of the dominant in any given plot appears essentially random. Eight different species dominate at least one plot. While it is impossible to rule out subtle microsite differences between plots, the homogeneity of the study site and the lack of apparent spatial correlation between dominants suggests chance inclusion or introduction, followed by competitive pre-emption, i.e., a "priority effect", of the kind described by Wilbur and Travis (1984). In this respect, the outcome mimics that predicted by "patch" models: in species with strong negative interactions, spatial fragmentation allows coexistence by increasing the probability that each species finds at least one location in which it is not displaced by biological enemies.

Another source of diversity in our more fragmented treatments is their greater perimeter-to-area ratio. Edge effects have attracted considerable attention in temperate (Pickett and Thompson 1979; Pickett and White 1985; Lande 1987) and tropical forest systems (Janzen 1983, 1986; Lovejoy *et al.* 1986). Thirteen of our experimental species were statistically associated with plot edges, and several were found only within a fraction of a meter of the edge. As with weedy species at the edges of disturbed forests, these species are overtopped or shaded by the interior canopy, yet are often removed outside the plots by grazing. Roughly half of the increased diversity in the more fragmented plots is attributable to edge effects. The causes of the similar patterns in the grassland arthropods are less certain. They do not seem to be responding to plant diversity alone, for once the effect of plot size is removed, arthropod diversity shows no correlation with plant diversity.

Other experiments that feature controlled degrees of fragmentation are needed to test the generality of these results. There is no reason that all communities should mimic the patterns in laboratory cultures, mangrove islands or annual grasslands. However, these studies clearly show that, when controlled for other confounding effects (in particular, habitat variability), fragmentation does not necessarily depress diversity as predicted in some of the current conservation literature.

Habitat subdivision and extinction

Not all species are equally vulnerable to natural and human disturbances. Consequently the maintenance of overall diversity in some reserve systems may be subordinate to the objectives of preventing the extinction of the most threatened species (see Soulé and Simberloff 1986 for a recent discussion). For this reason, general effects of habitat fragmentation on both local and global extinction rates are of special interest apart from effects on overall species diversity. As with species diversity, however, predictions about the effects of habitat subdivision *per se* on extinction probabilities have differed among various ecological traditions. In the laboratory experiments and patch models described above, spatial habitat variation maintains diversity by preventing or delaying regional extinctions of species that may disappear quickly in any given patch. In this tradition, it is almost axiomatic that habitat structure promotes ecological stability, operationally defined in most cases as lack of extinction (Ricklefs 1979; Begon *et al.* 1986).

Island studies have led to a very different expectation. Rates of extinction frequently decline with increasing island or habitat fragment size. Diamond (1984) provides an extensive review. The nature of the evidence varies. Most of the long-term estimates of extinction rates come from landbridge islands, such as those in the Sunda Shelf (Terborgh 1975; Diamond 1984), which were isolated by rising sea levels since the last glaciation. In most cases, these islands are assumed to have harbored a sizeable representation of the present mainland biota when isularization began. Island formation was then followed by a period of local population decay and extinctions, often referred to as a period of faunal "relaxation". Local species losses are assumed to have occurred more rapidly on smaller islands since they are now impoverished relative to the larger islands, and thus are thought to have lost a larger proportion of their hypothesized initial fauna. In a few cases, there is more direct evidence from the fossil record to support inferred post-Pleistocene extinctions (Hope 1973; Olson and James 1982; Case and Cody 1983). Over the shorter term, repeated census data may be used to estimate extinction rates (reviewed by Diamond 1984; Quinn and Hastings 1987).

For conservation purposes, extinction-area relations have been interpreted to mean that smaller parks are more vulnerable to extinction than larger ones. Soulé et al (1979) apply extinction rates inferred from landbridge islands to mammal populations of East African National Parks. The results of their analysis suggest that over 20 percent of large mammal species in the smallest parks will disappear over 50 years, and that over 70 percent will be lost over 5,000 years. They conclude that only very large parks have any possibility of providing effective mammal conservation. This conclusion has been challenged on both empirical (Western and Ssemakula 1981) and theoretical grounds (Boecklen and Simberloff 1986). Newmark (1987) performs a comparable analysis for large mammals in parks in western North America, and concludes that only the larger parks (those over 10,000 km^2) are adequate to protect large mammal populations. There is debate over these results as well, as most of the inferred extinctions are problematical (Quinn, van Riper and Salwasser 1990) and Newmark's analysis does not consider the effectiveness of multiple smaller parks in preventing extinctions (Quinn 1990).

The relationship between extinction in individual fragments and regional extinction depends upon whether the same species are vulnerable in different fragments. If some species disappear in deterministic fashion from smaller islands, the implication is that fragmented landscapes will experience higher overall extinction rates than unfragmented landscapes of comparable total area. Evidence for a deterministic order of extinctions is mixed, but there are cases in which this has been demonstrated. Dickerson and Robinson (1986) assembled a variety of freshwater microorganisms into artificial communities of laboratory beakers containing two different volumes of solution. Larger organisms were found to be more disposed to extinction in smaller volumes of solution. Soulé *et al.* (1988) find that bird species have disappeared in a very predictable order from canyons in and near San Diego, California, as these habitats were reduced and isolated by surrounding housing developments.

On a larger scale, Patterson and Atmar (1986; also Patterson 1988) find that mammal species on smaller montane habitat islands in the southern Rocky Mountains essentially represent nested subsets of those in the larger areas, suggesting a deterministic order of extinctions as montane habitat areas shrank with post-Pleistocene climate changes. In this case it is not clear what processes caused this pattern. The species missing from the smaller habitat islands are mostly smaller mammals, whereas larger-bodied mammals — which presumably require larger habitats (Belovsky 1987) — are found in most of the islands. In other studies, mechanisms are more understandable. Species depending upon spatially or temporally unpredictable resources are more likely to be susceptible to small fragment size. For example, Lovejoy *et al.* (1986) find that frugivorous saki monkeys are more likely than folivorous howler monkeys to disappear from experimental Amazon forest fragments.

Whatever the habitat in question, no theory is required to demonstrate that any given park or reserve will be more effective in preventing local extinctions if it is made larger. However, where resources or political capital are limited, larger reserves must eventually be traded off against increased numbers of reserves (e.g., Salwasser 1987). Consequently the question faced by decision makers is more often whether some quantity of land appended to the edge of a conservation land contributes more to protect a threatened species than does the same quantity of land in some other location. The answer to this question is not obvious from first principles, and even in the case of relatively well-studied species, there are a variety of different estimates of the efficacy of suggested conservation plans (Simberloff 1987; Lande 1988).

Theoretical considerations

Predicting extinction probabilities of populations broken into multiple, relatively independent subpopulations is in many ways analogous to the analysis of failures in mechanical or electronic systems with redundant backups. Extinction of a population in a park may be viewed as a failure of an individual unit of the conservation system. Multiple parks provide some redundancy, and extinction from all parks is an irrecoverable failure of the entire system. As with mechanical systems, total system failure may be made less likely either by increasing the reliability of individual subunits (e.g. enlarging parks), or by increasing the number of redundant subsystems (increasing the number of parks). The optimal strategy depends upon the relative costs of increasing subunit reliability versus replicating subunits.

Reliability theory (Polovko 1968) addresses this kind of problem. If failures are independent (i.e., parks are sufficiently widely distributed that they experience different environmental conditions), the expected persistence time of a system (population) rises roughly as the logarithm of the number of redundant subunits (subpopulations into which it is broken) (Quinn and Hastings 1987). The question then is whether, in a highly subdivided landscape, the increased extinction rates in the smaller individual fragments are enough to offset the increases in "system reliability" due to the larger number of "redundant" fragments. The answer obviously depends upon the relationship between fragment size and extinction risk.

In mathematical models, the dependence of extinction rate on area depends upon the extinction mechanism envisioned. Catastrophic extinction, such as those associated with major climatic shifts, may be essentially independent of population size or fragment area (Ewens *et al.* 1987). Against risks of this kind, it is obviously worthwhile to spread populations geographically. More usually, extinction results from a series of "stochastic" events. Shaffer (1981) has partitioned the stochasticity into demographic, genetic and environmental components.

The demographic component is most relevant when the populations are very small, and it represents the possibility that the random order of births and deaths will lead the population on a random walk to extinction. Assuming that the population has a tendency to grow when rare (average birth rate exceeds average death rate), demographic stochasticity models have the general feature that populations above some very small threshold sizes (N_e = 20) are likely to persist essentially indefinitely (Richter-Dyn and Goel 1972). Even in the limiting case where birth rate and death rate exactly balance, expected persistence time is proportional to population size (Hubbell 1979). If extinction processes approximate those envisioned in demographic stochasticity models, spatial subdivision should act to decrease the ability of a species to persist (Wright and Hubbell 1983; Quinn and Hastings 1987).

Genetic stochasticity refers to the potential for inbreeding depression and loss of the capacity for specialized adaptation (Franklin 1980). Both of these are the result of genetic drift, and are characteristic of very small, highly isolated populations. Although the processes themselves can be quite complex, extinction as a function of genetic stochasticity can be addressed mathematically in a rather straightforward manner. It is widely accepted that if habitat subdivision results in effective population sizes below 50, extinction can proceed rapidly, with the expression of deleterious recessive alleles, or gradually, with the erosion of heterozygosity through drift (Soulé 1980; Frankel and Soulé 1981). However the suggestion that there is a viability threshold (e.g., 50 breeding pairs) required for persistence of a population is oversimplified (see Lande 1988, for a review).

To a great extent, the expression of deleterious alleles depends on the history of a population (Ralls and Ballou 1983). Recently subdivided populations may contain more hidden recessives than populations that have been through one or more genetic "bottlenecks" as a result of previous subdivision. In some cases, fewer than 50 breeding pairs each may be sufficient to maintain a number of small, healthy populations, at least in the short term. For the longer term, it is argued that genetic drift can result in permanent losses of variation in local subpopulations with fewer than 500 breeding pairs (Franklin 1980). This estimate itself may be low, depending on the nature of inheritance of the traits of interest, as well as the kinds of selection acting on those traits (Lande and Barrowclough 1987).

Despite estimated differences in the effective population sizes required for persistence under demographic versus genetic constraints (20 versus 500 or more), there are reasons to expect that demographic stochasticity may be of greater concern in the conservation of rare species. Certain species which exist as metapopulations, infrequently sharing genes among numerous subpopulations, may critically depend on demographic factors (Lande 1987). When habitats occupied by such metapopulations are themselves subdivided, any locally-induced changes in demographic patterns can compound the population-wide influence of genetic stochasticity by reducing gene flow (Lande 1987; 1988).

The demographic stochasticity and genetic drift models assume a constant environment. In general, models of extinction due to environmental variability indicate that species persistence is less dependent on population or, by extension, habitat fragment size (Goodman 1987). If the environmental variation approximates white noise (i.e., frequent small perturbations), persistence times may be expected to change with roughly the logarithm of size (Leigh 1981). In this case, a population broken into a moderate number of independent subpopulations may be expected to last considerably longer than a single population of the same initial size (Quinn and Hastings 1987).

The relative contribution of each of these three stochastic components to natural extinctions is still not understood. As with our discussion of species diversity, we will turn to the empirical and experimental evidence on the effects of habitat fragmentation on extinction.

Fragment size and extinction rates

Averaged over a number of species, natural extinction rates vary with fragment or island size in a way consistent with simple environmental stochasticity models. In five of six studies reviewed by Diamond (1984) and six of six analyzed by Quinn and Hastings (1987), there is no statistical departure from the prediction that average extinction rates should vary (approximately) with logarithm of area. However these data could be interpreted in two ways. First, each individual species population might have probabilities of extinction rates weakly dependent upon area. In that case subdivision should reduce, or at least should not exacerbate, landscape-wide extinction (see Quinn and Hastings 1987). Alternatively, the weak relationship might represent the statistical blurring of different extinction versus fragment size thresholds for a number of species, each with a different minimum critical population size (as predicted by demographic stochasticity models). If this was the case, most species would be adversely affected by habitat subdivision, but with different species-specific sensitivities.

Very few studies examine extinction probabilities as a function of patch or island size for a single species, in part because a large number of replicate populations, and extinctions, are needed to make statistically reliable estimates. For species of the most concern in conservation, the experiment is obviously infeasible. Probably the clearest study of extinction rates in the field is an introduction experiment performed by Schoener and Schoener (1983), who introduced lizards (*Anolis sagrei* and *Leiocephalus punctatus*) onto small cays in the Bahamas. They found that populations below a threshold size (corresponding approximately to the critical size of 20 predicted by the demographic stochasticity model) disappeared within months. All of the larger populations persisted several years, beyond the end of the study. In laboratory tests, Forney and Gilpin (1989) compared large, small disconnected, and small connected populations of two *Drosophila*-species, and found that both were more extinction-prone in smaller populations, although one (*Drosophila pseudoobscura*) persisted nearly as well in small, connected populations as in larger ones.

Contrasting results are provided by an ongoing study with populations of a marine snail. We introduced the barnacle predator, *Nucella emarginata* in fixed densities onto 132 barnacle-encrusted plates of seven sizes, ranging from 1/64 m^2 to 1 m^2. Initial populations ranged from 2 to 128, spanning the range over which "minimum critical population sizes" have been predicted. Early results show that, 34 percent of the smallest populations persisted long enough to produce a second generation of eggs. The proportion of populations surviving increased gradually over 4 doublings in size, and the largest 2 sizes (2 more doublings) have not had any extinctions. The pattern is that expected for strong environmental stochasticity, and there is no sign of an abrupt threshold area for population survival (Quinn *et al.* 1989). Similar results are reported by Paine (1989) for patchy populations of an intertidal brown alga (*Postelsia palmaeformis*).

There is little to conclude about extinction mechanisms or the dependence of extinction rates on populations size from these contrasting studies, and we do not know how to explain the differences. It may be that the two intertidal species simply occupy a more variable environment than the lizards. This is quite likely the case in comparison with the fruitflies, which were grown in an essentially non-varying environment, and subject to nearly pure demographic stochasticity (Forney and Gilpin, 1989). Whatever the explanation, an archipelago of islands smaller than the threshold size will probably not support a population of island lizards (Schoener and Schoener 1983), whereas species behaving like the intertidal species may persist in a highly subdivided environment.

Extinction in experimentally fragmented communities

Few studies have examined species turnover in otherwise comparable habitats which are subdivided to different degrees. As with studies of fragmentation and species diversity, this comparison probably requires a controlled experiment, since area effects on species turnover in naturally-subdivided landscapes are generally inseparable from a variety of confounding variables.

As far as we know, our grassland experiment is the only field study to date which examines the effect of fragmentation on extinction directly — i.e., by comparing species losses from the same total area subdivided to different degrees. We have followed species turnover of vascular plants in habitat fragments for four growth seasons. Among the three subdivision treatments (two, eight, or 32 plots), there are no differences in annual extinctions, whether expressed as the absolute or relative number of species lost per treatment, or per plot. Thus, with area held constant, extinction rates were not affected by the extent to which habitat was spatially distributed.

Subdivision and stability

A more general question may be asked about the stability of ecological assemblages as a function of their spatial structure. Stability might be measured by extinction rates, but it also could include immigration, turnover, degree of population fluctuation, and tendency to rebound following perturbations. It has long been argued that increased spatial structure should tend to stabilize ecological communities (see Levin 1976; Caswell 1978; DeAngelis and Waterhouse 1987 for reviews). Contrary predictions may also be found, particularly for complex communities and coevolved associations, such as those represented by specialized tropical plant-pollinator interactions (Gilbert 1980).

There have been relatively few controlled field tests of the relationship between habitat subdivision and any form of ecological stability. Perhaps the most direct is that of Karieva (1987), who found, contrary to theoretical expectations, that the dynamics of an aphid-predator association was less stable in experimentally-fragmented stands of the host plant than in contiguous stands. In our grassland experiment, we find no relationship between stability of the species assemblages, as measured by the rate of species turnover, in either the plants (Robinson and Quinn 1988) or the arthropods (Quinn *et al.* unpublished manuscript).

Avenues for future research

Too little is known about the relationships among the spatial structure of fragmented habitats, species persistence, and the maintenance of overall biotic diversity. Clearly more experimentation with partitioned habitats is needed, including tests of the importance of fragment shape, as well as size. Furthermore, in future empirical studies the relationships between population size, habitat area, and extinction probabilities must be much better documented.

Two related problems seem especially in need of investigation. The first is the interaction between the kinds of population dynamics discussed here and the genetic structure of small and fragmented populations. Much work is under way on the genetics of small populations (see Soulé 1987; and the chapter by Hedrick, this volume). A recurring theme in many of these studies is the relationship between isolation and population size, which may have important implications for the future of many genomes. For some species, this relationship is of immediate concern, particularly those most vulnerable to inbreeding or outbreeding depression (Frankel 1974; Soulé 1980; Templeton 1986.)

The second issue is the role of migration between fragments, and, by implication, the value of maintaining "corridors" for migration between parks (see Simberloff and Cox 1987). No field experiment on fragmentation effects has addressed the role of migration specifically, but since recolonization affects both extinction and species diversity, future studies will no doubt include efforts to evaluate its importance. Although it is generally believed that corridors should increase persistence through reintroductions into failing or failed subpopulations [the "rescue effect" of Brown and Kodric-Brown (1977)], it must be noted that parasites, diseases, predators, competitors, and miscellaneous noxious exotics can also potentially move along corridors. It is not generally known whether the negative effects of movement of deleterious species will outweigh the positive impact of connecting threatened populations along corridors (Simberloff and Cox 1987).

In addition, a potentially useful approach in mathematical models would be more explicit incorporation of species reintroductions. In future conservation strategies, it is likely that purposeful reintroductions will be one of the more important tools in managing rare and extinction-prone species (Cade 1988).

Implications for reserve design

Effective species conservation has at least two components: incorporating biological diversity into reserve systems, and preventing subsequent extinction. It seems clear that at least some degree of fragmentation of habitats is frequently associated with increased diversity (Simberloff and Gotelli 1984; Soulé and Simberloff 1986; Quinn and Harrison 1988). This is particularly true in modern park systems, since the large number of small parks are specifically chosen to incorporate unique habitats and species. Island studies and controlled experiments suggest that other effects of subdivision are also important. In conservation terms, small reserves make extremely important contributions to the biological diversity incorporated in our protected lands.

The potential effects of habitat subdivision on extinction and community stability seems less clearcut. The variety of outcomes in the studies described in this review dictates that choosing optimal spatial arrangements of threatened populations or habitats is no simple matter. Theory suggests that in species subject to large, environmentally-driven population fluctuations, spreading risks in space, (i.e., increasing the number of reserves which contain subpopulations) may be at least as important as improving the chances for their persistence in single locations. For rare species subject to primarily demographic stochasticity, as well as those with very large home ranges or complex social structures, mean size of reserves may prove to be the more essential limitation.

In practice, the optimum nature reserve system for conservation of biotic diversity no doubt represents a mixed strategy — large parks for species with large area requirements and small parks to spread extinction risks for other species. Large parks may better maintain intact ecosystems and historical communities, while numerous small parks can better sample the complete range of habitat types. Present-day reserve systems consist of variably-sized parks and refuges, interspersed with multiple-use public and private lands. Nothing in the studies we review suggests that this arrangement is either ineffective or inappropriate, *a priori*, although contemporary strategies must incorporate more sophisticated efforts to pinpoint critical sites in need of protection (e.g., Margules, 1989). Many more conservation sites need to be selected in areas threatened by development, but we see little in ecological principles to suggest a change from traditional approaches to conservation, which emphasize an adequate representation of diverse natural communities coupled with species-specific management of particularly threatened populations.

Acknowledgements

Aaron Altura, Susan Harrison, Barbara Heinsch, Michael Judge, Richard Karban, Lisa Landsburger, Angela Lockwood, Roderick Thompson, and Carole Wolin have all contributed substantially to the experimental studies outlined in this review. Charles van Riper III and Peter Moyle made valuable comments on the manuscript. Funding was provided by the Hewlett Foundation through the University of California Public Research & Dissemination Program, the Sloan Foundation, the National Park Service, and the Committee on Research at the University of California at Davis. To all we are grateful.

Literature cited

Armstrong, R. A. 1976. Fugitive species: experiments with fungi and some theoretical considerations. Ecology, 57: 953—963.

Begon, M., J. L. Harper, and C. R. Townsend. 1986. *Ecology*. Sinauer, Sunderland, Massachusetts.

Belovsky, G. E. 1987. Extinction models and mammalian persistence. In: M. E. Soulé (ed.), *Viable populations for conservation*. Cambridge University Press, Cambridge, pp. 33—57.

Boecklen, W. J. 1986. Optimal design of nature reserves: consequences of genetic drift. Biol. Conserv. 38: 323—338.

Boecklen, W. J. and N. J. Gotelli. 1984. Island biogeography theory and conservation practice: species-area or specious-area relationships? Biol. Conserv. 29: 63—80.

Boecklen, W. J. and D. S. Simberloff. 1986. Area-based extinction models in conservation. In: D. K. Elliot (ed.), *Dynamics of Extinction*. Wiley-Interscience, New York, pp. 247—276.

Brown, J. H. and A. Kodric-Brown. 1977. Turnover rates in insular biogeography: Effect of immigration on extinction. Ecology 58: 445—449.

Cade, T. J. 1988. Using science and technology to reestablish species lost in nature. In: E. O. Wilson (ed.), *Biodiversity*. National Academy Press, Washington, D.C., pp. 279—288.

Case, T. J. and M. L. Cody. 1983. Synthesis: pattern and processes in island biogeography. In: T. J. Case and M. L. Cody (eds.) *Island Biogeography in the Sea of Cortez*. Univ. of California Press, Berkeley, pp. 307—341.

Caswell, H. 1978. Predator-mediated coexistence: A non-equilibrium model. Amer. Nat. 112: 127—154.
Connor, E. F. and E. D. McCoy. 1979. The statistics and biology of the species-area relationship. American Naturalist 113: 791—833.
Cook, R. R., J. F. Quinn, and C. van Riper III. 1990. A comparative analysis of mammalian inventory data for California's National Parks. In: C. van Riper III, T. J. Stohlgren, S. D. Viers, Jr. and S. C. Hillyer (eds.), *Examples of Resource Inventory and Monitoring in National Parks of California.* U.S. Dept. of the Interior, National Park Service, Washington, D.C., pp. 71—101.
Darwin, C. 1859. *The Origin of Species.* Harvard University Press, Cambridge. Mass. (1966 edition).
DeAngelis, D. L. and J. C. Waterhouse. 1987. Equilibrium and nonequilibrium concepts in ecological models. Ecol. Monogr. 57: 1—21.
Diamond, J. M. 1975a. The island dilemma: lessons of modern biogeographic studies for the design of natural reserves. Biol. Conserv. 7: 129—146.
Diamond, J. M. 1975b. Assembly of communities. In: Cody, M. L. and J. M. Diamond. (eds.), *Ecology and Evolution of Communities.* Harvard University Press, Cambridge, Massachusetts, pp. 342—444.
Diamond, J. M. 1976. Island biogeography and conservation: strategy and limitations. Science 193: 1027—1029.
Diamond, J. M. 1984. "Normal" extinction of isolated populations. In: M. H. Nitecki (ed.), *Extinctions.* University of Chicago Press, pp. 191—246.
Dickerson, J. E., Jr. and J. V. Robinson. 1986. The controlled assembly of microcosmic communities: The selective extinction hypothesis. Oecologia 71: 12—17.
Ewens, W. J., P. J. Brockwell, J. M. Gani and S. I. Resnick. 1987. Minimum viable population size in the presence of catastrophes. In: M.E. Soulé (ed.), *Viable Populations for Conservation.* Cambridge University Press, Cambridge, pp. 59—68.
Forney, K. A. and M. E. Gilpin. 1989. Spatial structure and population extinction: A study with *Drosophila* flies. Conservation Biology 3: 45—51.
Frankel, O. H. 1974. Genetic conservation: our evolutionary responsibility. Genetics 78: 53—65.
Frankel, O. H. and M. E. Soulé. 1981. *Conservation and Evolution.* Cambridge University Press, New York.
Franklin, I. A. 1980. Evolutionary change in small populations. In: M. E. Soulé and B. A. Wilcox (eds.) *Conservation Biology: An Evolutionay-Ecological Perspective.* Sinauer, Sunderland, Mass. pp. 135—149.
Game, M. and G. F. Peterkin. 1984. Nature reserve selection strategies in the woodlands of central Yorkshire. Biol. Conserv. 29: 157—181.
Gilbert, L. E. 1980. Food web organization and the conservation of neotropical diversity. In: M. E. Soulé and B. A. Wilcox (eds.), *Conservation Biology: An Evolutionary-Ecological Perspective.* Sinauer, Sunderland, Massachusetts, pp. 11—33.
Gilpin, M. E. and J. M. Diamond. 1980. Subdivision of nature reserves and the maintenance of species diversity. Nature 285: 567—568.
Goodman, D. 1987. The demography of chance extinction. In: M. E. Soulé (ed.), *Viable Populations for Conservation.* Cambridge University Press, Cambridge, pp. 11—34.
Harris, L. D. 1984. *The Fragmented forest.* University of Chicago Press, Chicago.
Hanski, I. and E. Ranta. 1983. Coexistence in a patchy environment: three species of *Daphnia* in rock pools. J. Animal Ecol. 52: 263—279.
Hastings, A. 1980. Disturbance, coexistence, history and competition for space. Theor. Pop. Biol. 18: 363—373.
Higgs, A. J. and M. B. Usher. 1980. Should nature reserves be large or small? Nature 120: 41—50.
Hope, J. H. 1973. Mammals of the Bass Straight Islands. Proc. Royal Soc. Vict. 85: 163—196.

Hubbell, S. P. 1979. Tree dispersion, abundance, and diversity in a tropical dry forest. Science 203: 1299—1309.
Huffaker, C. B. 1958. Experimental studies on predation: dispersion factors and predator-prey oscillation. Hilgardia 27: 243—283.
Huffaker, C. B., K. P. Shea and S. G. Herman. 1963. Experimental studies on predation. Hilgardia 34: 305—330.
IUCN. 1980. *World Conservation Strategy.* International Union for Conservation of Nature and Natural Resources. Gland, Switzerland.
Janzen, D. H. 1983. No park is an island: increase in interference from outside as park size decreases. Oikos: 41: 420—410.
Janzen, D. H. 1986. The eternal external threat. In: M. E. Soulé (ed.) *Conservation Biology: The Science of Scarcity and Diversity.* Sinauer, Sunderland, Massachusetts, pp. 286—303.
Jarvinen, O. 1982. Conservation of endangered plant populations: single large or several small reserves? Oikos 42: 396—398.
Karieva, P. 1987. Habitat fragmentation and the stability of predator-prey interactions. Nature 326: 388—390.
Lack, D. L. 1976. *Island Biogeography, Illustrated by the Land Birds of Jamaica.* Blackwell, Oxford, England.
Lahti, T. and E. Ranta. 1985. The SLOSS principle and conservation practice: an example. Oikos 44: 369—370.
Lande, R. 1987. Extinction thresholds in demographic models of territorial populations. Amer. Nat. 130: 624—635.
Lande, R. 1988. Genetics and demography in biological conservation. Science 241: 1455—1459.
Lande, R. and G. F. Barrowclough. 1987. Effective population size, genetic variation, and their use in population management. In: M. E. Soulé (ed.), *Viable Populations for Conservation.* Cambridge University Press, Cambridge, pp. 87—123.
Leigh, E. G., Jr. 1981. The average lifetime of a population. J Theor. Biol. 90: 213—239.
Levin, S. A. 1974. Dispersion and population interactions. Amer. Nat. 100: 207—225.
Levin, S. A. 1976. Population dynamic models in heterogeneous environments. Annual Review of Ecology and Systematics, 7: 287—310.
Lovejoy, T. E. and D. C. Oren. 1981. The minimum critical size of ecosystems. In: R. L. Burgess and D. M. Sharpe (eds.), *Forest Island Dynamics in Man-dominated Landscapes.* Springer-Verlag, New York, pp. 7—12.
Lovejoy, T. E., J. M. Rankin, R. O. Bierregard, Jr., K. S. Brown, Jr., L. A. Emmons and M. E. Van der Voort. 1984. Ecosystem decay of Amazon forest remnants. In: M. H. Nitecki, (ed.) *Extinctions.* University of Chicago Press, Chicago, pp. 295—326.
Lovejoy, T. E., R. O. Bierregard, Jr., A. B. Rylands, J. R. Malcolm, C. E. Quintela, L. H. Harper, K. S. Brown, Jr., A. H. Powell, G. V. N. Powell, H. O. R. Schubart and M. B. Hays. 1986. Edge and other effects of isolation on Amazon Forest fragments. In: M. E. Soulé (ed.) *Conservation Biology: The Science of Scarcity and Diversity.* Sinauer, Sunderland, Massachusetts, pp. 257—285.
MacArthur, R. H. and E. O. Wilson. 1967. *The Theory of Island Biogeography.* Princeton University Press, New York.
Margules, C. R. 1989. Introduction to some Australian developments in conservation evaluation. Biol. Conserv. 50: 1—11.
May, R. M. 1975. Island biogeography and the design of wildlife preserves. Nature 254: 177—178.
Mayr, E. 1963. *Animal Species and Evolution.* Harvard University Press, Cambridge, Mass.
McLellan, C. H., A. P. Dobson, D. S. Wilcove and J. F. Lynch. 1986. Effects of forest fragmentation on new- and old-world bird communities: empirical observations and theoretical implications. In: J. Verner, M. L. Morrison and C. J. Ralph (eds.), *Wildlife 2000.* Univ. of Wisconsin Press, Madison, pp. 305—313.

Newmark, W. D. 1986. Ph.D. Dissertation, School of Natural Resources, University of Michigan.
Newmark, W. D. 1987. A land-bridge island perspective on mammalian extinctions in Western North American parks. Nature 325: 430—432.
Nilsson, S. G. 1986. Are bird communities in small biotype patches random samples from communities in large patches? Biol. Conserv. 38: 179—204.
Osman, R. W. 1978. The establishment and development of a marine epifaunal community. Ecol. Monogr. 47: 37—63.
Olson, S. L. and James, H. F. 1982. Fossil birds from the Hawaiian Islands: evidence for wholesale extinctions by man before Western contact. Science 217: 633—635.
Paine, R. T. 1988. Habitat suitability and local population persistence of the sea palm *Postelsia palmaeformis.* Ecology 69: 1787—1794.
Patterson, B. D. and W. Atmar. 1986. Nested subsets and the structure of insular mammalian faunas in archipelagos. Biol. J. Linn. Soc. 28: 65—82.
Patterson, B. D. 1988. The principle of nested subsets and its implications for biological conservation. Conserv. Biol. 1: 323—334.
Pickett, S. T. A. and J. N. Thompson. 1978. Patch dynamics and the design of nature reserves. Biol. Conserv. 13: 27—37.
Pickett, S. T. A. and P. White. 1985. *The Ecology of Natural Disturbance and Patch Dynamics.* Academic Press, Orlando, Florida.
Polovko, A. M. 1968. *Fundamentals of Reliability Theory.* Academic Press, New York.
Preston, F. W. 1960. Time and space and the variation of species. Ecology 41: 611—627.
Quinn, J. F. 1990. Extinction rates and species richness of mammals in western North American parks. Biol. Conserv. (in press).
Quinn, J. F. and S. P. Harrison. 1988. Effects of habitat fragmentation and isolation on species richness: evidence from biogeographic patterns. Oecologia 75: 132—140.
Quinn, J. F. and A. Hastings. 1987. Extinction in subdivided habitats. Conserv. Biol. 1: 198—208.
Quinn, J. F., M. L. Judge and C. L. Wolin. 1989. An experimental analysis of patch size, habitat subdivision, and extinction in a marine intertidal snail. Conserv. Biol. 3: 242—251.
Quinn, J. F., C. van Riper, III and H. Salwasser. 1990. Mammalian extinctions from national parks in the western United States. Ecology (in press).
Quinn, J. F. and G. R. Robinson. 1987. The effects of experimental subdivision on flowering plant species diversity in a California annual grassland. J. Ecol. 75: 837—856.
Richter-Dyn, N. and N. S. Goel. 1972. On the extinction of a colonizing species. Theor. Pop. Biol. 3: 406—433.
Ricklefs, R. E. 1979. *Ecology.* Chiron Press, New York.
Roberts, L. 1988. Hard choices on biodiversity. Science 241: 1759—1761.
Robinson, G. R. and J. F. Quinn. 1988. Extinction, turnover and species diversity in an experimentally fragmented California annual grassland. Oecologia (Berlin) 76: 71—82.
Salwasser, H. 1987. Spotted owls: turning a battleground into a blueprint. Ecology 68: 776—779.
Schonewald-Cox, C. M., S. M. Chambers, F. MacBryde and L. Thomas (eds.) 1983. *Genetics and Conservation: A Reference for Managing Wild Animal and Plant Populations.* Benjamin/Cummings, Menlo Park, Calif.
Schoener, T. W. 1974. The species-area relationship within archipelagos: Models and evidence from land birds. Proc. 16th Int. Ornith. Congr., 629—642.
Schoener, T. W. and A. Schoener. 1983. The time to extinction of a colonizing propagule of lizards increases with island area. Nature 302: 332—334.
Shaffer, M. L. 1981. Minimum population sizes for species conservation. BioScience 31: 131—134.
Shaffer, M. L. and F. B. Samson. 1985. Population size and extinction: a note determining critical population size. Amer. Nat. 125: 144—152.

Simberloff, D. 1976. Experimental zoogeography of islands: Effects of island size. Ecology 57: 629—648.
Simberloff, D. 1987. The spotted owl fracas: Mixing academic, applied and political ecology. Ecology 68: 766—772.
Simberloff D. and L. G. Abele. 1976. Island Biogeography theory and conservation practice. Science 191: 285—286.
Simberloff, D. S. and L. G. Abele. 1982. Refuge design and island biogeography theory: Effects of fragmentation. Amer. Nat. 120: 41—50.
Simberloff, D. and J. Cox. 1987. Consequences and costs of conservation corridors. Conserv. Biol. 1: 63—71.
Simberloff, D. and N. Gotelli. 1984. Effects of insularization on plant species richness in the prairie-forest ecotone. Biol. Conserv. 29: 27—46.
Slatkin, M. 1974. Competition and regional coexistence. Ecology 55: 128—134.
Soulé, M. E. 1980. Thresholds for survival: maintaining fitness and evolutionary potential. In: M. E. Soulé and B. A. Wilcox (eds.), *Conservation Biology: An Evolutionary-Ecological Perspective.* Sinauer, Sunderland, Mass., pp. 151—169.
Soulé, M. E. (ed.) 1986. *Conservation Biology: The Science of Scarcity and Diversity.* Sinauer, Sunderland, Mass.
Soulé, M. E. (ed.) 1987. *Viable Populations for Conservation.* Cambridge University Press, Cambridge
Soulé, M. E. and D. Simberloff. 1986. What do genetics and ecology tell us about the design of nature reserves? Biol. Conserv. 29: 19—40.
Soulé, M. E. and B. A. Wilcox (eds.) 1980. *Conservation Biology: An Evolutionary-Ecological Perspective.* Sinauer, Sunderland, Mass.
Soulé, M. E., B. A. Wilcox and C. Holtby. 1979. Benign neglect: A model of faunal collapse in the game reserves of East Africa. Biol. Conserv. 15: 295—272.
Soulé, M. E., D. T. Bolger, A. G. Alberts, R. Sauvajot, J. Wright, M. Sorice, and S. Hill. 1988. Reconstructed dynamics of rapid extinctions of chaparral-requiring birds in urban habitat islands. Conserv. Biol. 2: 75—92.
Sullivan, A. L. and M. L. Shaffer. 1975. Biogeography of the megazoo. Science. 189: 13—17.
Sutherland, J. P. 1974. Multiple stable points in natural communities. Amer. Nat. 108: 859—873.
Templeton, A. R. 1986. Coadaptation and outbreeding depression. In: M. E. Soulé (ed.), *Conservation Biology: The Science of Scarcity and Diversity.* Sinauer, Sunderland, Mass, pp. 182—204.
Terborgh, J. 1975. Faunal equilibria and the design of wildlife preserves. In: F. B. Golley and E. Medina (eds.), *Tropical Ecological Systems: Trends in Terrestrial and Aquatic Research.* Springer, New York, pp. 369—380.
Terborgh, J. 1976. Island biogeography and conservation: strategy and limitations. Science 193: 1029—1030.
Terborgh, J., J. Faaborg, and H. F. Brockmann. 1978. Island colonization by Lesser Antillean birds. The Auk 95: 59—72.
Thiebodeau, F. R. 1983. Endangered species: Deciding which species to save. Environ. Mgmt. 7: 101—107.
Western, D. and J. Ssemakula. 1981. The future of savannah ecosystems: ecological islands or faunal enclaves? Afr. J. Ecol. 19: 7—19.
Whitcomb, R. F., J. F. Lynch, P. A. Opler and C. S. Robbins. 1976. Island biogeography theory and conservation: strategies and limitations. Science 193: 1030—1032.
Whitcomb, R. F., C. S. Robbins, J. F. Lynch, B. L. Whitcomb, M. K. Klimkiewicz and D. Bystrak. 1981. Effects of forest fragmentation on avifauna of the eastern deciduous forest. In: R. L. Burgess and D. M. Sharpe, (eds.) *Forest Island Dynamics in Man-dominated Landscapes.* Springer-Verlag, New York, pp. 125—205.
Wilbur, H. M. and J. T. Travis. 1984. An experimental approach to understanding pattern in

natural communities. In: D. R. Strong, D. Simberloff, L. G. Abele and A. B. Thistle (eds.), *Ecological Communities: Conceptual Issues and the Evidence.* Princeton University Press, Princeton, New Jersey, pp. 113—122.

Wilcox, B. A. 1980. Insular ecology and conservation. In: M. E. Soulé and B. A. Wilcox (eds.), *Conservation Biology: An Evolutionary-Ecological Perspective.* Sinauer, Sunderland, Mass, pp. 95—117.

Wilcox, B. A. and D. D. Murph. 1985. Conservation strategy: The effects of fragmentation on extinction. Amer. Nat. 125: 879—887.

Wilson, E. O. and E. O. Willis. 1975. Applied biogeography. In: M. L. Cody and J. M. Diamond (eds.), *Ecology and Evolution of Communities.* Harvard University Press, Cambridge, Massachusetts, pp. 522—534.

Wright, S. 1970. Random drift and the shifting balance theory of evolution. In: K. Kojima (ed.), *Mathematical Topics in Population Genetics. Springer-Verlag.* Berlin, pp. 1—31.

Wright, S. J. and S. P. Hubbell. 1983. Stochastic extinction and reserve size: A focal species approach. Oikos 41: 466—476.

Zimmerman, B. L. and R. O. Bierregaard. 1986. Relevance of the equilibrium theory of island biogeography and species-area relations to conservation with a case from Amazonia. J. Biogeog. 13: 133—143.

11. Perspectives on adaptive policy design in fisheries management

CARL J. WALTERS

Abstract

Fisheries management is necessarily an adaptive process, where decision makers can obtain some guidance from basic biological research and population dynamics theory, but must ultimately rely on direct management experience to test the validity of that guidance. The adaptive or learning process can be either "passive" or "active". A basic issue for decision makers is whether to treat scientific advice as correct until it proves untenable in practice (a "passively adaptive" or "evolutionary" strategy), or instead to deliberately experiment with policy choices so as to reveal the better one more quickly (an "actively adaptive" or "probing" strategy). Passive strategies may fail to reveal opportunities for improved management in several common situations, including (1) harvest policies for stocks that may have been severely depleted before much population data were gathered; (2) harvest policies for stocks that show "cyclic" behavior; (3) resource enhancement programs; and (4) management policies for multispecies "assemblage" fisheries. In these situations it is argued that basic research and theory cannot in principle provide the decision-maker with confidence about correct choices in advance, and conservative ("safe", risk averse) decisions would not be informative; the decision maker must either accept continuing uncertainty, or conduct a risky management experiment.

Introduction

Among the assortment of topics that are loosely referred to as applied population biology, fisheries population dynamics and management are usually thought of as "mature" subject areas, with a relatively long historical record of practical experience, empirical research, and theoretical development in terms of both biological and economic modeling. The basic theoretical concerns of fisheries population dynamics were identified early in this century (Hjort 1914; Baranov 1918), and a very extensive literature has accumulated about the processes (growth, mortality, recruitment) that contribute to fish production. The literature is characterized by a relatively healthy balance and interplay between theoretical and empirical studies, reinforcing one another to give apparently clear signals of considerable progress over the years.

S. K. Jain and L. W. Botsford (eds), Applied Population Biology, 249—262.

However, in spite of this extensive literature the practical management of fisheries and other renewable resources is generally viewed as a complicated art rather than a science, involving the integration of scientific information with social and economic concerns (objectives, constraints) so as to produce sustained benefits over time. There is often fierce debate about the reliability and relevance of whatever scientific data happen to be available, about appropriate objectives, and about the various models that are used (both intuitive and formal) for relating parts of the data and for predicting responses to particular policy measures. However, there has been little debate about one very basic aspect of the viewpoint, namely the assumption that decision making can and should be functionally separated from scientific information gathering. This separation is widely reflected in the structure of fisheries organizations (research versus management divisions, etc.), in publication guidelines for fisheries journals, and in cooperative arrangements between academic researchers and fisheries agencies. However, management decisions are a potentially valuable source of perturbations (disturbances) that could be used to gain scientific understanding of fisheries systems. The existing functional separation of research and management activities may be weakening both the science and the decision making.

Recently there has been considerable interest in the idea of "adaptive management" (Holling 1978; Ludwig and Waters 1982; Walters and Hilborn 1976, 1978; Walters 1981, 1986, 1987a), where decision making in the face of uncertainty is seen as a sequence of "experiments" that may be more or less informative about system potentials. This paper reviews our thinking about three aspects of the idea: (1) weaknesses in research approaches that propose to develop useful information independent of decision making; (2) the issue of passive versus active adaptation (i.e., when to deliberately undertake risky management experiments); and (3) prototypical situations where it appears that experimental management approaches may be necessary to eventually gain scientific understanding. Readers familiar with discussions in the general ecological literature about the need for field experimentation (Diamond 1986) will recognize some of the arguments mentioned below; an added ingredient not present in the general literature is a concern with the social and economic valuation of experimental policy options.

We shall illustrate a prototypical situation where an adaptive approach to management might produce substantial economic benefits. The Fraser River sockeye salmon stock supports one of Canada's most valuable fisheries. Like most fisheries, a key problem in managing this stock is to identify the best number of fish to allow past the fisheries to spawn each year. To identify that best number, we use historical data (Walters 1986) to estimate an empirical relationship between spawners and the number of subsequent recruits that these spawners produce to future fisheries. The "stock-recruit" relation shows the net effect of all reproduction and survival factors that lead from spawners to recruitment. But we may note there is a serious problem: over the range of available historical data, it appears that more spawners lead to proportionally more recruits (no density dependence evident in reproduction/survival rates). This means that more spawners might produce more recruits to catch. However, managers in the system have hesitated to allow more spawners, because to do so would involve immediate reduction in catches (and loud complaints from fishermen) and might produce "overspawning" (poor recruitment from too many spawners). If we end up with gross extrapolations beyond historical experience, there is no way to decide which is correct (or how the population would actually respond) without actually allowing more fish to spawn. Allowing more spawners could trigger a variety of biological responses (crowding, dispersal, colonization) that cannot be predicted by just studying the fish under the spawning conditions that they face today. To increase spawning stocks would be a gamble: the fishermen would have to give up some catch in the short term in order to allow more spawners, with no guarantee of bigger returns in the future. The key problem in adaptive management is to decide under what conditions such gambles are worth taking.

Research approaches in fisheries

Biologists have taken three basic approaches to gathering information that they presume will eventually be valuable in fisheries management: reductionist/functional, empirical/correlative, and experimental/comparative. In this section I will argue that the first two of these, which comprise the bulk of current research activity and publication in fisheries, are of much more limited value (and future potential) than is usually supposed. They can provide the manager with innovative ideas about how management might be improved, but not adequately dependable predictions about what will happen if those ideas are implemented. Basic concepts and examples are reviewed with an emphasis on population management, to argue in favor of an experimental/comparative approach in fisheries.

Reductionist/functional approach

The reductionist/functional approach involves trying to "understand" a managed system by studying the behavior of its parts, or functional components, while using mathematical devices (models) to assemble the components into overall predictions. Early fisheries work identified three key components of behavior in exploited populations (growth, natural mortality, and recruitment), and several researchers (Ricker 1975; Beverton and Holt 1957) provided models for putting these components together; much of fisheries research is still plugging away at these components, seeking to measure them more precisely and understand them in more detail. As fisheries management concerns have broadened, particularly in relation to pollution and aqua-culture, there has also been explosive growth in the literature on a number of detailed topics at the individual organism level (fish diseases, toxicology, impacts of chemical changes on all sorts of aquatic organisms, etc.)

In the last 20 years we have gained much experience with trying to synthesize the results of reductionist research into mathematical models. This experience has been extremely frustrating, for three reasons: (1) fisheries "systems" have no natural logical boundaries, making it impossible to decide in advance what breadth of components (just biology or biology plus uncontrolled economic behavior, etc.) need to be included (in models) to successfully predict responses to management; (2) some key biological processes are only expressed at large (expensive to study) space-time scales in the field, and hence have not been practical to study experimentally (especially recruitment, and processes like predation that define species interactions); (3) attempting to avoid the scale problem by modeling large processes in more detail, where the details are hopefully easier to study, has led to explosions in model complexity with only fragmentary data available on many of the details. These frustrations may be symptoms of fundamental theoretical limits on how well we can ever deal with complex systems by reductionist approaches, as claimed by system theorists (e.g. Rosen 1978). But whether or not there are theoretical limits, it has certainly not been practically possible to construct functional models that scientist would trust without direct field tests or that a resource manager would dare rely on and could fully justify as a basis for making decisions.

Empirical/correlative approach

The empirical/correlative approach involves taking space/time data series (usually of population level variables) that have been gathered without any particular experimental design, then seeking useful patterns in the data. The main fisheries examples are: attempts to describe the relationship between stock and recruitment estimation of mortality and recruitment rates from changes in age composition associated with changes in fishing effort, analyses of patterns in productivity in relation to environmental factors such as nutrient loading, and the search for environmental variables that are correlated with recruitment.

Beyond the standard scientific warning that "correlation does not imply causality", a few more pitfalls in the empirical/correlative approach are worth noting: (1) natural variability and measurement errors result in a nonrepresentative time series of system responses, and hence biased estimates of model parameters (Walters 1985, 1986); (2) many environmental data series are available to fisheries scientists, so a diligent search will always reveal some good (but probably spurious) correlation patterns; and (3) biological and physical linkages (over space, time, and among processes) are complex enough to guarantee that a plausible functional explanation (a particular set of causal linkages) can be found for any correlation that happens to exist. Each of these pitfalls should be sufficient to make an intelligent resource manager suspicious of predictive models constructed from correlative data; taken together they imply that the manager should treat such models only as suggestive of hypotheses that may deserve further research (but certainly not as reliable tools for decision making).

Experimental/comparative approach

The direct experimental/comparative approach to study of how fisheries systems behave, as intact units in the field, is gaining a renewed popularity. It actually has quite a long history, for example, in the development of management strategies for farm pond fish communities, in the laboratory population studies of Silliman (1968), and in experiments aimed at comparing alternative regulatory policies (size limits, gear restrictions, etc.) for freshwater sport fisheries (Roedel 1975). The main recent examples involve seeing how lake ecosystems respond to disturbances such as nutrient addition and acidification (Schindler 1975; Schindler and Turner 1982), and how salmonid populations are influenced by watershed disturbances such as logging (Hartman 1982; Murphy *et al.* 1986). These examples have been a humbling experience for the reductionist model builders, by revealing various "surprising" dynamic responses in the field (unexpected buffering capacity in lake acidification experiments, big fish production changes from modest fertilization, lack of effect or even enhancement of coho salmon production by logging activities that seemed to cause deterioration in water quality).

Unfortunately, most fisheries systems are defined on space/time scales that are too large for conventional experimental study using the basic devices of replication, randomization, and factorial arrangement of treatments (including controls). Even where replication is possible and the logistics of disturbance and monitoring are practical, large scale experiments usually take so long to provide definitive results that they are not attractive to the brightest, most ambitious scientists. Who is silly enough to work hard at setting up an experiment that will not bear fruit for 20 years, when it is perfectly acceptable to engage instead in reductionist studies that can be published next year? While the scientist's response to this question is obvious, it is important to recognize that the resource manager should not even be asking the question; unfortunately, many managers do ask it, but with the word "published" replaced by the phrase "used to justify whatever action he feels intuitively is best".

Active versus passive adaptation

Considering that predictive management models constructed from functional or correlative data cannot be trusted, resource managers would be wise to view each of their decisions as an "experiment" with an uncertain outcome but the potential to provide more experience upon which to base future decisions. In the language of control system theory, each decision has a "dual effect" (Fel'dbaum 1960—61): it impacts immediately on system performance, and it impacts in the longer term on the uncertainty that future decision makers will have to face. Generally there is a tradeoff between these effects: more dramatic decisions today are more costly (and risky), but provide stronger signals (clearer contrasts among alternative hypotheses) for future data analysis. Thus the statement that we should view decisions as experiments is in fact not a simple recommendation at all: it implies a very difficult decision problem involving risks and tradeoffs between short and long term values.

There have been various attempts to develop formal optimization models for the tradeoff between short term costs and long term benefits of experimental decisions (reviews in Walters 1986, 1987a), but no clear and general guidelines have emerged from these attempts. Two main difficulties have been encountered: (1) the modeling/optimization formulations are very complex, since one must model not only alternative system responses to each decision (i.e. predictions from alternative models), but also how future data will be gathered and used (i.e. how future data will be analyzed statistically in relation to the alternative models); and (2) the expected value of risky decisions often differs only slightly from the expected value of more cautious, conservative decisions.

Instead of viewing decisions as experiments, a "passively adaptive" alternative would be to assume at each decision point that some particular "best" model (based on fit to historical experience up to that decision point) is in fact correct. This strategy can result in quite an informative sequence of decisions and data over time, due to changing assessments about which model is best and also due to informative variation imposed on the system by natural disturbances outside the manager's control.

A common outcome in our optimization calculations has been to find that passive adaptation is substantially better (30—50%) than policies involving no monitoring and learning from experience, while the additional gain due to active adaptation (with deliberate experimental "probing" through variation in policy choices) is quite modest (5—15%). What happens in such calculations is not that deliberate experiments have little potential value, but rather that their expected value across possible outcomes is pulled down by bad possibilities. For example, suppose a passive choice has an expected value of 100; the best experimental choice might then have a best outcome of 150, but a worst one of 60, so that its expected value if both outcomes are equally likely is only 105.

Although these results indicate that passive adaptation might work well in principle for many fisheries systems, we have grave doubts about its application in practice. We have encountered a number of situations where analysis of historical data indicated quite clearly (to us) that historical management practice had been faulty; for example stock-recruitment analysis indicates that higher spawning stock levels would result in higher productivity from the major sockeye salmon stocks of Bristol Bay and the Fraser River. When discussing such analyses with local fisheries managers, we have found that they can generally articulate quite clearly (and vigorously defend) some alternative model that rationalizes the historical policy choices (i.e. a model for which the choices have been optimal). Notice that any recommendation for change in management practice must involve an extrapolation of what will happen for choices outside the range of historical experience; what any manager can validly argue is that his rationalizing model (which says to stay in the range of past experience) is as good an extrapolation as the model which leads to the recommendation for change. In other words, there is always a "scientific" defense (my extrapolation is just as good as yours) for avoiding policy changes and hence for avoiding even passive learning. To make matters worse, there are obvious psychological reasons for resource managers to defend their rationalizing models. The bottom line is that passively adaptive strategies may work fine in automatic control systems and computer simulations, but it is naive to expect that they will work efficiently or effectively in fisheries decision making.

Situations where actively adaptive policies are needed

Even when decision making is fully objective, so that there are no strong attempts to rationalize or cling to historical models and policies, there are some important situations where passively adaptive management will likely fail to reveal opportunities for improved management. In these situations, behaving cautiously (or according to the predictions of a single "best" model based on historical data) will lead to continued collection of data that are no more informative than the historical information already available, unless the manager is lucky enough to experience a drastic natural disturbance. This section reviews four such situations that are of considerable current interest in fisheries management.

Depleted stocks

Many of our major fish stocks, ranging from the large sockeye salmon stocks of Alaska and British Columbia to Pacific halibut to herring, may have been reduced to less than the most productive stock levels by harvesting before effective monitoring systems became available. In more recent times, these populations have generally been managed by stabilizing conservation policies that prevent further depletion but do not promote recoveries. Due to biases associated with (1) random errors in stock size measurement and (2) interaction between random environmental effects and stock size, analyses of recent data for these stocks will likely lead to the conclusion that productivity (recruitment rate, surplus production rate) is independent of stock size (Walters and Ludwig 1981; Waters 1985; Walters and Staley 1987), even if much higher stock sizes would in fact be much more productive.

This situation generates considerable debate and confusion. On the one hand, the recent historical data give a "clear" signal that it would not be beneficial to reduce harvest rates. On the other hand, there are often hints from older data that stock sizes could be much larger, and there may appear to be habitat potential (spawning area, etc.) for stock increase. One can construct endless functional models, based on food chain or habitat requirements data, that suggest the possibility of considerable increase. But these models are easily criticized (correctly) on the grounds that they may omit some key limiting factor (such as disease outbreaks) that has gone unnoticed in recent historical times.

It is important to recognize that there is no way, even in principle, to avoid the criticism about hidden limiting factors by doing more careful, reductionist biological research on a stock while it remains at low levels. To any experimental demonstration on a local scale (laboratory, field substock) that factor X is absent or does not prevent recovery, one may simply rejoin either that there is another factor Y or that factor X does not express its impacts on the local space-time scale studied so far. In other words the effects of factor X may be an "emergent property" of the overall stock's behavior, and will only be visible if the overall stock is manipulated (i.e., allowed to recover).

Experimental recovery policies have been considered for a number of salmon populations in the Pacific northwest, and have been implemented for one major British Columbia stock (rivers Inlet sockeye); so far the Rivers inlet experience has been disappointing (stock has not recovered as expected). Higher spawning stocks are being allowed in a few other British Columbia salmon stocks, such as the Adams River sockeye, when circumstances (unusually good natural survival) permit. Impetus for further experiments has been provided by the results of a dramatic "natural" experiment in Bristol Bay, Alaska, where increased sockeye salmon escapements in the 1970's were followed by the largest returns in history (Eggers and Rogers 1987).

Cyclic stocks

A number of major fish and invertebrate stocks have displayed more or less regular abundance "cycles"; the most dramatic examples along the Pacific coast are the Dungeness crab off California, Pacific cod off British Columbia, and the Pacific halibut off British Columbia and Alaska. There have been two basic types of models or explanations for these cycles (Botsford 1986; Skud 1982; Walters *et al.* 1986; Deriso 1987): "density dependent recruitment" and "environmental forcing". The density dependent recruitment model proposes that there is a dome-shaped relationship between parental stock size and recruitment, such that low stock sizes produce peak recruitments which then result in higher stock sizes and lower recruitments. The environmental forcing model proposes that recruitment variation is driven not by stock size but by some environmental factor that varies in a cyclic pattern over time (or at least displays extreme values at fairly regular intervals). Botsford (1986) has argued that both effects may be present, such that the environmental forcing leads to exaggeration of any cyclic tendencies that may be present due to density dependent recruitment.

The two models imply strikingly different policies for maximizing long term yield. If recruitment is strongly density dependent, then fishing rates should be kept very high so as to hold stock size down (and recruitment high) and break up the cyclic pattern. If recruitment is mainly dependent on environment, then either fishing rates do not matter or else should be kept quite low, particularly following periods of low recruitment, to insure that stock size is not dangerously depleted. The conservative, safe management choice is to behave as thought he environmental forcing model is correct (use low harvest rates, accept cyclic variation). Unregulated fishing effort, where effort moves up and down in response to stock size (usually with at least a one year lag), may exaggerate the cycles but is also "safe" in the sense that lower efforts at low stock sizes will tend to prevent gross depletion.

Both models predict essentially the same stock behavior under conservative management policies: continued cycling and an inverse correlation between stock size and recruitment rate. As noted earlier, with the large oceanographic data bases that are available it will always be possible to find environmental measurements that are well correlated with the recruitment variation, and to concoct causal explanations for such correlations. Should any particular correlation fail over time, one may always argue that the environmental measurement used was just not quite the right "index" for the environmental syndrome that causes recruitment variation, and may follow up this argument by doing a new (and likely successful) search for a correlated index.

A key question is whether the environmental forcing models can be "tested" while a conservative policy is maintained, by doing reductionist/functional studies on the detailed links between environmental factors and recruitment. For example, one might hope to demonstrate that water temperature affects larval growth rate, which in turn influences how long the larvae are exposed to particular predators. I believe that the answer is no, for two reasons: (1) detailed field monitoring can only show whether the correlations present at gross levels of measurement (e.g. total recruitment) are also present at lower levels (e.g. growth or survival through short stages), which would only confirm that gross behaviors are predictable from lower levels (rather than confirming causality in relation to the lower level correlations); and (2) any particular detailed model for linkages can always be criticized by pointing out how linkages that it does not consider (there are always some of these) may, in the future, cancel or reverse any patterns that have been evident to date.

The critical field test of density dependence versus environmental forcing models would be a management experiment: keep fishing hard enough to hold stock size down, then see if recruitment remains high. This is obviously a risky experiment; if the environmental forcing model is correct (and recruitment collapses or remains cyclic), lost yields to allow stock recovery may be greater than the short term gains obtained during the initial period of hard fishing. Calculations by R. B. Deriso and Anna Parma (Deriso 1987) suggest that the balance of gains and losses will favor doing the experiment only if a high prior probability is assigned to the density dependence model.

Resource enhancement programs

Major investments are being made around the world, for a great variety of fish species, to increase production through habitat improvements and hatchery programs. Such investments have been widely viewed as grand management experiments, and adaptive optimization calculations (outcomes under alternative hypotheses, times required to detect correct hypothesis under various monitoring programs, etc.) have indicated that they can be very good gambles (Walters 1977). The $300 million Canadian Salmonid Enhancement Program was explicitly designed and funded as a two-phase experiment, with Phase I (1977—present) being a 10-year test of various technologies (hatcheries, spawning channels, lake fertilization, etc.) and Phase II to involve larger scale implementation and continuation of whatever technologies prove most effective. Though it has turned out that a 10 year test period will not be sufficient (20 years would have been more realistic but politically difficult to sell), the Canadian government has kept its promise to engage in careful evaluation and program revision over time. A similar experimental program is currently being developed for future investments in mitigation of hydropower impacts on salmon production in the Columbia River Basin, through the Northwest Power Planning Council (Webb 1986).

A striking and unexpected pattern has appeared in chinook and coho salmon production from some Canadian and American hatcheries: beginning in the early to mid-1970s, survival rates (measured as catch plus escapement per smolt) began to decline rapidly, so that total production has not increased in recent years in spite of substantial increases in number of smolts released. A variety of hypotheses have been advanced to explain the declines (review in Walters 1987b), and these involve two general alternatives: (1) increasing smolt production has stimulated some agent(s) of mortality (such as disease) either within the hatcheries or in their receiving environments, so that survival would improve again if smolt releases were reduced; or (2) there has been a change in ocean rearing conditions (upwelling and temperature changes have been correlated with the declines). These alternative hypotheses are strikingly similar to the density dependence and environmental forcing models for cyclic stocks.

Since there are many hatcheries dumping fish into essentially the same ocean environment, there is a unique opportunity to do a controlled experiment to test between the smolt production and ocean rearing hypotheses. The idea would be to reduce smolt releases from a few (replicate) hatcheries, while maintaining current operating practices in the others (controls). A few new hatcheries should be added to the treatment group each year over several years, to control for time-treatment interactions. The value of controls in this experiment is obvious: if rearing practices were simultaneously changed in all hatcheries (which has unfortunately happened in relation to some practices), it would be impossible to say whether a positive response was due to the change or to a reversal of environmental conditions; we would be left with the haunting concern that the environmental problem might reappear.

Multispecies assemblage fisheries

Many fisheries are directed not at single stocks but at whole assemblages of species; prominent examples are most trawl fisheries and coral reef fisheries. There has been much agonizing by fisheries scientists about how to deal with these situations (Pauly 1979); analysis on a stock-by-stock basis is practically impossible in many cases (monitoring and functional data requirements too great), it is difficult to gather data for modeling for the various species interactions that may be important, and it is unclear how to handle conflicting objectives (protection of less productive species versus catching more of the productive ones). In short, all of the problems and pitfalls of reductionist/functional and empirical/correlative research are multiplied many-fold over the single stock case. One attempt to avoid these problems has been to look at empirical relationships between yield and fishing effort, either over time or among contrasting fishing areas (examples in Murphy and Pauly 1983); a difficulty with this approach is that the point estimates of yield may not represent sustainable or equilibrium values.

Fortunately most assemblage fisheries are spatially structured into a number of similar, more or less isolated "experimental units" that could be used as the basis for planned comparisons among alternative fishing policies. Two types of comparisons are much needed today, both as a basis for management and to provide insights about profitable directions for more detailed research. First, there should be comparisons of temporal behavior for assemblage units that are subject to different, fixed (over time) exploitation rates; such comparisons are needed to establish whether, and over what time scale, equilibrium is reached in species composition and total production. Second, there should be comparisons of steady exploitation versus "rotational" harvesting, where the assemblages are fished heavily only once every several years and are allowed "fallow periods" between harvests. Rotational policies may represent a more natural (and hence sustainable) disturbance pattern than annual harvesting, helping to preserve diversity that is due to natural successional dynamics.

Such experiments would obviously be rather expensive to monitor and maintain over time (enforce against poaching, etc.), and would require substantially better cooperation from fishermen than most management agencies have managed to muster. But if the experiments are not done, we will be stuck with trying to construct multispecies models without any clear idea of even the general phenomenological patterns that the models ought to predict; the transient, highly variable data that are currently available from most assemblage fisheries can be interpreted in as many different ways (i.e., fit as many different models) as there are analysts to look at them.

We do not yet know how to properly evaluate the difficult decision choices associated with situations such as depleted and cyclic stocks, where informative decision choices would involve immediate loss in yields or risks of longer term collapse. Informative choices in these situations are essentially gambles on untested opportunities for improved yields, and there is no coherent theory about gambling in the context of public decision making: should public decision makers behave in a risk averse manner (as would most individuals), and how should they measure the social cost of hardships felt by fishermen that are displaced by changing harvest rates? These are not scientific questions, and until they are resolved it would be imprudent to suggest that actively adaptive strategies are better than accepting continuing uncertainty over time.

However, there are no good excuses for inaction in situations like salmon enhancement and coral reef fisheries, where the risks and costs of experimentation can be isolated upon relatively few spatial units while the benefits of improved policies would accrue from many more units. The main challenge in these situations is to our imagination: to identify innovative policy options, to find experimental designs that provide for adequate replication and control, and to find clever ways of reducing monitoring costs across the experimental units.

References

Baranov, F. I. 1919. On the question of the biological basis of fisheries. Nauch. Isssled. Ikhtiologicheskii Inst. Izv. 1: 81—128 (In Russian).

Beverton, R. J. and S. J. Holt. 1957. On the dynamics of exploited fish populations. U.K. Min. Agric. Fish., Fish. Invest. (Ser. 2) 19: 533 pp.

Botsford, L. W. 1986. Population dynamics of the Dungeness crab (Cancer magister). In: Jamieson, G. S., N. Bourne (eds.), North Pacific Workshop on stock assessment and management of invertebrates. Can. Spec. Pub. Fish. Aquat. Sci. No. 92.
Deriso, R. B. 1987. Pacific halibut: biology, fishery, and management. Northwest Environmental Journal 3: 129—144.
Diamond, J. 1986. Overview: laboratory experiments, field experiments, and natural experiments. pp. 3—22. In: Diamond, J. and T. J. Case (eds.). Community Ecology. Harper and Row, New York. 665 pp.
Eggers, D. and D. Rogers. 1987. The cycle of returns of sockeye salmon (Oncorhynchus nerka Walbaum) to the Kvichak River, Bristol Bay, Alaska: cyclic dominance or depensatory fishing? Can. J. Fish. Aquat. Sci. (In press). [has this one made it to press yet?? dmd]
Fel'dbaum, A. A. 1960—61. Theory of dual control I—IV. Automatic Remote Control USSR 21: 1240—49, 1453—65; 22: 3—16, 129—43 (In Russian).
Hartman, C. F., ed. 1982. Proceedings of the Carnation Creek Workshop: a ten-year review. Pacific Biological Station, Nanaimo, B.C.
Hjort, J. 1914. Fluctuations in the great fisheries of northern Europe, viewed in the light of biological research. Rapp, P.-V. Reun. Cons. Perm. Int. Explor. Mer. 20: 1—228.
Holling, C. S., ed. 1978. Adaptive environmental assessment and management. John Wiley, New York.
Ludwig, D. and C. J. Walters. 1982. Optimal harvesting with imprecise parameter estimates. Ecological Modelling 14: 273—92.
Murphy, G. and D. Pauly, eds. 1983. Management of tropical fisheries, theory and practice.
Murphy, M. L., J. Heifetz, S. W. Johnson, K. V. Koski and J. F. Thedinga. 1986. Effects of clear-cut logging with and without buffer strips on juvenile salmonids in Alaskan streams. Can. J. Fish. Aquatic Sci. 43: 1521—1533.
Pauly, D. 1979. Theory and management of tropical multispecies fisheries. ICLARM Studies and Reviews, No. 1, Manila. 35 pp.
Ricker, W. E. 1975. Computation and interpretation of biological statistics of fish populations. Bull. Fish. Res. Bd. Canada, No 191. 382 pp.
Roedel, P. M., ed. 1975. Optimum sustainable yield as a concept in fisheries management. Amer. Fish. Soc. Spec. Publ., No. 9.
Rosen, R. 1978. Principles of measurement and representation of natural systems. Elsevier/ North Holland, Inc., New York.
Skud, B. E. 1982. Dominance in fishes: the relation between environment and abundance. Science 46: 144—49.
Walters, C. J. 1977. Design of experimental salmon enhancement policies. In: D. V. Ellis (ed.), Pacific salmon: management for people. Chapter 10. University Victoria Press, Western Geographical Series, Vol. 13. Victoria, B.C.
Walters, C. J. 1981. Optimum escapements in the face of alternative recruitment hypotheses. Can. J. Fish. Aquat. Sci. 38: 678—89.
Walters, C. J. 1985. Bias in the estimation of functional relationships from time series data. Can. J. Fish. Aquat. Sci. 42: 147—49.
Walters, C. J. 1986. Adaptive management of renewable resources. Mcmillan Pub. Co., New York.
Walters, C. J. 1987a. Approaches to adaptive policy design for harvest management. In: T. Vincent (ed.), Proceedings 2nd US/Australia workshop on applied control theory in natural resource management. Springer-Verlag, Lecture Notes in Biomathematics, Berlin.
Walters, C. J. 1987b. Mixed stock fisheries and the sustainability of enhancement production for chinook and coho salmon. In: R. McNeill (ed.), Proc. 1st Intl. Salmonid Conference. Oregon State Univ. Press.
Walters, C. J. and R. Hilborn. 1976. Adaptive control of fishing systems. J. Fish. Res. Bd. Canada 33: 145—59.
Walters, C. J. and R. Hilborn. 1978. Ecological optimization and adaptive management. Ann. Rev. Ecol. Syst. 9: 157—88.

Walters, C. J. and D. Ludwig. 1981. Effects of measurement errors on the assessment of stock-recruitment relationships. Can. J. Fish. Aquat. Sci. 38: 704—10.

Walters, C. J. and M. J. Staley. 1987. Evidence against the existence of cyclic dominance in Fraser River sockeye salmon (*Oncorhynchus nerka*). Can. J. Fish. Aquat. Sci.

Walters, C. J., M. Stocker, A. V. Tyler, and S. J. Estrherim. 1986. Interaction between Pacific Cod (Gadus macrocephalus) and Herring (*Clupea harengus pallasi*) in the Hecate Strait, British Columbia. Can. J. Fish. Aquat. Sci. 43: 830—37.

Webb, T. 1986. Approaches to the application of adaptive management principles to Columbia Basin planning. Final Report by ESSA Ltd. to Northwest Power Planning Council, Portland, Oregon.

12. Applying the principles of population biology: assessment and recommendations

LOUIS W. BOTSFORD and SUBODH K. JAIN

Abstract

In addition to describing recent developments in specific fields, the previous chapters demonstrate a common theme, that the solutions to practical problems have not involved the ready application of a set of firm principles of population biology. Here we (a) explore several possible reasons why this is so, (b) evaluate several weaknesses in the ways in which theoretical principles are currently developed, and (c) suggest several modifications of our approach to both the development of theory and the application of the resulting principles of population biology to practical problems. There is a large amount of uncertainty inherent in the problems of population biology, due to both natural variability on a wide range of temporal and spatial scales, and limited knowledge of mechanisms. The former represents an inherent limit on predictability, and the latter is unique to population biology because of the inherent heterogeneity of mechanisms and behavior. Both should be more widely appreciated among decision makers, as they limit reasonable expectations. They deserve wider appreciation among population biologists as well, as they set a requirement for a special approach tailored to the specific nature of this uncertainty. We recommend an approach in which results from applications are monitored so that they can form an empirical extension of the process of developing theoretical principles. This approach would both foster development of a theory that could be more useful in the solution of practical problems in applied population biology and provide additional empirical support for the theory itself. In the development of a "general, predictive" theory, we recommend (1) closer adherence to the more restrictive definition of the word "general" (i.e., holds for many specific cases) and (2) an appreciation for the weak implications of correct predictions. The former will require synthesis from a larger number of examples, which can be supplied from practical problems. The latter suggests a shift to more of a hypothetico-deductive scheme, which can guide the incorporation of results from applied problems into the empirical process. We point out several existing trends in directions consistent with these recommendations, as well as the points of view of a number of prominent population biologists who argue against a closer relationship between theory and the real world.

An overview of examples

Although these ten chapters (2 through 11) do not provide comprehensive coverage of all the fields in which population biology is applied, they include examples of most of the inherent problems typically encountered in applica-

S. K. Jain and L. W. Botsford (eds), Applied Population Biology, 263–286.

tions. They give us, therefore, some idea of how the principles of population biology are applied, as well as potential problems in the current state of that art. Perhaps the most important observation in the context of this book is that in no instance is a set of established principles taken "off the shelf" and used to provide a predictive solution to a practical problem. In many cases the specific details of natural history and related facts from fields such as soil physics or chemistry are of more value (e.g., in Chapter 2, information on plant nutrient requirements was of more direct use than theories from community ecology). However, in some instances, particularly those involving the genetic basis of plant breeding and the role of genetic variability in weed control, theoretical principles provide useful tactics, though not details (e.g., Chapters 3 through 6). Also, in a few other instances, they provide *post hoc* explanations of the current state of the population or community being managed, based on past events, again more so in problems in which genetics, rather than population dynamics, is the major focus (e.g., the explanation of current patterns of variation in terms of a genetic bottleneck in Chapters 3 and 4; and the selection response in a single large versus several small subdivided populations in Chapter 6).

In Chapters 3 and 4, both Hedrick and Kesseli discuss practical implications of a useful construct from genetic principles, the effective population size (N_e). Information on certain reproductive features is used to convert actual number of individuals in a specific population to the effective number of randomly mating individuals by adjusting for the effects of various factors such as sex ratio and breeding system on the expected retention rates of total genetic variance. In this case, theory provides a valuable means for comparing different species on a common basis. Hedrick outlines the potential for loss of fitness through inbreeding depression in small populations of normally outbred species. Many Species Survival Plans, developed mainly for the so-called charismatic megavertebrates, center on these simple genetic predictions and some simple solutions (e.g., Lande 1988; Western and Pearl 1989). There are also several detailed studies in rare plant species which seek minimum viable population size, based on both genetic and demographic data. Population geneticists have become actively involved in conservation biology (Schonewald-Cox *et al.* 1983; Soulè 1987). However, as Hedrick points out in Chapter 3, many simplifying assumptions are made in estimating N_e and predicting the consequences of small N_e, and much more needs to be learned from the long-term records of various conservation programs.

In Chapter 4, Kesseli focusses on problems specific to rare and endangered species of plants. In a plant community context, he describes how habitat fragmentation, losses of seed output (due to lack of pollinators), and losses of recruitment niche (poorer dispersal or germination) might play a larger role than genetic variation. It appears from surveys of population genetic data that different *Limnanthes* species might survive these extinction threats in quite different ways. Some species might evolve high inbreeding and shorter life cycles, others seem to have greater reliance on phenotypic plasticity, and one taxon (*L. douglasii* var. *rosea*) shows an increased role of gynodioecy, and outbreeding system originating locally from intervarietal hybrids. Here too, genetic principles provide useful guidelines for estimating the effective size of populations, but confirmation of their accuracy and the ability to make specific recommendations for management will require more examples, either through experiments or monitoring applications (Jain 1991).

In Chapters 5 and 6, genetic principles provide valuable *post hoc* insights into the dynamics of a less controlled case (weeds) and valuable guidance in dealing with the agricultural species. In both, the effects of large amounts of genetic variability, either as a goal of selection or a constraint in biological control, are not fully understood. On the other hand, several new plant breeding methods fully utilize the theory of selection response, role of recombination and varying mating designs. Host-pathogen coevolution, now increasingly analyzed with models from population biology (Jayakar and Zonta 1990), will undoubtedly play a major role in new methods of weed control and the management of crop diseases.

In Chapter 7 by Dobson and Hudson on red grouse and Chapter 8 by Smith on parasites in cattle and sheep, mathematical models were applied in two different ways. The former involves the use of models to determine a causal mechanism in a population problem, and the latter uses models to predict results, albeit only as an indicator of relative behavior, not an absolute predictor. Although both of these chapters made use of a recently developed approach to modeling parasites (Anderson and May 1978), neither really used previously established theoretical results. General principles regarding specific life history characteristics that lead to cycles, or general statements regarding vulnerability of parasites to treatment would have been valuable. These would require further observations and analysis.

In the two chapters in which existing population dynamic theory was instrumental, Chapter 9 on pest control and Chapter 10 on reserve size, it was found lacking. Previous results from simple, models were not consistent with observations, and more detailed analysis of more realistic models was recommended. In Chapter 10, Murdoch reviews the theoretical conclusion that biological control must involve the controlling species driving the controlled species to a low, stable level and maintaining it there. Exceptions to this are more the rule, and in actual cases dispersal between subpopulations and aggregation on hosts are more important (cf., Strong 1988a, 1988b; May and Hassel 1988a, 1988b). In Chapter 10, Robinson and Quinn critically analyze data from studies that have been used to support earlier conclusions from island biogeography. Again, the earlier conclusions, in this case that single large reserves always provide for greater diversity, do not hold up.

In Chapter 11, Walters outlines some of the ways of specifically addressing the uncertainty involved in practical problems in population biology. Based on long experience with practical problems, Walters and his colleagues have developed a formal, general approach to solving environmental problems, most of which involve some aspect of population biology (Holling 1978; Walters 1986). This approach, termed adaptive management, takes advantage of the fact that prescribed solutions to practical problems can be used to gain more information about the system being managed, in addition to accomplishing whatever other objective is desired. In his chapter, Walters points out that the procedures to use in adaptive management are neither simple nor obvious. Rather comprehensive analysis of data-gathering and modeling is required to iteratively increase our understanding of various methods.

Although none of the other chapters deals with adaptive approaches explicitly, some of them hint at a continuing, iterative interaction between theory and application to establish a better understanding of the system of interest. For example, in Chapter 2, Bradshaw proposes we use our current understanding of succession to reconstruct communities, but more as a test of that understanding than a reliable guide to the outcome. Also, plant and animal breeders continually adapt their breeding methods in the light of previous results from selection, and evolutionary responses of genetic stocks dictate some of these adaptive choices. As another example, in their prescription of ways of reducing the effects of the parasite on red grouse, Dobson and Hudson point out those that facilitate the accumulation of information regarding the dynamics of this system.

Sources of problems in applications

One of the goals of this book is to address the problem of the perceived lack of success in solutions to practical problems in population biology. The solutions to practical problems addressed in the previous ten chapters rarely involve the application of any principles of population biology. We are, therefore, led to the questions of why population biology may not be providing adequate principles on which applied solutions can be based, and whether there are any ways in which this situation could be improved. In Chapter 1 we mentioned three possible causes for the rarity of application of principles to solve practical problems in population biology: (1) limitations of theory, (2) inherent natural variability, and (3) a flawed approach to applications. The first is the possibility that the methods by which we are developing the principles of population biology do not lead to the kinds of principles that are useful in practical problems. The second is the possibility that the degree of uncertainty and randomness present in populations is so great that we can not possibly achieve any greater success rate. The last possibility is that our approach to solutions of practical problems in population biology is inherently flawed; the idea that we can develop principles, then use them to confidently prescribe solutions to practical problems, with little further involvement in each problem, may itself be wrong. We evaluate each of these here, before going on to suggest a program for improved success.

Limitations of current theory

One of the difficulties encountered in attempting to critically evaluate the theory of population biology is that, as outlined in Chapter 1, there are several branches of theory. Theoretical population ecology is distinct from systems ecology (Levin 1976; Levin *et al.* 1989), and there are also several other theories of limited scope (e.g., optimal foraging theory, life history theory, etc.; see Chapter 1). A second consideration is that existing theory was not developed solely for the solution of practical problems. Some models are merely metaphors developed for pedagogical reasons. In fact, theoretical ecology and applied ecology are often viewed as quite independent pursuits. Some theoretical ecologists maintain their distance from practitioners, referring to practical problems only in the justification section on the last page of their grant applications, while some applied ecologists eschew theoretical ecology, observing that it has little relevance to real problems. The latter has probably been responsible for the largely independent development by some fields (e.g., fisheries) of their own theory (McIntosh 1985, p.158 and references therein) and the implementation of solutions independent of theory in other fields, such as integrated pest management (Kogan 1986). Also, some of the existing theory in population genetics had its origins in applied problems (e.g., Wright's interest in the theory of mating systems arose from his involvement in dairy pedigree analysis). These difficulties notwithstanding, we will analyze the body of knowledge that has resulted from the pursuit of a "general, predictive" theory of population biology, by first simply focusing on the implications of those two words.

Some of the limitations of the theory of population ecology are possibly inherent in the ambiguity in the stated goal of a *general, predictive* theory. While few would question having these two qualities as final goals, part of the inadequacy of existing theory for the solution of practical problems may actually arise from wrongheaded pursuit of them. Taking the former first, as mentioned in Chapter 1 the quest for generality often involves ignoring important details. This does not really contradict the meaning(s) of the word general; in most dictionaries it has at least two definitions: (1) applying to all, or almost all, cases, and (2) involving only main features, not detailed or specific. The former would obviously be much more useful in practical applications. Unfortunately, the latter is the definition most often implied in theoretical population biology. In the construction of theory, population biologists rarely investigate more than one or two cases before suggesting a general law, and there is much written in the theoretical literature to justify simplification of models (see Chapter 1). A theoretical construct, or law, that is general in this sense (i.e., is based on only a few observations or a vague model with details removed) has no guaranteed value in practical applications. There is little reason to believe that it applies to the situation of immediate interest, and it may have been simplified to such a degree that it says little beyond the obvious.

The logistic equation is a good example of the latter definition. It states basically that populations grow rapidly when small, but at some point cease growing. There is no correspondence between the functional reason why growth rate changes in the model as biomass or numbers change, and the actual population mechanisms (e.g., decreasing reproductive or survival rate) (but see Schoener 1975). Generality in this second, looser sense is, however, not without some value. It can serve to promote understanding and communicate mechanisms when used as a metaphorical analogue. When explaining typical population behavior (e.g., populations grow rapidly when small, but eventually reach a level at which they remain roughly constant), it is often useful to strip it of unessential details. This pedagogical value is of little use, however, when the model is being used to attempt to understand why a population behaves in a certain way, or to project actual behavior of a specific population.

Adoption of the second goal, a *predictive* theory, is based on the widely held belief that prediction implies understanding (e.g., Paine 1981), or is in fact the supreme test of our understanding. This belief however, may not necessarily be true. Some of the ambiguity arises out of the fact that prediction and understanding are associated with different logical bases (cf., Toulmin 1961). Because deductive systems are nonampliative (i.e., the information in the conclusions is contained in the premises), the word predictive implies an argument that is essentially inductive (see, e.g., Salmon 1966, 1973). On the other hand, most agree that our theory or understanding of population biology is arrived at, through methods that are in large part deductive (the degree of dependence on deductive arguments varies, with Popper's version being the extreme, depending completely on falsifying hypotheses). Whether one follows a strict falsificationist program or not, in virtually all versions of the hypothetico-deductive method (e.g., Platt 1964; Salmon 1966), correct predictions are weaker results (in terms of increasing our understanding) than incorrect predictions.

Confusion between one of the *goals* of theory construction (being able to predict) and the *methods* of construction (that weaker conclusions can be drawn from correct predictions) have led to weaker theory. Concluding that a hypothesis or theory is true because a prediction derived from it is true is an example of the error in deductive logic, "denying the consequent". Although ecologists are frequently cautioned against this practice (e.g., Dayton 1973; Koehl 1989), this argument is often used to justify both verbal theory and models (e.g., Werner and Mittelbach 1981; May 1981). Because predictiveness is a goal of the theory eventually constructed, it is often prematurely inserted as a step in the process of theory construction. (See Loehle 1983 for further discussion of prediction versus theory, and Lehman 1986c regarding the differences between knowledge and understanding in ecology.)

Prediction can actually be based on no understanding at all. In population biology, predictions are often based on a simple regression with no requirement for understanding of underlying mechanisms. Thus, prediction does not necessarily imply understanding, and furthermore, the frequent failure of such predictive relationships, for example in predicting recruitment to marine populations, is cause for concern (e.g., Sissenwine 1984; Walters and Collie 1988).

It would be comforting to know that the converse held; that understanding implied prediction; however, even that is not straightforward, for several reasons. We may have a good understanding of a process, and yet not be able to predict outcomes because of either: (a) inherent randomness or (b) particular kinds of behavior of nonlinear systems that make them almost independent of initial conditions, hence virtually unpredictable (see the section below on Inherent Unpredictability). As Ehrlich (1989) and others point out, this may be a constraint on our field, not unlike, but of greater magnitude than similar constraints in physics.

Is the fact that prediction does not imply understanding necessarily bad for theory or practical application? Some would argue that prediction is our paramount goal, and that whether we understand or can explain the predictions is unimportant. However, this view ignores not only the value of better understanding in making better predictions, but also other benefits, such as the facilitation of solutions to other problems through contribution to general theory. To illustrate these benefits consider the well-known example from the history of astronomy. Predictions of eclipses and other celestial events from empirical formulas based on Ptolemy's epicycle theory were, for some time (even after the acceptance of the Copernican system), much more accurate than predictions based on the Copernican view of our solar system. The latter mechanisms eventually led to better predictions, but the important point here is that the Ptolemaic view was a theoretical dead end. It could not have led to the later contributions of Newton and Kepler, and our current view of the universe. In population biology the same may be true of simple models which are used to provide predictions, but do not explicitly incorporate mechanisms.

To summarize, the goals of making our theory or set of principles both general and predictive contain often unforeseen, inherent difficulties and require considerable care. Principles or laws which are general in the sense that they are vague, are not likely to be useful in specific applications. Establishing generality in the more useful sense will require more examples and a closer relationship between theory and observation. With regard to prediction, population biologists must keep in mind that a correct prediction does not necessarily imply understanding of underlying principles, nor is an incorrect prediction without value. In most interpretations of the hypothetico-deductive method (or strong inference) correct predictions (positive outcomes) are weaker results. (As applied ecologists, we should be pleased that predictions that turn out to be false are the most valuable in such a scheme, since in applied ecology that is the usual result.) However, the important point to note is that these "failures" have future value *only* if they are monitored and are incorporated as part of the empirical process.

Inherent unpredictability

With regard to the second potential cause of problems in applications, inherent natural variability, there is little doubt that there is considerable variability in the systems dealt with by population biologists, and that this leads to a high degree of uncertainty in the solutions to population problems. In that respect, population biology or biological ecology may be slightly worse off than other fields such as physics (cf., Ehrlich 1989). The question at issue here is whether we could deal more effectively with the existing uncertainty. The answer depends on the kinds of variability present and their sources.

The relatively large amount of inherent biological variability due to random events such a weather is responsible for only part of the uncertainty present. Another significant source of variability is inherent in identifiable structure within populations (e.g., genetic, age, size, etc.). A third source of variability in population data is the errors incurred in observation and measurement. The remaining variability is simply due to a lack of adequate understanding of the processes with which we are dealing (cf. Walters and Hilborn 1978; Walters 1986, p. 162; Getz and Bergh 1988, for further discussion of kinds of uncertainty).

There is no remedy that allows removal of the first type, meteorological or oceanographic variability, and, to the degree that it is present in any specific system, applied population biologists cannot be expected to be any better at prediction than meteorologists. However, an understanding of how environmental variability influences a population, and statistical characterization of the random influence can be useful in management. The second type, intra-population variability, due to structure, can be handled, at least conceptually, through incorporation in structured models, if required (e.g., Botsford 1991). There are well-developed techniques for dealing with errors of the third type, errors in measurement or observation, but when applied to population biology they often suggest observation over prohibitively large time periods or numbers of replicates in space.

The last type of uncertainty (that due to inadequate understanding) continues to exist for several reasons and is a primary focus here. One is the inherent variability and uniqueness of natural histories. To paraphrase Slobodkin (1988), large amounts of effort on the frontiers of lizard ecology may bear little fruit for ornithology. Note that population or community dynamics may differ from genetics on this point, in that there is more similarity between genetic mechanisms in birds and those in lizards than there is in population mechanisms.

A particularly perplexing aspect of variability in ecology that is beginning to receive attention is that it occurs in varying amounts at different spatial and temporal scales (Levin 1989; Powell 1989). The variability on slow temporal scales and long spatial scales that occurs in many of our problems suggests that sampling over broad spatial scales and long temporal scales will be required for adequate understanding. Difficulties due to variability are more obstinate in applied, as opposed to basic, research. It is reasonably easy to deal with uncertainty in basic research, because one has more leeway in choice of research topic (i.e., scale, species, system), hence greater opportunity to match bite to chewing ability. In basic research one can choose a system that is easily measured, and relatively isolated from exogenous randomness. In practical problems, however, we are faced with a proposed action or an existing problem that may involve considerable uncertainty on a specific spatial scale, and we do not have the luxury of studying only part of it or studying it for a long time before making a decision.

Methods of dealing with uncertainty are a broad topic, and beyond the scope of this chapter. We simply note here that historically, applied biological ecology can be fairly described as having progressed from: (1) believing that it could be (or had been) conquered (i.e., we had, or would eventually be capable of almost perfect predictions), to (2) ignoring it in solutions to problems then apologizing for it *post hoc,* to (3) admitting that it existed and trying to explain it qualitatively to users, to (4) attempting to better quantify it, and we may now finally be at a stage where we (5) actually attempt to devise ways of accounting for it and reducing it. Our answer to the question, are we doing as well as we can, given the nature of the problem, is therefore in the negative. More can probably be done, we have only recently begun to address the problem specifically. We agree that biological ecology, which is a relatively young and under-appreciated science (e.g., in terms of funding) and is operating in an environment of considerable uncertainty, has made substantial contributions in some applied fields, however we emphasize the potential for improving our performance.

Value of current theory

Our last potential contributor to the high failure rate in applications is that perhaps we are taking the wrong approach in assuming that theory in population biology provides us valuable principles on which to base solutions to practical problems. Before evaluating potential flaws and suggesting changes to the ways in which practical problems are approached and theory is developed, we explore the nature of the usefulness of current theory. Rather than laws or principles on which to base policy, current theory provides us a set of suggestions of what to look for, what to expect, an idea of how the systems of interest *might* behave, or a *post hoc* explanation of observations.

To date, the most valuable use of theory has been in terms of identifying possibilities, rather than accurate prediction. As an example, consider the problem faced by water managers in California (and elsewhere) who wish to provide adequate flows to assure a viable environment for fish and wildlife species influenced by flows. Without the benefit of theoretical principles of population biology, they might design regulations so that a model species of recreational interest, such as striped bass (*Morone saxatilus*), could persist. Such a scheme, might allow an occasional year of flows that were low enough to prevent successful recruitment, on the basis that loss of a single years' reproduction would not lead to local extinction of an iteroparous species. However, theoretical principles would tell us that this would be a problem for other semelparous species. In California, the delta smelt (*Hypomesus transpacificus*), a native semelparous species which is under consideration for endangered status, would be jeopardized by such regulations (Moyle *et al.* 1989). It is doubtful that this fact would even be considered if there were not a theory or set of principles upon which to rely.

Genetic aspects of population theory have a similar value. In Chapter 5 on weed control, without a genetic theory describing the consequences of colonizing episodes and the effects of different mating systems, we would not have known to look for genetic bottlenecks and the possibility that all individuals of certain species might be genetically related and almost identical. Such theoretical predictions and subsequent discoveries of variational patterns in weedy species also formed the basis for some control programs. In Chapter 3 on zoos and reserves, genetic principles led to quantification of various effects such as inbreeding and loss of fitness due to genetic load. They also formed the basis for design of genetic management of viable populations. Theory in all of these cases points to the serious gaps in our data bases on the genetic and ecological processes regulating populations.

Another characteristic of current theory is that it is more useful in a comparative, rather than an absolute sense. An example is the development of theories dealing with minimum viable populations (c.f., Ehrlich 1989; Soulè 1987). The work on population dynamics has been mapping the relationship between probability of extinction and population size, but we do not have a theory that enables us to accurately forecast probabilities of extinction from a specified population size. This situation is similar to the use of the concept of effective population size (Chapters 3 and 4). In many cases the theory gives us the direction of change in N_e, as we change population structure, but not the magnitude.

In instances in which theoretical constructs have been taken to imply more than a comparative suggestion, they have often failed, primarily because they were not adequately justified to begin with. For example, the notion that "compensation" (i.e., density-dependence) must exist in populations has been used to argue that the effects of larval fish mortalities induced by power plants are minimal (i.e., mortalities will be compensated for by the inherent density-dependent mechanisms in a population). As outlined in Barnthouse *et al.* (1984), use of this "law" served only to confuse the issues and draw out the arguments in the courts. Another example is the idea that the American lobster acted as a keystone predator in the urchin-kelp-lobster system in the northwest Atlantic, a notion that even made its way into several textbooks, in spite of the limited empirical support (Elner and Vadas 1990). Other examples are the earlier recommendations regarding pest management and reserve design which were found to be inadequate in Chapters 9 and 10. Similarly, a model of losses in genetic variability due to drift in a small population led to rather superficial guidelines in genetic conservation (Frankel and Soulè 1981). It was noted, for example, that an effective size of $N_e = 50$ would control drift per generation and that $N_e = 500$ would keep even the cumulative inbreeding levels below 5 or 10 percent. However, neither the plans to achieve specific values of N_e without adequate data on population substructure or reproductive biology, nor attempts to measure the evolutionary or demography links between variation and fitness parameters have been widely successful. In fact, Lande (1988) argues for a reexamination of population genetic arguments, since genetic effects may be of much lesser importance than demographic effects.

Flawed approach

These examples reflect a belief that underlies a standard approach to practical problems, that we have a set of principles which we can use "off the shelf" to confidently determine solutions to practical problems. As mentioned in Chapter 1 this belief probably results from attempts by ecologists to measure up to the physical sciences (and thus is another form of physics envy [Cohen 1971]). However, a cursory examination of early physical applications shows that, even for physics, this model does not always hold. For example, most of us have seen pictures of the bridge at Tacoma Narrows flapping in the breeze and eventually destroying itself when winds excited its natural resonant frequency. Other major bridges also failed in the early days of large metal structures. That this does not occur today is probably due to accumulated experience and "rules of thumb", rather than better theory. A second example in which uncertainty prevented adequate design is the airplane. In the early days of development it was not clear whether aircraft should be designed to be inherently stable or unstable. The first successful airplane and other early designs were unstable, probably due in part to the earlier experience of the inventors, the Wright brothers, with the bicycle, which is inherently unstable (at least in yaw) (Vincenti 1988).

Thus, our approach to practical problems in population biology may be falsely colored by a mistaken notion of the way the "harder" sciences operate. While it would no doubt be to our advantage if this model were true for population biology, we may be doing ourselves a disservice by continuing to operate as though it were true. There are several flaws inherent in such an approach. One is that it does not explicitly acknowledge the uncertainties in solutions prescribed on the basis of existing knowledge. Our principles are not "general" enough to support the optimism often implied. By not acknowledging uncertainty we incur a second disadvantage, we are unable to take advantage of any knowledge of the uncertainty, statistical or other. We could fashion approaches designed to progressively reduce uncertainty as we gain experience. A third, related disadvantage of such a procedure is that the involvement of the population biologist ends after the first solution is proposed. Failure to monitor the outcome of our proposed action represents a wasted opportunity to improve the chances of future success. It may be to our advantage to fashion solutions that involve further monitoring. A fourth disadvantage closely related to the last two, is that each practical application is to some degree treated as an isolated case. There is typically little attempt to incorporate outcomes of solutions to problems which are only loosely related (e.g., involve different species or recur in different applied contexts) into a general body of knowledge.

In parallel to flaws in our approach to practical problems, there are potential problems with the theoretical principles themselves, brought about by the nature of our field of study. The entities of interest in population biology, populations of plants and animals and their natural (and often not so natural) environment, are of such great complexity and heterogeneity and vary on such large spatial scales and slow temporal scales that they make the essential problem of this field unique among other applied fields. Scientific progress is slower, and generalizations harder won. Spatially, we cannot scale our problems down to representative physical models. Temporally, we cannot wait until we have enough information to design a solution to a problem (to use the airplane analogy again, we are inventing the airplane as we fly).

These difficulties in gaining information create tremendous pressure to base our conclusions on lesser amounts of information in the development of theoretical principles, but especially where applications are concerned. The latter is currently acute in certain "crisis" questions, such as the effects of genetically engineered releases or how to design reserves. As outlined above and in Chapter 1, there is a marked tendency to draw "general" conclusions prematurely, and to put greater stock in a few positive outcomes, rather than take the time to pursue a hypothetico-deductive program. Although other fields such as medicine have similar levels of complexity, there is not nearly the impetus to solve ecological problems that there is to solve medical problems (cf., Slobodkin 1988). The problems of applied population biology are unique, and they therefore require a unique approach to their solution.

Recommended approach

In applied population biology, because the consequences of our understanding of a population system are going to be translated into a management action, we must demand closer adherence to rules that will provide rapid approach to a model that is accurate for practical purposes. We can not afford the luxury of readily accepting mathematically attractive models or dramatic theories based on little comparison to real data. Continued use of such models and theories in applications are not giving the right answers in applications.

We propose that both the problems involved in our current approach to practical problems, and the inherent constraints to development of useful theory can be reduced by treating applications in population biology as a continuous part of the empirical process. Considering applications as experiments, designing them as such, interpreting them as such, and selling them as such, can circumvent the above outlined flaws in our current approach. Use of outcomes of applications as experimental tests of theory could supplement the short supply of observations which are difficult to obtain on the appropriate scales. This proposal will involve changes in both our approach to practical problems, and the development of theory, leading to a much closer interdependence between them. Rather than being just a result of our current understanding of population biology, solutions to practical problems will also be a test of our current understanding of population biology. Learning from these tests will require monitoring the results of management actions. Whatever is then learned can then both contribute to theory and be useful in future applications.

Our earlier discussion of the pursuit of generality suggests that a shift in emphasis would improve our approach to the construction of theory so that it will be more useful in practical situations. Instead of tending to proclaim general results, based on analysis of a simple model that may not realistically portray even a single population, we should consider adopting the other definition of generality, establishing it on the basis of realistic models of many situations. Instead of rendering models so vague that they cannot be compared to the real situation, we should stick with realistic models of each situation, then search for general aspects of behavior among them. The advantage of this shift is that we will be constructing a theory that will apply to a greater number of real situations, hence will actually be useful in practical problems. In terms of existing approaches, this would, in a sense, involve observation of the results of many applications of the tactical approach to modeling (*sensu* Holling 1964), and greater emphasis on special theories (*sensu* Oster 1981; Levin *et al.* 1989). In demanding greater realism and less vagueness in population models, we will be increasing the empirical content (i.e., how falsifiable a model is) (*sensu* Popper 1935) (also see Chapter 2 in Rothman 1986).

This approach to achieving a general theory has not been more closely followed because it requires more specific examples which are expensive and difficult to obtain. However, in drawing this conclusion, the framers of theory have often overlooked the plethora of ecological observations and potential for "experiments", in applied ecology (e.g., applied ecology is not mentioned in Diamond's [1988] otherwise thorough review of the various types of experiments used in community ecology, nor is this possibility mentioned in Kareiva's [1989] list of empirical work currently needed to better develop theory). In our current approach to practical problems, the opportunities presented by these are rarely fully utilized for learning. Our recommended approach of incorporating the solutions of practical problems into the empirical process will provide a larger number of specific examples from which to argue the generality of theoretical principles, in addition to providing a theory with greater probability of success in practical applications.

Development of a general theory that is more useful in the solution to practical problems can also benefit from a different philosophical approach with a more skeptical evaluation of observations before drawing conclusions, and a more tentative view in our understanding of how populations work. These would be manifested in a more selective approach to accepting hypotheses as true. Acceptance of theoretical constructs or specific population models on the basis of one or a few observations increases the risk of incorrect predictions, hence incorrect management prescriptions. Because the acceptance of theoretical ideas or models as being true involves the philosophical issue of confirmation, it is useful to examine current thinking on that issue.

As noted in Chapter 1, different versions of the hypothetico-deductive method vary in the significance they attach to positive outcomes, with Popper's strict falsificationist view being the extreme. Some prominent ecologists have pointed out that (1) Popper's views are passé among philosophers of science and (2) a falsificationist doctrine does not jibe well with the way science is actually done. Both of these are essentially true, but some elaboration is required for their complete understanding. Philosophers of science distinguish between two aspects of the scientific process, discovery and justification. The former involves the creative processes by which scientists arrive at scientific theories, and the latter involves the more formal, sometimes *post hoc* process by which they decide whether they are going to incorporate them into existing theory (see e.g., Suppe 1977 , pp. 125, 233). In the early part of this century, philosophers generally focused on the latter to the exclusion of the former. A second, related distinction has to do with how philosophers view their role in science. This distinction is between a simple, uncritical description of what scientists do and a methodological prescription of what scientists should be doing (see, for example Hull 1974, p. 88). Earlier in this century, philosophers of science confined themselves to the latter, but in the early 1960s, in part in response to criticism that they were not describing the way science was actually done (cf., Roughgarden 1984; May 1981 for similar comments regarding ecology specifically), they shifted their approach more to the former (see e.g. Cohen 1977). Thus, one of the reasons why Popper is passé is that philosophers of science are not addressing the same question that they addressed earlier; they have shifted from a methodological prescription regarding justification to more of a passive description that includes the process of discovery as well as justification. Also, although it is true that Popper is passé among philosophers, the philosophers whose ideas we usually quote instead built closely on Popper's ideas (see, for example, quote from Feyerabend in Suppe 1977, p. 166 [also see footnote on that page], Kuhn 1977).

In our suggestions as to how population biologists should go about constructing a theory that will be more useful in the solution of practical problems, we focus more on justification, rather than discovery, and, since we seek a better way of doing things, we seek a methodological prescription, rather than a passive description. Because, presumably, both the construction of theory and the solution of practical problems depend on selecting a correct model of how the population system behaves, development of an efficient methodology hinges on development of a specific view of confirmation (i.e., how we decide to accept explanations of behavior as being true, based on observations). As noted above and in Chapter 1, virtually all views of the hypothetico-deductive method value instances of negative outcomes in testing predictions over positive outcomes. This is founded on the basis that only one instance of an exception can disprove a generality (*sensu stricto*), whereas even many instances of agreement do not definitely prove it.

As discussed earlier in this chapter, in both theoretical and applied population biology, there is a predominant tendency to accept models or verbal theoretical constructs on the basis of a single (or few) positive outcome(s) (cf., Dayton 1973; Koehl 1989; Loehle 1987). A greater presence of the mode of acceptance suggested by a hypothetico-deductive scheme in the approach of population biologists would reduce the tendency to generalize prematurely. We would tend to structure our beliefs more on instances of negative outcomes from which we could disprove models or hypotheses. One problem with this approach (addressed below) is that it is more comforting to be able to say "I know how it works", rather than "I know how it doesn't work." An additional benefit of attention to the rigors of a hypothetico-deductive scheme is that rejection of a model or idea sharpens the focus on just exactly what is or is not being rejected, hence leads to more realistic (or mechanistic) models.

In referring to a hypothetico-deductive method, with Popper's view as an extreme, we do not mean to necessarily prescribe a Popperian program for progress in population biology. For one, we are describing justification, not discovery, hence we are saying nothing about a large part of a research program. Also, as pointed out in Chapter 1, we recognize problems with that approach, such as multiple causes (Hilborn and Stearns 1982; Quinn and Dunham 1984) (though even that one is not insurmountable, e.g., see Rothman 1986). Our remarks regarding greater reliance on disproving hypotheses apply equally well whether we are actually talking about distinct hypotheses or a continuum of combinations of causes.

Our recommended approach, with its stricter adherence to a hypothetico-deductive method in the acceptance of hypotheses presents its own substantial problem for applied ecologists. The problem stems from the urgency of most practical problems. We do not have the luxury of being able to wait while we perform additional experiments or take additional data to completely determine how a population or system of populations works. An answer is usually required immediately. There is an urgency in applications that is not present in purely theoretical work. We must usually make a decision or take management action when we have only an incomplete view of how the population system "works" (i.e., more than one unfalsified hypothesis).

Because the approach we have argued for thus far involves greater dependence on negative outcomes, from which we are able to reject hypotheses, at the time when an urgent decision needs to be made we may still have several possible (not yet rejected) ways in which the population might work. Because of this quandary, in most practical applications a choice is made of the "most correct" hypothesis. However, we need not arbitrarily (or probabalistically) adopt a single view of how the system works and manage on that basis. Rather we can base management on the consequences of *all* unfalsified hypotheses. In practical applications we can (at least conceptually) compute, on a probabalistic basis, the consequences of each of the various hypotheses being true, and manage accordingly.

For example, if there were several unfalsified hypotheses regarding a specific dynamic aspect of a certain population's behavior about which a management decision had to be made, rather than deciding to adopt only one of the hypotheses and following the implied management strategy, we could base management on the consequences (with the appropriate probabilities) of all of the hypotheses being true. Thus, the optimal strategy would not necessarily require choosing a specific hypothesis, but rather would be to choose a policy that optimized the desired cost or benefit functional over the complete range of possible outcomes, using a probabalistic description of the various hypotheses. This approach accounts for existing uncertainty, and avoids the waste of energy involved in the more common approach, each biologist choosing a different hypothesis, then debating to choose the most "correct" one (see Chapter 11).

Practical examples demonstrate how this problem arises in applications. Most of the time a population biologist's recommendation for a specific problem requires first determining a critical aspect of population dynamics of a population. One example is the cause of cyclic fluctuations in red grouse in Chapter 7. Another example is the work on demersal fisheries in Australia, where it is not obvious which model of community ecology is appropriate (Sainsbury 1988). A second example from fisheries, is the general problem which stems from the fact that recommendations regarding fishery policy involve knowing what controls recruitment, at least whether recruitment is density-dependent or environmentally forced (e.g., see Discussion in Botsford 1986). Taking this last example a bit further, if evaluation of the various recruitment hypotheses is conducted according to a falsificationist interpretation of the hypothetico-deductive method, a number of them may be rejected, others may not be rejected, and still other potential causes (possibly the true cause) may not even occur to population biologists. One is left with a few mechanisms that are known not to be controlling recruitment, and possibly several that may be. The question then is how to formulate a management policy based on knowledge of what *does* not control population behavior. This presents a fundamental problem that is usually avoided in practice. Even if one has gone this far with the hypothetico-deductive method (and that is not likely in current fisheries management), there is strong motivation at this point (even much earlier) to choose one's favorite hypothesis, formulate management policy on the basis of that hypothesis, and spend much of your time defending your hypothesis (cf. Walters 1986; Walters and Collie 1988, Chapter 11). The approach we suggest here is to choose the management policy that optimizes the criteria (e.g., long-term yield, economic rent, etc.) "averaged" over all possible hypotheses.

The procedure presented thus far does not yet consider how we should design this procedure to incorporate learning (i.e., differentiating between hypotheses) as a criterion in addition to the primary criterion. Neither have we evaluated how policy should be changed after we had seen its initial effects. Evaluation of these is the core of modern adaptive management (see Walters 1986, Chapter 11).

Views of others – pro and con

The approach recommended here is not completely new. The basic idea of designing management policies with the goal of extracting further information in addition to other criterion, is closely related to the idea of adaptive environmental management that has been under development by C. S. Holling, C. Walters, and R. Yorque since the 1970s (Holling 1978; Walters 1986; Walters and Hilborn 1976). We argue here for an adaptive approach, but with a greater role for process-oriented research on mechanisms, rather than relying solely on management experimentation. We suggest a much closer relationship between applications and ecological theory. The adaptive approach put forth here is much more than a technique to be used only by the more mathematically oriented practitioners. Rather it should be regarded as the accepted approach to all practical problems in population biology, it should be appreciated by users (i.e., decision makers) as well as practitioners, and (perhaps even more importantly) theorists should be aware of the advantages it can hold for them.

Our ideas are similar to some of the conclusions drawn in NRC (1986). That study of how ecological knowledge is applied to practical problems, also recommends treating practical projects as experiments and stresses the importance of developing and testing hypotheses (predictions) as well as monitoring to determine their outcome (NRC 1986, p. 15). Further, it emphasizes several important practical implications of this approach, the most important being that it will require looking for close analogues and attempting to do generic experiments. The book contains several recommendable examples in which the experimental approach has been used: a study of the impact of hydroelectric development on caribou (Chapter 16, Kiell *et al.* 1986), a study of the water quality of Lake Washington (Chapter

20, Lehman 1986a), and a study of raising the water level in a subarctic lake (Chapter 21, Lehman 1986b).

Slobodkin (1988) questions the practical value of simple models from theoretical population biology. May (1984, p. 14) defends the use of simple models on the basis of expediency by accurately characterizing the urgency associated with practical problems, "The choice is not between perfect and imperfect advice to managers, but between crudely imperfect advice and no advice at all" (which insists on reminding us of the actor Ronald Reagan's defense of the low quality of his films: "They didn't want 'em good, they wanted 'em Thursday"). Slobodkin (1988) answers that in economics as in ecology, "it is by no means obvious that 'crudely imperfect advice' from a supposed expert is more or less valuable than 'no advice at all'. Slobodkin also recommends closer ties between theory and practice.

> Ecology in the absence of practical questions is in danger of deluding itself by a vain hunt for generality, answering only easy questions that it poses for itself, and becoming irrelevant to anyone outside its own academic village. Practical questions of environmental management in the absence of ecology are likely to receive misleading and even dangerous answers.

Oster (1981) presents a similar view of the limits of simple, general models. He refers to two kinds of theory, *general* theory composed of models that describe general phenomena, but not any single specific case, and *special* theory composed of models that answer questions regarding a specific situation. With regard to the former he notes that:

> Thousands of papers have been published on competition and predation theory, most of which are simply devoted to examining the mathematical properties of a set of differential equations that purport to model interactions between real populations. Remarkably, for a long time most ecologists took these equations quite seriously — as if there were hidden in them some great, but subtle, truth about nature. What was lost in the proliferation of paper was that the subtlety was mostly mathematical, and the truth they contained mostly allegorical.

He proposes, as we do, that the general principles of ecology will suggest themselves only after many special cases are understood.

In Kogan's (1986) book, his view that integrated pest management has become detached from theory, and sometimes pushed on ahead of theory is cited as cause for grave concern. The chapters on theory provide only marginal hope for change, and the chapters on applications lament the "lack of guidance from theory" (Kennedy 1986). Perhaps new directions with more frequent comparisons between modeling results and applications, as outlined in Chapter 11 of this book will lead to changes in that situation.

Loehle (1987) has evaluated the issue of confirmation and the philosophical approach of ecologists. He notes first that both a bias toward confirmation and tenacious adherence to ones own theory have been shown to be prevalent in current science. Although he finds them necessary for theory maturation, which is in turn necessary for valid testing of the theory, he points to ecological theories adopted on that basis that show no progress toward maturation. He also notes the impediment to progress that long time scales present, and argues for an increase in empirical content.

There are other indications of leanings in the directions specified here. For example, most of the papers in the recent issue of *Oikos* whose purpose was to evaluate the impact of theory on research into the causes of microtine population fluctuations, recommended closer ties between theory and practice (Lomnicki 1988; Krebs 1988; Hansson 1988; Hansson and Stenseth 1988). As another example, Levin *et al.* (1989) have recently argued for a more empirical approach to ecotoxicology, with less dependence on bold predictions.

Our views also appear to be consistent with Ehrlich's (1989) recent assessment of whether ecological theory is any good in practice, though he has a much more positive view of the benefits of current theory. He notes, as we do, that where applications are concerned, empirical studies are far behind where they should be, because they are expensive and difficult. Also, he uses the concept of minimum viable population to demonstrate a view stated here, that theory does not provide us with specific predictions, but rather alerts managers to the dangers of extinction and inbreeding in certain situations. Perhaps the most illuminating example of how his positive view of the practical value of current theory can be consistent with his appreciation of its lack of data base and inherent limitations, is his description of why competition and niche theory cannot explain the decline of the California sardine, "here again a lack of data on guild structure and community relationships is probably more of a barrier than a failure of theory".

Our ideas are also consistent with those expressed by Elner and Vadas (1990) in their recent critique of sea urchin phenomena in the northwestern Atlantic. They criticize the affirmative oriented approach used in arguments to establish the lobster as a keystone predator and the similar subjective attempts to continue to support that idea, as well as the relative paucity (and subjective interpretation) of experimental or other data. Their recommendation of a more rigorous experimental approach based on stronger natural history observations parallels our arguments for a more skeptical, hypothetico-deductive approach fueled by a greater number of observations or experiments through applied problems.

To seek a balanced presentation, we attempt to outline some recent writings that seem to go against our point of view, though this may be due to a different definition of theory, rather than a disagreement over the role that must be played by a specific type of endeavor, whether or not one calls it theory (also see Loehle 1983 for definition and justification of different types of models). There are many ecologists who would reserve the right to develop theory with little dependence on data. They would hold that theory and empirical work (and most assuredly applications) are different, independent fields of endeavor. Levin (1981) provides an example in his argument that the traditional view of testing models against the real world relegates theory to a subordinate role, and that that is not the true meaning of theory, but rather a derivative. Part of the problem in these arguments is a different interpretation of the concept of a theory. He appears to argue for an ecological theory that is much more akin to a mathematical theory, rather than a scientific theory. In mathematics, a theory is essentially a number of deductions from relatively few axioms. The results may or may not be related to the real world or be useful in a practical sense.

This view of theory differs from the view of theory established via some version of the hypothetico-deductive method. To test causal mechanisms one usually derives predictions (implications) from them, then compares these to real data to prove or disprove the mechanisms. This notion of deducing predictions from a set of hypotheses (or proposed theory) then comparing the predictions to the real world to draw inferences about the hypotheses, is not the same as the development of theory in mathematics, since, in mathematics, theories are not developed to test axioms, but rather to elaborate their consequences. (Also see Loehle 1987, p. 406.)

Levin (1981) further argues that theory and the real world should remain separate, especially where generalizations are concerned. In his view, the Lotka-Volterra equations "have had their value seriously undermined by overuse and by misguided attempts to parameterize them on the basis of data". He suggests that these equations "should not be taken literally, but as guides to theory, thought, and experimentation", arguing for a role of theory as metaphorical. As noted earlier in this chapter, this role of theory can serve a valuable pedagogical function. The practical problem this presents lies in keeping the two kinds of theory separate. We are frequently faced with occurrences such as the impact of harvest on marine species in the Southern Ocean being explained using the Lotka-Volterra equations (May *et al.* 1979), and most of the fisheries in the world being managed using the logistic equation. Although they have some use, it is doubtful that we should manage with metaphors.

Another point of view allows comparison between the real world and theory, but differs from the recommendations made here in that a match between model and data is sought, and the model is accepted once the match is found. This mode of operation is more in the spirit of ready confirmation, rather than a program of falsification. May's (1981) view is representative of the views of many others. He proposes that simple models can clarify the essential features in a complicated natural situation. To do so, he says, applied mathematicians try to identify the essential features using common sense or physical intuition, incorporate them in models, then make predictions. "If the predictions accord with reality, our understanding is advanced; if not, we try to find what necessary ingredient was omitted" (May 1981).

As a final point of possible disagreement, throughout our arguments for a shift in emphasis we have made the reasonable (to us) assumption that we seek the model that accurately describes the functioning of a population (or populations) of interest as best as we can understand it. It seems that both theorists, who are, after all, concerned with describing the way populations actually work, and practitioners will be more successful if they use such a realistic model. Although we have assumed that the procedure that allows the most rapid approach to that goal is the one we seek, we realize that that view is not universally held. Several prominent ecologists have argued that we need not have a realistic description for applied problems, and in fact in some cases a model that is known not to be the true description can give better answers (c.f. Ludwig and Walters 1985; Walters 1987 (Chapter 5), and Botsford's 1987 comments).

In summary, based on conclusions drawn from discussions with chapter authors, we have presented practical real world arguments as well as a more formal, philosophical development of how and why we think that applications of population biology should be incorporated into the empirical process of theory construction. Without taking this further here, we simply observe that a large part of the field of population biology appears to be isolated from the real world and engaged in affirmative arguments based on little long term or comparative (i.e., between species, systems, etc.) data, while there is a plethora of applied problems that provide not only the basis for a more comprehensive theory, but the motivation for development of a new methodology. We hope that current trends in population biology, such as realization of the importance of manipulation (Murdoch 1970; Bender, *et al.* 1984) and long term ecological research, the turn to more realistic models that incorporate mechanisms (Schoener 1986; Tilman 1987), and reminders of philosophical problems (Loehle 1987) can be melded into a new approach to practical problems. We have also mentioned areas in need of further work. While some ecological modelers have begun to include environmental variability in their models and analyses, very few (e.g., Carl Walters; Don Ludwig, and associates) have begun to analyze the consequences of new methods that incorporate structural uncertainty and it consequences. Applied population biologists would be well advised to re-evaluate their approach to practical problems, and there are several problems as to how this should be done, that are in urgent need of solution.

Acknowledgements

We thank James F. Quinn for his comments on an earlier version of this chapter.

Literature cited

Anderson, R. M. and R. M. May. 1978. Regulation and stability of host-parasite population interactions: I regulatory processes. J. Anim. Ecol. 47: 219—249.

Barnthouse, L. W., J. Boreman, S. W. Christensen, C. P. Goodyear, W. VanWinkle, and D. S.

Vaughn. 1984. Population biology in the courtroom: the Hudson River controversy. BioScience 34: 14—19.

Bender, E. A., T. J. Case, and M. E. Gilpin. 1984. Perturbation experiments in community ecology: theory and practice. Ecology 65: 1—13.

Botsford, L. W. 1986. Population dynamics of Dungeness crab (*Cancer magister*). In G. S. Jamieson and N. Bourne. Can. Special Pub. Fish. Aquat. Sci. 92: 140—153.

Botsford, L. W. 1987. Particiant's Comment. Pp. 137—138 in T. L. Vincent, Y. Cohen, W. J. Grantham, G. P. Kirkwood, and J. M. Skowronski (eds.), *Modeling and Management of Resources under Uncertainty, Proceedings of the Second U.S.—Australia Workshop on Renewable Resource Management.* Springer-Verlag, 318 pp.

Botsford, L. W. 1991. Individual state structure in population models. In Populations, Communities and Ecosystems: an Individual Perspective. (in press).

Cohen, I. B. 1977. History and the philosopher of science. Pp. 308—349 in Suppe, F. (Ed.), *The Structure of Scientific Theories.* University of Illinois Press, Urbana, IL. 818 pp.

Cohen, J. 1971. Mathematics as a metaphor. Science 172: 674—675.

Dayton, P. K. 1973. Two cases of resource partitioning in an intertidal community: making the right prediction for the wrong reason. Amer. Naturalist 107: 662—670.

Diamond, J. 1988. Overview: Laboratory experiments, field experiments, and natural experiments. Pp 3—22 in J. Diamond and T. J. Case (eds.), *Community Ecology*, Harper and Row, New York 665 pp.

Ehrlich, P. R. 1989. Discussion: Ecology and resource management — Is ecological theory any good in practice? Pp. 306—318 in Roughgarden, J., R. M. May, and S. A. Levin (Eds.), *Perspectives in Ecological Theory.* Princeton University Press., Princeton, NJ. 394 pp.

Frankel, O. and M. Soule. 1981. Genetics and Conservation. Cambridge Univ. Press, Cambridge.

Getz, W. M. and M. O. Bergh. 1988. Quota setting in stochastic fisheries. Pp. 259—274 in W. S. Wooster (ed.) *Fishery Science and Management.* Springer-Verlag, New York.

Hansson, L. 1988. Empiricism and modelling in small rodent research: how to partition efforts. Oikos 52: 150—155.

Hansson, L. and N. C. Stenseth. 1988. Modelling small rodent population dynamics: suggestions to empiricists, theoreticians and editors. Oikos 52: 227—229.

Hilborn, R. and S. C. Stearns. 1982. On inference in ecology and evolutionary biology: the problem of multiple causes. Acta Biotheoretica 31: 145—164.

Holling, C. S. 1964. The analysis of complex population processes. Can. Ent. 96: 335—347.

Holling, C. S. (Ed.) 1978. *Adaptive Environmental Assessment and Management.* Wiley International Series on Applied Systems Analysis, Vol. 3. Chichester, U.K.

Hull, D. L. *Philosophy of Biological Science.* Prentice-Hall, Inc. Englewood Cliffs, NJ.

Jain, S. 1991a. Genetics and demography of rare annual plants. *In* P. Fiedler and S. Jain (eds.), *Conservation Biology in Theory and Practice*, Chapman and Hall, New York.

Jain, S. 1991b. Restoration; ecology. *In* P. Fiedler and S. Jain (eds.), *Conservation Biology in Theory and Practice*, Chapman and Hall, New York.

Jayakar, S. D. and L. Zonta. 1990. Coevolution at two trophic levels. Pp. 349—366 in K. Wohrmann and S. Jain (eds.), *Population Biology: Ecological and Evolutionary Viewpoints.* Springer-Verlag, Berlin.

Kareiva, P. Renewing the dialogue between theory and experiments in population ecology. Pp 68—88 in Roughgarden, J., R. M. May, and S. A. Levin (Eds.), *Perspectives in Ecological Theory.* Princeton University Press., Princeton, NJ. 394 pp.

Kennedy, G. G. 1986. Plant-pathogen-insect interactions. Pp 203—216 in Kogan, M. (Ed.) 1986. *Ecological Theory and Integrated Pest Management.* Wiley Inter-Interscience, New York. 362 pp.

Kiell, D. J., E. L. Hill, and S. P. Mahoney. 1986. Protecting caribou during hydroelectric development in Newfoundland. Pp. 205—226 in National Research Council. 1986. *Ecological Knowledge and Environmental Problem Solving: Concepts and Case Studies.* National Academy Press, Washington, D.C. 388 p.

Koehl, M. A. R. 1989. Discussion: From individuals to populations. Pp 39—53 in Roughgarden, J., R. M. May, and S. A. Levin (Eds.), *Perspectives in Ecological Theory.* Princeton University Press., Princeton, NJ. 394 pp.
Kogan, M. (Ed.) 1986. *Ecological Theory and Integrated Pest Management.* Wiley Inter-Interscience, New York. 362 pp.
Krebs, C. J. 1988. The experimental approach to rodent population dynamics. Oikos 52: 143—149.
Kuhn, T. S. 1977. Second thoughts on paradigms. Pp 459—482 in Suppe, F. (Ed.), *The Structure of Scientific Theories.* University of Illinois Press, Urbana, IL. 818 pp.
Lande, R. 1988. Genetics and demography in biological conservation. Science 241: 1455—1460.
Moyle, P. B., J. E. Williams, and E. D. Wikramanayake. 1989. Fish species of special concern in California. Inland Fisheries, California Department of Fish and Game.
Lehman, J. T. 1986a. Control of eutrophication in Lake Washington. Pp. 301—316 in National Research Council. 1986. *Ecological Knowledge and Environmental Problem Solving: Concepts and Case Studies.* National Academy Press, Washington, D.C. 388 p.
Lehman, J. T. 1986b. Raising the level of a subarctic lake. Pp. 317—344 in National Research Council. 1986. *Ecological Knowledge and Environmental Problem Solving: Concepts and Case Studies.* National Academy Press, Washington, D.C. 388 p.
Lehman, J. T. 1986c. The goal of understanding in limnology. Limnol. Oceanogr. 31: 1160—1166.
Levin, S. A. (Ed.) 1976. *Ecosystem Analysis and Prediction.* Society for Industrial and Applied Mathematics. 337 pp.
Levin, S. A. 1981. The role of theoretical ecology in the description and understanding of populations in heterogeneous environments. Amer. Zool. 21: 865—875.
Levin, S. A., M. A. Harwell, J. R. Kelly, and K. D. Kimball. 1989. Ecotoxicology: problems and approaches. Pp 3—7 in Levin, S. A., M. A. Harwell, J. R. Kelly, and K. D. Kimball (Eds.) Ecotoxicology: Problems and Approaches. Springer-Verlag, New York.
Loehle, C. 1983. Evaluation of theories and calculation tools in ecology. Ecological Modelling 19: 239—247.
Loehle, C. 1987. Hypothesis testing in ecology: psychological aspects and the importance of theory maturation. Quart. Rev. Biol. 62: 397—409.
Lomnicki, A. 1988. The place of modelling in ecology. Oikos 52: 139—142.
Ludwig, D. and C. J. Walters. 1985. Are age-structured models appropriate for catch-effort data? Can. J. Fish. Aquat. Sci., 42: 1066—1072.
May, R. M. 1981 The role of theory in ecology. Amer. Zool. 21: 903—910.
May, R. M. 1984. An overview: real and apparent patterns in community structure. Pages 3—16 in D. R. Strong, D. Simberloff, L. G. Agele, and A. D. Thistle, eds. *Ecological Communities: Conceptual Issues and the Evidence.* Princeton University Press, Princeton.
May, R. M. and M. P. Hassel. 1988a. Population dynamics and biological control. Phil. Trans. R. Soc. Lond. B 318: 129—169.
May, R. M. and M. P. Hassel. 1988b. Parasitoid theory: against Manichaeism. Trends in Ecology and Evolution.
May, R. M., J. R. Beddington, C. W. Clark, S. J. Holt, and R. M. Laws. 1979. Management of multispecies fisheries. Science 205: 267—277.
McIntosh, R. P. 1985. *The Background of Ecology: Concept and Theory.* Cambridge University Press, 383 pp.
Murdoch, W. W. 1970. Population regulation and population inertia. Ecology 51: 497—502.
National Research Council. 1986. *Ecological Knowledge and Environmental Problem Solving: Concepts and Case Studies.* National Academy Press, Washington, D.C. 388 p.
Oster, G. 1981. Predicting populations. Amer. Zool. 21: 831—844.
Paine, R. T. 1981. Truth in ecology. Bulletin of the Ecological Society of America 62: 256—258.
Platt, J. R. 1964. Strong inference. Science 146: 347—353.

Popper, K. 1935. *The Logic of Scientific Discovery.* [A translation of *Logik der Forschung,* 1935] Harper and Row 1959, 480 pp.
Powell, T. M. 1989. Physical and biological scales of variability in lakes, estuaries, and the coastal ocean. Pp. 157—176 in Roughgarden, J., R. M. May, and S. A. Levin (Eds.), *Perspectives in Ecological Theory.* Princeton University Press., Princeton, NJ. 394 pp.
Quinn, J. F. and A. E. Dunham. On hypothesis testing in ecology and evolution. Pp. 22—37 in G. W. Salt (ed.) *Ecology and Evolutionary Biology: A Round Table on Research.* U. of Chicago Press. 130 pp.
Rothman, K. J. 1986. *Modern Epidemiology.* Little, Brown and Company, Boston. 358 pp.
Roughgarden, J. 1984. Competition and theory in community ecology. Pp. 3—21 in G. W. Salt (ed.) *Ecology and Evolutionary Biology: A Round Table on Research.* U. of Chicago Press. 130 pp.
Sainsbury, K. J. 1988. The ecological basis of multispecies fisheries, and management of a demersal fishery in tropical Australia. In J. A. Gulland (ed.) *Fish Population Dynamics* (Second Ed.) John Wiley and Sons, Ltd. 422 pp.
Salmon, W. C. 1967. *The Foundations of Scientific Inference.* University of Pittsburgh Press, Pittsburgh.
Salmon, W. C. 1973. *Logic.* Prentice-Hall, Inc. 150 pp.
Schoener, T. W. 1975. Population growth regulated by intraspecific competition for energy or time: some simple representations. Theoret. Pop. Biol. 4: 56—84.
Schoener, T. W. 1986. Mechanistic approaches to community ecology: a new reductionism. Amer. Zool. 26: 81—106.
Schoenwald-Cox, C., S. M. Chamber, B. MacBryde and L. Thomas (eds.). 1983. *Genetics and Conservation.* Benjamin-Cummings.
Sissenwine, M. P. 1984. Why do fish populations vary? Pp. 59—94 in R. M. May (ed.) Exploitation of Marine Communities. Springer-Verlag, Berlin, 366pp.
Slobodkin, L. B. 1988. Intellectual problems of applied ecology. Bioscience 38: 337—342.
Strong, D. R. 1988a. Parasitoid theory: from aggregation to dispersal. Trends in Ecology and Evolution 3: 277—280.
Strong, D. R. 1988b. Reply from Donald Strong. Trends in Ecology and Evolution
Soulè, M. E. (ed.) 1987. *Viable Populations.* Cambridge U. Press.
Suppe, F. (Ed.), *The Structure of Scientific Theories.* University of Illinois Press, Urbana, IL. 818 pp.
Thompson, J. D. 1991. The biology of an invasive plant. Bioscience 41: 393—401.
Tilman, D. 1987. The importance of the mechanisms of interspecific competition. Am. Nat. 129: 769—774.
Toulmin, S. 1961. *Foresight and Understanding.* Harper and Row, New York. 117 pp.
Vincenti, W. G. 1988. How did it become "obvious" that an airplane should be inherently stable? Invention and Technology, Spring/Summer 1988: 52—56.
Walters, C. J. 1986. *Adaptive Management of Renewable Resources.* Macmillan, New York. 374 pp.
Walters, C. J. and J. Collie. 1988. Is research on environmental factors useful to fisheries management? Can. J. Fish. Aquat. Sci. 45:1848—1854.
Walters, C. J. and R. Hilborn. 1976. Adaptive control of fishing systems. J. Fish. Res. Board. Can. 33: 145—159.
Werner, E. E. and G. C. Mittelbach. 1981. Optimal foraging: Field tests of diet choice and habitat switching. Amer. Zool. 21: 813—830.
Western, D. and M. Pearl (eds.). 1989. *Conservation in the Twenty First Century.* Oxford Univ. Press, Oxford.

List of contributors

S. C. H. Barrett, Department of Botany, University of Toronto, Toronto, Ontario, Canada M5S 1A1
L. W. Botsford, Department of Wildlife and Fisheries Biology, University of California, Davis, CA 95616, U.S.A.
A. D. Bradshaw, Department of Botany, University of Liverpool, Liverpool, U.K
A. P. Dobson, Department of Biology, Princeton University, Princeton, New Jersey 08544, U.S.A.
P. W. Hedrick, Department of Biology, Pennsylvania State University, University Park, PA 16802, U.S.A.
P. J. Hudson, The Game Conservancy, Grouse Research Project, Crubenmore Lodge, Newtonmore, Iverness-shire, PH20 1BE, Scotland
S. K. Jain, Department of Agronomy and Range Science, University of California, Davis, CA 95616, U.S.A.
R. V. Kesseli, Department of Vegetable Crops, University of California Davis, CA 95616, U.S.A.
W. W. Murdoch, Department of Biological Sciences, University of California, Santa Barbara, CA 93106, U.S.A.
G. R. Robinson, Department of Botany, University of California, Davis, CA 95616, U.S.A.
J. F. Quinn, Division of Environmental Studies, University of California, Davis, CA 95616, U.S.A.
G. Smith, Section of Animal Health Economics, University of Pennsylvania, School of Veterinary Medicine, Kennett Square, PA 19348, U.S.A.
C. J. Walters, Institute of Animal Resource Ecology, University of British Columbia, Vancouver, B.C. V6T1W5 Canada

S. K. Jain and L. W. Botsford (eds), Applied Population Biology, 287.

Index